Lecture Notes in Computer Science 2428

Edited by G. Goos, J. Hartmanis, an

Springer
Berlin
Heidelberg
New York
Hong Kong
London
Milan
Paris
Tokyo

Holger Hermanns

Interactive Markov Chains

And the Quest for Quantified Quality

Springer

Series Editors

Gerhard Goos, Karlsruhe University, Germany
Juris Hartmanis, Cornell University, NY, USA
Jan van Leeuwen, Utrecht University, The Netherlands

Author

Holger Hermanns
University of Twente, Faculty of Computer Science
Formal Methods and Tools Group
P.O. Box 217, 7500 AE Enschede, The Netherlands
E-mail: hermanns@cs.utwente.nl

Cataloging-in-Publication Data applied for

Die Deutsche Bibliothek - CIP-Einheitsaufnahme

Hermanns, Holger:
Interactive Markov chains : and the quest for quantified quality / Holger Hermanns. - Berlin ; Heidelberg ; New York ; Hong Kong ; London ; Milan ; Paris ; Tokyo : Springer, 2002
(Lecture notes in computer science ; 2428)
ISBN 3-540-44261-8

CR Subject Classification (1998): D.2.4, F.3.2, F.1.2, C.4, D.2.2, G.3

ISSN 0302-9743
ISBN 3-540-44261-8 Springer-Verlag Berlin Heidelberg New York

Springer-Verlag Berlin Heidelberg New York
a member of BertelsmannSpringer Science+Business Media GmbH

http://www.springer.de

© Springer-Verlag Berlin Heidelberg 2002
Printed in Germany

Typesetting: Camera-ready by author, data conversion by Christian Grosche, Hamburg
Printed on acid-free paper SPIN 10873887 06/3142 5 4 3 2 1 0

Foreword

To devise methods for the construction of high quality information processing systems is a major challenge of computer science. In most contexts, however, the definition of what constitutes (high) quality in a more concrete sense is problematic, as invariably any definition seems to fall short of its essence. Computer science proves no exception to the rule, and its quest for quality in relation to the analysis of system designs has given birth to two main interpretations: quality as correctness, and quality as performance.

The first interpretation assesses quality by showing formally that (a model of) a system satisfies the functional requirements of its formal specification. Its methods are rooted in logic and discrete mathematics, and are based on the all-or-nothing game imposed by the Boolean lattice: unless satisfaction has been demonstrated completely, nothing can be said. This is both the strength and the weakness of the approach: results have the utmost precision, but are hard to obtain.

The second interpretation aims to assess quality on a continuous scale that allows for quantification: using stochastic system models one tries to calculate system properties in terms of mathematical expectation, variation, probability, etc. The strong point of this approach is that it allows for quality in other than absolute terms, e.g. a message loss of less than 0.01%, service availability of more than 99.99%, etc. Its weaker side is that it cannot handle very well system properties that are not directly related to repeatable events, including many functional system properties, such as e.g. absence of deadlock, reachability of desirable system states, etc.

It is clear that the analysis of the quality of system designs must ultimately encompass both of the approaches above. A first step in this direction was the development of stochastic Petri net models, which combine a classical functional model for (concurrent) systems with stochastic features. The latter allow the derivation of performance models in the form of continuous-time Markov chains directly from a system description using such nets. Thus the formalism in principle allows functional and performance analysis of systems in terms of an integrated model and perspective.

This potentially great leap forward from the existing practice of studying correctness and performance through unrelated models (and by different scientific communities) proved harder to materialise than was initially hoped for. One of the main causes was the infamous state space explosion: the fact that the number of global states of a system grows exponentially with the number of components of the system. Because of this, non-trivial system designs give rise to large Petri net models, which in turn yield huge Markov models that can no longer be effectively manipulated, even with the aid of computers.

In the early 1990s this observation motivated the study of what is now referred to as *stochastic process algebras.* In the preceding decade process algebras had proven an effective means for the modelling and analysis of the functionality of concurrent systems. They address the problem of state space explosion by a powerful formalisation of system composition by process algebraic operators, combined with the study of *observational congruence* of behaviours. The latter allows for a *compositional* control of state space complexity: replacing components with observationally congruent but simpler components the state space can be reduced without explicitly generating it first.

The study of stochastic process algebra has for a considerable part been driven by the non-trivial question of how best to add stochastic features to process algebra, combining sufficient stochastic expressivity with compatibility with existing process algebraic theory. The present LNCS volume by Holger Hermanns contains his answer to this question for Markovian process algebra, i.e., where the stochastic model of interest is that of continuous-time Markov chains. Written in a clear and refreshing style it demonstrates that it is not only Hermanns' answer, but really 'the' answer.

Where others before him treated stochastic delay as attributes of system actions, Hermanns saw the enormous advantages of a completely different approach: treating delays as actions in their own right that silently consume exponentially distributed amounts of time, and treating system actions as instantaneous actions. This separation of concerns bears all the signs of a great idea: it is (retrospectively) simple and leads to very elegant results. A complication that mars the other approaches, viz. the synchronisation of delays as a by-product of synchronising actions, is completely avoided. Only system actions are subject to synchronisation, and delays in different components of a system are interleaved. Due to the memoryless nature of exponential distributions this yields a perfectly natural interpretation of the passage of (stochastic) time. It is the Platonic discovery that interleaving process algebra and Markov chains are a perfect couple. Another advantage of the decoupling of system actions and delays is that there can be more than one delay preceding an action. This extends the class of (implicit) action delays far beyond that of the exponential distribution, viz. to the (dense) class of phase-type distributions.

The author must be commended for the technical skills with which he has reaped the full benefits of this idea. In addition to defining and applying his formalism, he has also firmly embedded it in standard process algebraic theory by providing full axiomatisations for the stochastic varieties of observation congruence which are conservative extensions of the non-stochastic cases. Also, the link between the concepts of lumpability in Markov chains and bisimilarity in process algebra that was first observed by Hillston, comes to full fruition in the hands of Hermanns. Based on standard algorithms for bisimulation a low complexity algorithm is devised for lumping (Markov chain) states that can be applied compositionally. The latter is a fine example of the advantages of interdisciplinary research, as such an algorithm was not available in the standard theory of Markov chains.

We believe that this monograph by Holger Hermanns represents an important step in the quest for the integrated modelling and analysis of functional and performance properties of information processing systems. It is also written in a very accessible and, where appropriate, tutorial style, with great effort to explain the intuition behind the ideas that are presented. With a growing number of researchers in the performance analysis and formal methods communities that are interested in combining their methods, we think that this monograph may serve both as a source of inspiration and a work of reference that captures some vital ingredients of quality.

May 2002 Ed Brinskma, Ulrich Herzog

Preface

Markov chains are widely used as stochastic models to study and estimate a broad spectrum of *performance and dependability characteristics*. In this monograph we address the issue of *compositional specification* and analysis of Markov chains. Based on principles known from *process algebra*, we develop an algebra of *Interactive Markov Chains* (IMC) arising as an *orthogonal* extension of both continuous-time Markov chains and process algebra. In this algebra the interrelation of delays and actions is governed by the notion of *maximal progress*: Internal actions are executed without letting time pass, while external actions are potentially delayed. IMC is more than 'yet another' formalism to describe Markov chains. This claim is substantiated by a number of distinguishing results of both theoretical and practical nature. Among others, we develop an *algebraic theory* of IMC, devise *algorithms* to mechanise *compositional aggregation* of IMC, and successfully apply these ingredients to analyse state spaces of several million states, resulting from a study of an ordinary telephone system.

The contents of this monograph is a revised version of my PhD thesis manuscript [96] which I completed in spring 1998 at the University of Erlangen, Germany. I am deeply indebted to Ulrich Herzog and Ed Brinksma for their enthusiastic support when preparing its contents, and when finalising this revision at the University of Twente, The Netherlands.

Many researchers had inspiring influence on this piece, or on myself in a broader context, and I take the opportunity to express my gratitude to all of them. I am particularly happy to acknowledge enjoyable joint research efforts with Christel Baier, Salem Derisavi, Joost-Pieter Katoen, Markus Lohrey, Michael Rettelbach, Marina Ribaudo, William H. Sanders, and Markus Siegle which have led to various cornerstones of this book. Henrik Bohnenkamp, Salem Derisavi, and Marielle Stoelinga read the manuscript carefully enough to spot many flaws, and gave me the chance to iron them out in this monograph. Cordial thanks go to Alfred Hofmann at Springer-Verlag for his support in the process of making the manuscript a part of the LNCS series. And finally, there is Sabine and the tiny crowd. Those who know her are able to assess how perfectly happy I account myself.

June 2002 Holger Hermanns

Contents

1. Introduction

1.1 Performance and Dependability Estimation with Markov Chains

The purpose of this book is to provide a *compositional* methodology of modelling and analysis with *Markov chains.* Markov chains are widely used as simple and yet adequate models in many diverse areas, not only in mathematics and computer science but also in other disciplines such as operations research, industrial engineering, biology, demographics, and so on. Markov chains can be used to estimate performance and dependability characteristics of various nature, for instance to quantify throughputs of manufacturing systems, locate bottlenecks in communication systems, or to estimate reliability in aerospace systems.

It is often possible to represent the behaviour of a system by specifying a discrete number of states it can occupy and by describing how the system moves from one state to another as time progresses. If the future evolution of the system only depends on its present state, the system may be represented as a (time homogeneous) Markov chain. If the future evolution depends in addition on some *external* influence, we fall into the basic model class considered within this monograph. We take the view that the evolution of a system can be the result of *interaction* among different parts of the system. We provide means to specify these parts, as well as combinators to compose parts. In this way, complex Markov models can be built in a compositional, hierarchical way. Since the inherent structure of nowadays and tomorrows systems is becoming more and more complex, the possibility to specify Markov chains in a compositional way is a significant advantage.

During the last two decades *process algebra* has emerged as *the* mathematical framework to achieve compositionality. Process algebra provides a formal apparatus for reasoning about structure and behaviour of systems in a compositional way. The theoretical basis developed in this monograph will therefore be a process algebraic one. It will turn out that compositionality is not only favourable to specify complex situations but also facilitates the analysis of such models.

1.2 The Challenge of Compositional Performance and Dependability Estimation

It is worth to have a look at the historical development of performance and dependability evaluation methodology. From the very beginning, *queueing systems* have been used as intuitive means for describing system and analysing their performance [60, 129]. However, in the late 1970-ies it has been recognised that different real world phenomena could not be expressed satisfactorily by means of queueing systems. In particular the need to model *synchronisation* and *resource contention* was recognised, as a consequence of the (still ongoing) trend towards distributed systems [45]. Thus, a bunch of extensions has been proposed for queueing systems in order to reflect these issues in an intuitive way. The unfortunate result was that the exact semantics of such extensions was unclear, due to a lack of formal meaning of the queueing approach. Even more severe, the interference among different extensions was confusing.

Instead of adding more and more symbols to a more and more ambiguous notation, the feeling grew that (extended) queueing systems, developed from an engineer's perspective, could benefit a lot from a scientific analysis of the core concepts inherent to distributed systems [110].

Petri nets turned out to be rather close to an abstract view on (extended) queueing systems [45]. In contrast to queueing systems, Petri nets are very parsimonious with respect to their basic ingredients [162]. This is beneficial in order to develop a precise theory. Nowadays, at least in the setting of Markov chains, there is a common agreement that many kinds of extended queueing system can be represented as a generalised stochastic Petri net (GSPN), an extension of Petri nets with exponentially timed and immediate transitions [4, 3]. In particular, various add-ons to queueing systems can be concisely expressed in terms of Petri nets.

Queueing and *scheduling disciplines* are exceptions. They have been incorporated into the Petri net terminology in the same informal way as in queueing systems, namely by adding a remark, say 'LIFO' or 'JSQ', to the respective entity of the net. ('LIFO' stands for 'last in first out' scheduling while 'JSQ' describes 'join the shortest queue' queueing strategy.)

The problem of this informality is more severe than it appears to be at first glance. Consider, for instance, a small system with a Markovian arrival stream and two queues, each having a few places and each connected to a private server. Assume further that the two servers have drastically different (exponentially distributed) service times and 'JSQ' queueing strategy. This system is depicted in Figure 1.1. Nevertheless, we point out that such a description, does not at all give rise to an *unambiguous* Markov chain.

Ambiguity arises whenever both queues are equally occupied. Then, the remark 'JSQ' does *not determine* where the next arrival will be scheduled. This phenomenon, known as *nondeterminism* or *underspecification*, has an important impact on performance estimates of such systems, when service

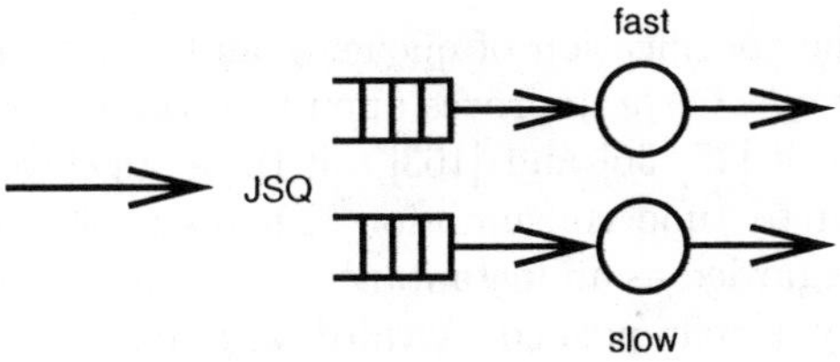

Figure 1.1. A Queueing System with 'Join the Shortest Queue' Strategy

times differ drastically. In such a scenario, if always the slower server will be selected in case of nondeterminism, the system will behave less productive than if always the faster server would be chosen in case of equally filled queues. Thus the performance of the system depends on the actual implementation of the 'JSQ' strategy, that unfortunately is not precisely fixed in neither queueing system nor Petri net notation.

With the rapidly increasing complexity of distributed system designs, another drawback of these traditional approaches became obvious. The nice visual representation of Petri nets (and queueing systems) is not really helpful for complex system designs. The main characteristics of distributed systems is that they consist of a hierarchy of autonomous (however interacting) components. These components aim to fulfil a common task transparent for the user. But since Petri nets and queueing systems are *flat* models, this inherent structure cannot be adequately represented [20].

Research on process algebra [116, 145, 14] has inspired efforts to introduce *compositionality* into this framework [110], an approach where complex models are constructed in a stepwise fashion out of smaller building blocks. An *abstraction* mechanism provides means to treat the components as *black boxes*, making their internal structure invisible. Nowadays, there is a growing awareness that if there is hope that industrial-size designs can be handled by a performance estimation methodology, *this methodology must be based on the two premises of compositionality and abstraction.*

In order to meet this *challenge of compositional performance estimation* the combination of process algebra and Markov chains has been brought up in the pioneering tutorial of Götz *et al.* [79]. This work has opened the floodgates for a variety of approaches that are nowadays subsumed as *stochastic process algebras* [42, 107, 114, 163, 25]. Significantly different from queueing systems and Petri nets, these algebras provide *composition operators* to define complex Markov chains stepwise. The practicability of this approach has been illustrated in several case studies (see e.g. [106, 71, 99, 177]), and important progress has been made in exploiting compositionality also for analysis purposes [115, 99, 114, 140, 139].

However, these approaches are still lacking important conceptual features. First, the idea of abstraction (and its benefits) are less developed for stochastic process algebras, compared to their non-stochastic ancestors. In addition, with respect to the specification of queueing and scheduling strategies none of the approaches *effectively* improves upon traditional techniques. With the notable exception of [25, 36] and [163], all these algebras treat interaction inseparably linked to time consumption. Therefore the scheduling of jobs which is usually regarded as an instantaneous decision, can only be modelled if compositionality is relinquished. A third, severe problem is indeed also a consequence of this strict link between time consumption and interaction: There is no obvious interpretation of the nature of the time elapsing during interactions between components [113]. We will argue that none of the approaches addresses this issue in an entirely satisfactory manner.

The calculus of *Interactive Markov Chains* developed in this monograph takes a drastically different view. Interaction and time consumption are *strictly separated*: each interaction takes place instantaneously. We will show that this separation is the key to improve upon all the above mentioned shortcomings. This possibility is indeed just a side result of a more general distinguishing property of the calculus. Interactive Markov Chains are a proper, conservative extension of both, non-stochastic process algebra as well as Markov chains, continuous-time Markov chains to be precise. In particular, and different from the above mentioned approaches, the concept of nondeterminism is retained. Nondeterminism is a valuable tool in a compositional specification methodology. It is useful to reflect that many system designs behave nondeterministically at a certain level of abstraction (as in the above example), unless all relevant information is specified. Furthermore, nondeterminism allows one to concisely model implementation freedom as well as the impact of different influences of an external environment [41, 145, 40]. On the other hand, since nondeterminism is absent in Markov chains, we shall put emphasis on the issue how to transform Interactive Markov Chains into Markov chains in order to estimate performance.

To represent the basic ideas of our calculus, we shall find it convenient to use a graphical notation that resembles Petri nets. We use this notation for the basic components of Interactive Markov Chains, and introduce operators to compose components. In the beginning we just work with two composition operators, ***parallel composition*** and ***abstraction*** that appear as the main operators allowing for a hierarchical description of complex Interactive Markov Chains. We will later introduce means to specify also the leaves of such a hierarchy of interacting components with the help of some additional operators. In this way, we obtain an entirely textual way of describing a system. This textual notation, a formal language, sets the ground for an algebraic theory of Interactive Markov Chains, that will be developed. In addition, we exemplify how the textual notation can be enriched with further operators useful to specify difficult dependencies that occur frequently.

The algebra developed in this monograph is mainly concerned with the question how to decide whether two Interactive Markov Chains can be regarded to be equivalent. It will be an important observation that ***bisimulation*** *equivalences*, fundamental in non-stochastic process algebra, are of great value for Interactive Markov Chains as well. In particular, *weak bisimilarity* turns out to be crucial in order to exploit abstraction and to enable performance estimation of a given chain by transformation into a Markov chain.

Due to their importance, we shall put emphasis on the algebraic properties enjoyed by strong and weak bisimilarity, as well as *weak congruence*, a slight variant of weak bisimilarity. Specifically, we develop *sets of equational laws* that are *sound* with respect to strong bisimilarity, respectively weak congruence. This means that the laws can be applied to transform Interactive Markov Chains into equivalent ones. Furthermore, we shall prove that the set of laws is *complete* for arbitrary expressions of the formal language. 'Completeness' refers to the fact that any two equivalent Interactive Markov Chains can be transformed into each other by application of the laws. Hence, strong bisimilarity and weak congruence can be decided by purely syntactic transformation. It is worth to mention that this completeness result is unique, not only in the area of stochastic process algebra. In fact, it also solves an open problem for a class of non-stochastic process calculi with priorities.

Apart from this algebraic theory we concentrate on algorithmic issues related to the computation of bisimilarities. We develop strategies that efficiently factor the state space of an Interactive Markov Chain into equivalence classes of bisimilar states. The fact that the algorithms return equivalence classes of states has an interesting side effect. By essentially representing each class by a single state, it is possible to construct a quotient, an *aggregated* Interactive Markov Chain that is equivalent to the original one. In practice, this 'equivalence preserving state space aggregation' is highly favourable. Models of *real* systems, consisting of a hierarchy of interacting components, tend to have very large state spaces that are often even too large to be stored in memory. For such systems, any kind of analysis already fails because the state space to be analysed remains unknown. However, *substitutivity*, one of the central properties we shall demand for bisimilarities, allows one to replace arbitrary interacting components by their aggregated representatives, since the above aggregation is bisimilarity preserving. This implies that the size of the state space of the hierarchical system is reduced as well, but with the additional gain that it avoids the construction of the original state space which would be prohibitively large. This general technique is known as *compositional aggregation* and has successfully been applied to a variety of very large non-stochastic systems in order to verify behavioural properties [29, 82, 83, 44, 132]. In the context of Interactive Markov Chains we will apply compositional aggregation in order to study the stochastic behaviour of systems with more than 10 million states. Since continuous-time Markov chains are a proper subset of Interactive Markov Chains, this strategy is of potential interest for anyone aiming to solve large Markov chains.

1.3 Roadmap

The monograph is divided in 7 chapters, each of them concluded by a discussion of the contribution of the respective chapter in the view of related material.

- In Chapter 2 we give an intuitive introduction into the basic concepts of process algebra to set the ground for understanding the following chapters. We introduce a rudimentary calculus of *interactive processes* and discuss useful equivalences on this model, essentially recalling the motivation for strong and weak bisimilarity. We also sketch efficient algorithms to compute these equivalences. A small case study is used to illustrate the benefits of these ingredients, in particular for compositional aggregation techniques.
- Chapter 3 focuses on *Markov chains* of both discrete and continuous nature. Based on insight gained in Chapter 2 we discuss useful equivalences on either of these models, touching on the notion of *lumpability*, and provide efficient algorithms for them.
- In Chapter 4 we introduce the central formalism of this monograph, *Interactive Markov Chains*, that arise as an integration of interactive processes and *continuous-time* Markov chains. We justify our decision to treat both ingredients as orthogonal as possible by evaluating other approaches that have appeared in the literature. We develop strong and weak bisimilarity on our model and provide algorithms to compute these relations. The compositional application of these algorithms is discussed by means of a small example.
- Chapter 5 develops an *algebra* of Interactive Markov Chains. In particular, we present a language to generate Interactive Markov Chains and derive a sound and complete algebraic theory for arbitrary expressions of the language. In addition we introduce further operators that are useful to enhance modelling and analysis convenience for real applications.
- Chapter 6 discusses the achievements of the previous chapters by means of some small and medium size *case studies*. In particular we focus on different ways to exploit compositionality, specifically using either compositional aggregation or applying the algebraic theory. We will also highlight the limitations of our concept with respect to performance estimation in the presence of nondeterminism.
- Chapter 7 concludes by addressing the question whether the *challenge of compositional performance and dependability estimation* has been met, and to what extent it improves on previous work addressing similar challenges. Furthermore, we point out relevant directions for further work.

2. Interactive Processes

In this chapter we introduce the basic framework that will be used throughout the whole monograph. We introduce labelled transition systems together with two composition operators, abstraction and parallel composition. We then proceed discussing useful equivalences on this model, essentially recalling the motivation for strong and weak bisimilarity. We also sketch efficient algorithms to compute these equivalences. The contents of this chapter is a collection from [145, 27, 154, 125, 86, 76, 73, 22]. Its purpose is to give an intuitive understanding of the key features of process algebra. A reader familiar this features is invited to fleetingly glance through this chapter in order to catch a glimpse of the notations used.

2.1 Transition Systems and Interactive Processes

State-transition diagrams, automata and similar models are widely used to describe the dynamic behaviour of systems. In the context of process algebras, they usually appear in a specific form, labelled transition systems. A transition system consists of a set of states S and a set of possible state changes. The latter set is given as a binary relation on states, i.e., a subset of the crossproduct $S \times S$. Intuitively a pair of states (P, Q) is contained in this relation if it is possible to change from state P to Q in a single step. It is convenient to denote such a transition relation with an arrow (e.g. $\longrightarrow$), because $(P, Q) \in \longrightarrow$ can be written in infix notation, $P \longrightarrow Q$, nicely representing the possible state change between P and Q.

Labelled transition systems are a particular variant where state changes are conditioned on occurrences of actions drawn from a set *Act*. A state change between P and Q then implies the occurrence of a specific action. Therefore, the transition relation $\longrightarrow$ is a subset of $S \times Act \times S$ rather than a binary relation on S. Again a kind of infix notation, $P \xrightarrow{\mathsf{a}} Q$, is convenient to denote $(P, \mathsf{a}, Q) \in \longrightarrow$. Here the action appears as the label of the transition explaining the term 'labelled transition' system.

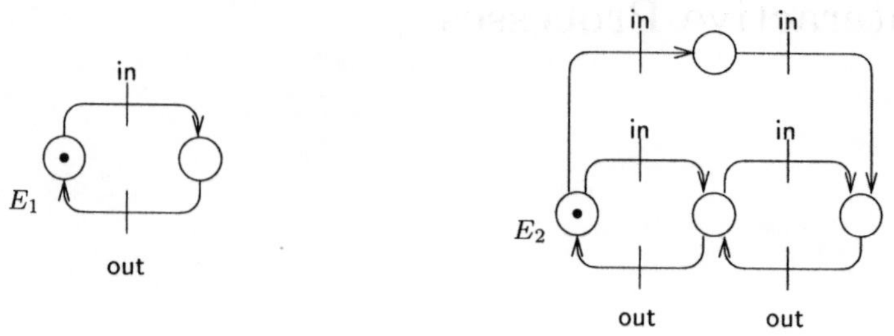

Figure 2.1. Two interactive processes

Definition 2.1.1. *A labelled transition system is a triple $(S, Act, \longrightarrow\!\!\!\!\!\rightarrow)$, where*
- *S is a nonempty set of states,*
- *Act is a set of actions, and*
- *$\longrightarrow\!\!\!\!\!\rightarrow \subset S \times Act \times S$ is a set of labelled transitions, sometimes also called* interactive *transitions.*

In order to use labelled transition systems as an operational model of systems it is common practice to identify a specific initial state P in the transition system where operation starts. A transition system with an initial state is called an interactive process.

Definition 2.1.2. *An interactive process is a quadruple $(S, Act, \longrightarrow\!\!\!\!\!\rightarrow, P)$, where $(S, Act, \longrightarrow\!\!\!\!\!\rightarrow)$ is a labelled transition system and $P \in S$ is the initial state.*

Example 2.1.1. Figure 2.1 contains two examples of interactive processes. States are represented as circles labelled with identifiers from S. We adopt the convention to omit labels of states if they are not required to understand the example. We use $\odot$ to denote the initial state. The first interactive process, with initial state E_1, is a simple one-place buffer. It accepts data via the action in *and releases it upon the action* out*. The right process is able to buffer two values but with a slightly unusual restriction. From its initial state E_2 there is an implicit* nondeterministic *decision between two actions* in *that is taken upon the acceptance of the first datum. The lower branch leads to a usual two place buffer behaviour, whereas on the upper branch a datum cannot be released before two data have been accepted.*

To avoid to overburden our ideas with mathematical notation we do not insist on the distinction between the initial state of an interactive process and the process itself. Unless stated otherwise we assume that all transition systems that appear in the remainder of the monograph are contained somewhere in an immense transition system $(\mathcal{S}^{\text{all}}, Act^{\text{all}}, \longrightarrow\!\!\!\!\!\rightarrow)$ that is (at least) the union of all those we will discuss. $\mathcal{S}^{\text{all}}$ is a superset of all appearing states and Act^{all} comprises all appearing actions. We shall use this immense transition system and its elements in the sequel. In fact, we will give its definition in Chapter 5 (on page 108).

We are able to exactly identify an interactive process $(S, Act, \twoheadrightarrow, P)$ by its initial state P in this immense transition system, because S can be defined as the set of states reachable from P, $\twoheadrightarrow$ is the transition relation restricted to S and Act contains all actions labelling this relation. S is called the state space of P.

We use the term 'interactive' process in order to indicate that a process may interact with other processes to exchange data, or, more general, to cooperate in order to achieve a common goal. Several different shapes of process interaction have been studied in the literature of process algebra [141, 145, 41, 117]. The most distinctive features are asynchronous versus synchronous and binary versus multi-party interaction. Our way of interaction is synchronous multi-party interaction as used, for instance, in LOTOS [27, 122], the specification language standardised by the ISO.

In order to give processes the potential of interaction, we introduce a parallel composition operator that is equipped with those actions $\mathsf{a}_1 \ldots \mathsf{a}_n$ that both processes have to synchronise on. All other actions can be performed independently by either of the processes. If P and Q are two processes, such synchronous parallel composition is denoted $P \mathrel{\|_{\mathsf{a}_1 \ldots \mathsf{a}_n}} Q$. By varying the set of synchronising actions, parallel composition ranges from full synchrony when the set comprises all the possible actions, to arbitrary interleaving when the set is the empty (in this case we use the concise notation $P \,|||\, Q$). The intuition of this operator can be made more precise by the following (still informal) properties.

- A state change of $P \mathrel{\|_{\mathsf{a}_1 \ldots \mathsf{a}_n}} Q$ is possible if P may change to, say P', on the occurrence of an action a that is not contained in $\{\mathsf{a}_1 \ldots \mathsf{a}_n\}$. The result of the state change is $P' \mathrel{\|_{\mathsf{a}_1 \ldots \mathsf{a}_n}} Q$, since only P has changed state.
- Symmetrically, a state change of $P \mathrel{\|_{\mathsf{a}_1 \ldots \mathsf{a}_n}} Q$ is also possible if Q may change to some Q', on the occurrence of an action a that is not contained in $\{\mathsf{a}_1 \ldots \mathsf{a}_n\}$, resulting in $P \mathrel{\|_{\mathsf{a}_1 \ldots \mathsf{a}_n}} Q'$.
- In order to be able to interact on an action a contained in $\{\mathsf{a}_1 \ldots \mathsf{a}_n\}$, both P and Q have to be able to perform a and thereby evolve to some P' and Q'. If this prerequisite is fulfilled, $P \mathrel{\|_{\mathsf{a}_1 \ldots \mathsf{a}_n}} Q$ may in a single step change state to $P' \mathrel{\|_{\mathsf{a}_1 \ldots \mathsf{a}_n}} Q'$.
- No other transitions are possible for $P \mathrel{\|_{\mathsf{a}_1 \ldots \mathsf{a}_n}} Q$.

We formalise these requirements in the definition of a transition relation for the process $P \mathrel{\|_{\mathsf{a}_1 \ldots \mathsf{a}_n}} Q$. The first three requirements will be used to define the respective parts of the transition relation of $P \mathrel{\|_{\mathsf{a}_1 \ldots \mathsf{a}_n}} Q$. Each part will allow to derive a certain transition of $P \mathrel{\|_{\mathsf{a}_1 \ldots \mathsf{a}_n}} Q$ if P and/or Q exhibit a certain transition. The last property will be reflected by requiring that the relation itself is the least relation satisfying the definition, i.e., it does not possess non-derivable transitions.

Definition 2.1.3. *Let P and Q be two interactive processes with state spaces S_P and S_Q. Parallel composition of P and Q on actions $a_1 \ldots a_n$ is (again) an interactive process $(S, Act, \longrightarrow, P \overline{\underline{a_1 \ldots a_n}} Q)$, where*

- $S := \{P' \overline{\underline{a_1 \ldots a_n}} Q' \mid P' \in S_P \wedge Q' \in S_Q\}$,
- $\longrightarrow$ *is the least relation satisfying that*
  if $a \notin \{a_1 \ldots a_n\}$, then

$$P' \xrightarrow{a} P'' \text{ implies } P' \overline{\underline{a_1 \ldots a_n}} Q' \xrightarrow{a} P'' \overline{\underline{a_1 \ldots a_n}} Q',$$

$$Q' \xrightarrow{a} Q'' \text{ implies } P' \overline{\underline{a_1 \ldots a_n}} Q' \xrightarrow{a} P' \overline{\underline{a_1 \ldots a_n}} Q'',$$

if $a \in \{a_1 \ldots a_n\}$, then

$$P' \xrightarrow{a} P'' \wedge Q' \xrightarrow{a} Q'' \text{ implies } P' \overline{\underline{a_1 \ldots a_n}} Q' \xrightarrow{a} P'' \overline{\underline{a_1 \ldots a_n}} Q''.$$

The style of this definition goes back to Plotkin [157], it is usually called structured operational semantics since it defines the operational behaviour of a process inductively over its structure. Note that each state of a parallel composition possesses a specific syntactic structure, since it contains the syntactic operator $\overline{\underline{}}$.

It should be obvious that this definition reflects the informal requirement that we have stated before. Indeed, it is more precise than the informal definition. There, nothing has been said about the concrete actions that are exhibited by a parallel composition, we have only focused on the question when a state change should be possible, but not discussed the resulting action label. In Definition 2.1.3 we have decided to reuse the label a occuring in the preconditions. In other words, actions exhibited by parallel composition are those exhibited by their components regardless of the fact if they are obtained from synchronisation or not. The uninitiated reader may find this solution straightforward, but other solution are equally reasonable; [143] contains a discussion of this topic. Our choice (borrowed from [153, 27]) is the key to enable multi-party synchronisation, where the same action a can be used in further parallel composition contexts.

Example 2.1.2. Figure 2.2 shows parallel composition of two interactive processes E_3 and E_5. The resulting interactive process $E_3 \overline{\underline{mid}} E_5$, obtained by applying Definition 2.1.3, is depicted underneath. Recall that we blur the distinction between an interactive process and its initial state.

In general, the possibility that multiple processes can synchronise on the same action has been proven to be convenient from the system engineering point of view. But, with the concept we have introduced so far, arbitrary actions of an interactive process can be used for synchronisation at an arbitrary level of specification. This is undesirable, since modern top-down (or bottom-up) system design methodology tries to use a black box view on components,

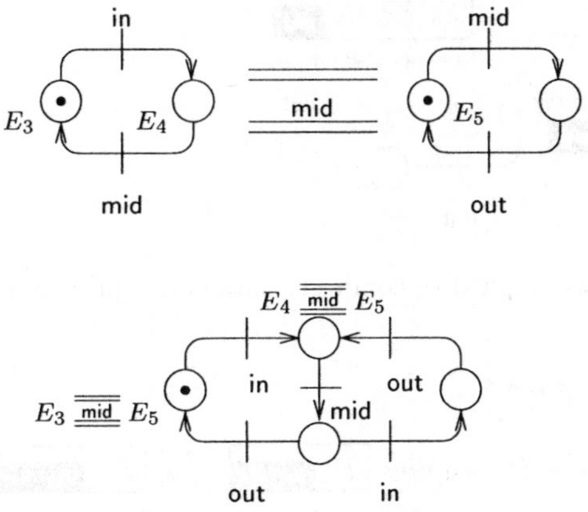

Figure 2.2. Parallel composition of two interactive processes

in order to abstract from internal details that are irrelevant at a certain level of design.

The concept of abstraction arises natural in the framework of process algebra. The key to this feature is a distinguished action, usually named τ, that symbolises an internal activity, e.g. an internal state change that only depends on local variables. Actions different from τ are called external actions. In the context of parallel composition $P \overline{\overline{\underline{\underline{\,\mathsf{a}_1 \ldots \mathsf{a}_n\,}}}} Q$, internal activities of P should be definitely independent from internal actions of Q. Hence, synchronisation on internal actions has to be strictly ruled out. In other words, it is not allowed that τ occurs in $\{\mathsf{a}_1 \ldots \mathsf{a}_n\}$.

This already ensures that internal actions remain internal in the context of arbitrary levels of parallel composition. However, when building a large system it is desirable to internalise actions in a stepwise fashion after they have been used for (multi-party) interaction. This ensures the above mentioned black box view on a process, where only those actions are accessible for synchronisation that are still needed and therefore not internalised. A specific operator, called ***abstraction operator***, is dedicated to this purpose. For a given process P and actions $\mathsf{a}_1 \ldots \mathsf{a}_n$, it internalises those actions by simply renaming them into τ. We use $\boxed{P\ \mathsf{a}_1 \ldots \mathsf{a}_n}$ to denote this abstraction.

Definition 2.1.4. *Let P be an interactive process with state space S_P. Abstraction of actions $\mathsf{a}_1 \ldots \mathsf{a}_n$ in P is an interactive process $\big(S, Act \longrightarrow,$ $\boxed{P\ \mathsf{a}_1 \ldots \mathsf{a}_n}\big)$, where*

- $S := \left\{ \boxed{P'\ \mathsf{a}_1 \ldots \mathsf{a}_n} \,\middle|\, P' \in S_P \right\}$
- $\longrightarrow$ *is the least relation satisfying that*
  if $\mathsf{a} \notin \{\mathsf{a}_1 \ldots \mathsf{a}_n\}$, *then*

$$P' \xrightarrow{\mathsf{a}} P'' \text{ implies } \boxed{P'\ \mathsf{a}_1 \ldots \mathsf{a}_n} \xrightarrow{\mathsf{a}} \boxed{P''\ \mathsf{a}_1 \ldots \mathsf{a}_n},$$

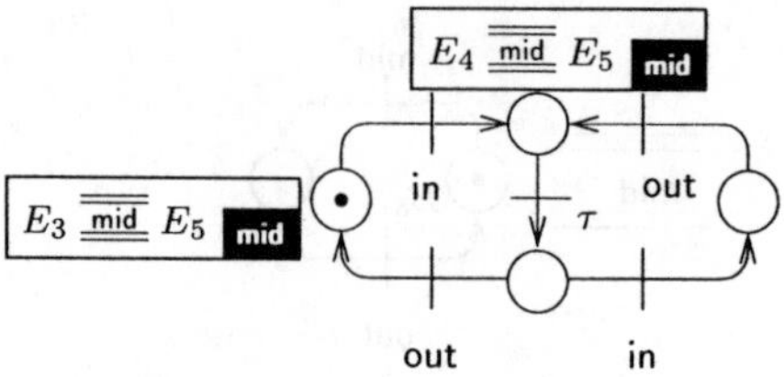

Figure 2.3. Abstraction applied to composed interactive processes

if $\mathsf{a} \in \{\mathsf{a}_1 \ldots \mathsf{a}_n\}$, *then*

$$P' \xrightarrow{\mathsf{a}} P'' \textit{ implies } \boxed{P' \;\; \mathsf{a}_1 \ldots \mathsf{a}_n} \xrightarrow{\tau} \boxed{P'' \;\; \mathsf{a}_1 \ldots \mathsf{a}_n}\,.$$

Example 2.1.3. In Figure 2.3 we have depicted the result of internalising the action mid *in the process* $E_3 \mathbin{\underline{\overline{\mathsf{mid}}}} E_5$ *by means of abstraction. The behaviour of the resulting process can be best described as a two place buffer, composed out of two one-place buffers,* E_3 *and* E_5*. The first accepts data with action* in*. These are* internally *passed to the process* E_5 *which is now ready to output them.* E_3 *may accept a second datum, but before it can pass the datum to* E_5 *this process has to get rid of the datum accepted earlier. So, this process appears as a two-place buffer but is internally realized as a serial connection of two one-place buffers.*

We will later rely on a more compact notation for structural operational definitions. Instead of using the form

if A*, then* B *implies* C we simply write $\dfrac{B}{C}\ A,$

and we call the latter a structural operational rule. Applying this abbreviation to Definition 2.1.3 and 2.1.4 leads to the rules given in Table 2.1. When the syntactic structure of an interactive process is becoming more complex, this rule scheme is conveniently used in 'proof trees' for transitions.

Example 2.1.4. In order to prove that $\boxed{E_3 \mathbin{\underline{\overline{\mathsf{mid}}}} E_5 \;\; \mathsf{mid}}$ *possesses an outgoing transition labelled with action* in *we construct the following proof tree:*

$$\dfrac{\dfrac{E_3 \xrightarrow{\mathsf{in}} E_4}{E_3 \mathbin{\underline{\overline{\mathsf{mid}}}} E_5 \xrightarrow{\mathsf{in}} E_4 \mathbin{\underline{\overline{\mathsf{mid}}}} E_5}}{\boxed{E_3 \mathbin{\underline{\overline{\mathsf{mid}}}} E_5 \;\; \mathsf{mid}} \xrightarrow{\mathsf{in}} \boxed{E_4 \mathbin{\underline{\overline{\mathsf{mid}}}} E_5 \;\; \mathsf{mid}}}$$

Table 2.1. Structural operational rules for interactive processes

$$\frac{P \xrightarrow{\mathsf{a}} P'}{P \overset{\mathsf{a}_1\ldots\mathsf{a}_n}{\equiv} Q \xrightarrow{\mathsf{a}} P' \overset{\mathsf{a}_1\ldots\mathsf{a}_n}{\equiv} Q} \qquad \mathsf{a} \notin \{\mathsf{a}_1\ldots\mathsf{a}_n\}$$

$$\frac{Q \xrightarrow{\mathsf{a}} Q'}{P \overset{\mathsf{a}_1\ldots\mathsf{a}_n}{\equiv} Q \xrightarrow{\mathsf{a}} P \overset{\mathsf{a}_1\ldots\mathsf{a}_n}{\equiv} Q'} \qquad \mathsf{a} \notin \{\mathsf{a}_1\ldots\mathsf{a}_n\}$$

$$\frac{P \xrightarrow{\mathsf{a}} P' \qquad Q \xrightarrow{\mathsf{a}} Q'}{P \overset{\mathsf{a}_1\ldots\mathsf{a}_n}{\equiv} Q \xrightarrow{\mathsf{a}} P' \overset{\mathsf{a}_1\ldots\mathsf{a}_n}{\equiv} Q'} \qquad \mathsf{a} \in \{\mathsf{a}_1\ldots\mathsf{a}_n\}$$

$$\frac{P \xrightarrow{\mathsf{a}} P'}{\boxed{P\ \blacksquare_{\mathsf{a}_1\ldots\mathsf{a}_n}} \xrightarrow{\mathsf{a}} \boxed{P'\ \blacksquare_{\mathsf{a}_1\ldots\mathsf{a}_n}}} \qquad \mathsf{a} \notin \{\mathsf{a}_1\ldots\mathsf{a}_n\}$$

$$\frac{P \xrightarrow{\mathsf{a}} P'}{\boxed{P\ \blacksquare_{\mathsf{a}_1\ldots\mathsf{a}_n}} \xrightarrow{\tau} \boxed{P'\ \blacksquare_{\mathsf{a}_1\ldots\mathsf{a}_n}}} \qquad \mathsf{a} \in \{\mathsf{a}_1\ldots\mathsf{a}_n\}$$

2.2 Equivalences on Interactive Processes

From the very beginning an essential part of process algebra theory has been devoted to the development of equivalence notions. The starting point of all process algebraic equivalences is the observation that different processes may exhibit the same behaviour. This behaviour-oriented (as opposed to state-oriented) point of view directly implies that the labelling of states is inherently negligible whereas the labelling of transitions is not. This is common for all process algebraic equivalences. The differences among the variety of equivalences are caused by different ideas about what is a distinguishable part of the behaviour. R.J. van Glabbeek has extensively studied different notions of an experimenter that interacts with an interactive process in order to determine its behaviour [75, 74]. Instead of recapitulating his work we sketch here some commonly used equivalences and try to address the question what makes an equivalence a good equivalence. This question will be raised again in later chapters, where probabilities and probability distributions come into play. First, we consider so called 'strong' equivalences, where internal and external actions are treated in the same way. Afterwards we discuss 'weak' equivalences, that aim to abstract from internal state changes as much as possible.

2.2.1 Strong Equivalences

Since a transition system is essentially an automaton, language equivalence of automata is surely an important notion. Two automata are language equiva-

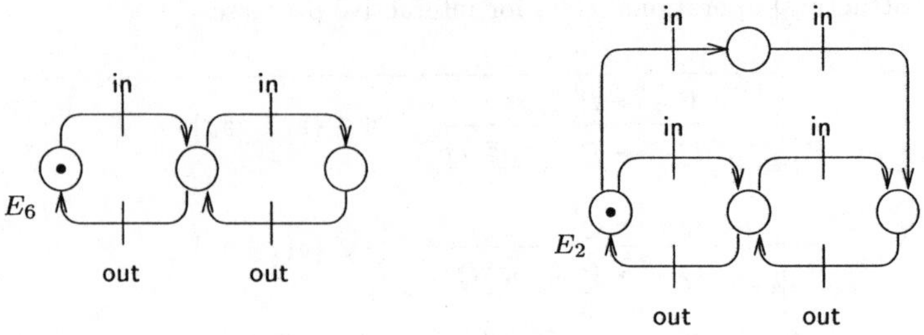

Figure 2.4. Two interactive processes with equivalent traces

lent if they accept the same language, i.e., the same set of finite sequences over Act^{all}. In the context of process algebra this relation is called trace equivalence. It is defined using $P \xrightarrow{a_1} \xrightarrow{a_2} \cdots \xrightarrow{a_n}$ to denote that there are some processes $P_1, P_2, \ldots, P_n$ such that $P \xrightarrow{a_1} P_1 \xrightarrow{a_2} P_2 \cdots P_{n-1} \xrightarrow{a_n} P_n$.

Definition 2.2.1. *Let P and Q be two interactive processes. P and Q are strong trace equivalent, written $P \sim_{tr} Q$, if*

$$P \xrightarrow{a_1} \xrightarrow{a_2} \cdots \xrightarrow{a_n} \quad \textit{if and only if} \quad Q \xrightarrow{a_1} \xrightarrow{a_2} \cdots \xrightarrow{a_n}.$$

Example 2.2.1. Consider the process depicted in Figure 2.4. E_6 describes a usual two place buffer. E_2 has already appeared in Figure 2.1. Depending on the in *branch taken E_2 may loose the possibility to output the first datum before a second has been accepted. However, both interactive processes are trace equivalent according to Definition 2.2.1.*

This example gives some insight into the weaknesses of trace equivalence. The process E_2 is not always able to release a datum after it has accepted one whereas E_6 always is. This is problematic, if E_2 is put in a context where an output is required for synchronisation after every input, (with E_1 of Figure 2.1, for instance, synchronising on action out and in). E_6 would be able to interact on action out after action in has occured, whereas E_2 may be not, forcing the synchronising partner to deadlock. In other words, trace equivalence does not preserve deadlocks.

The main reason why trace equivalence is suitable for automata theory, while it does not fit with the process algebraic theory of processes, is the difference between their models of interaction. Automata theory assumes complete control of the automaton over its transitions. In the process algebraic view of processes all observable actions are under the *joint* control of the process and its environment. In this context, automata can be seen as processes with only internal actions (but not necessarily labelled with τ), or alternatively, as a process with a completely cooperative environment, i.e. one that is always capable of synchronising on the action of the automatons' choice. The standard interaction between process algebraic processes, however, assumes an interactive resolution of choices, at least between observable transitions. This means

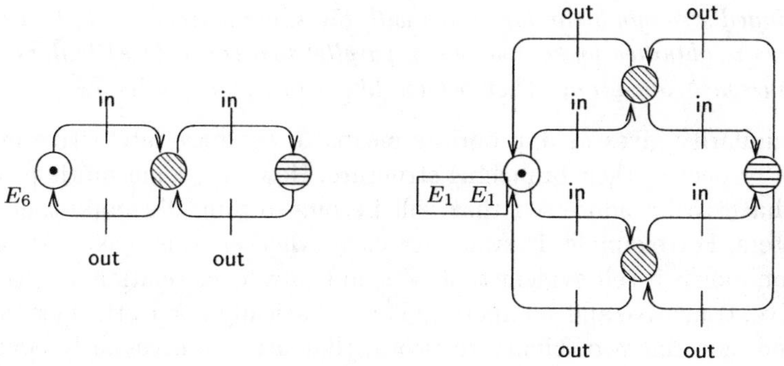

Figure 2.5. Two strongly bisimilar interactive processes

that a process cannot select a transition labelled with an observable action if this action is not also enabled by the environment. If several such jointly enabled transitions exist, then the choice is made nondeterministically.

In the presence of concurrent composition it is natural that two transition systems should be equivalent if and only if they interact in the same way with arbitrary environments. In view of the above, that means the way in which they constrain the choices between different actions is relevant. This is also referred to as the *branching (time) structure* of processes, as opposed to the *linear (time) structure* of classical automata.

Milner and Park [145, 155] have introduced the most important class of equivalence relations that respect the branching structure of a process and hence *are* deadlock preserving. It is the class of *bisimulation equivalences* or *bisimilarities*. Two processes are bisimilar if they can simulate each other stepwise. The word 'stepwise' describes an essential difference to trace equivalence, because bisimulation is defined inductively, only considering a single step.

Definition 2.2.2. *A binary relation $\mathcal{B}$ on $\mathcal{S}^{\text{all}}$ is a strong bisimulation if $(P,Q) \in \mathcal{B}$ implies for all* $\mathsf{a} \in Act$:

– $P \xrightarrow{\mathsf{a}} P'$ *implies* $Q \xrightarrow{\mathsf{a}} Q'$ *for some* Q' *such that* $(P',Q') \in \mathcal{B}$,
– $Q \xrightarrow{\mathsf{a}} Q'$ *implies* $P \xrightarrow{\mathsf{a}} P'$ *for some* P' *such that* $(P',Q') \in \mathcal{B}$.

Two interactive processes, P and Q, are strongly bisimilar, written $P \sim Q$, if they are contained in some strong bisimulation $\mathcal{B}$,i.e., $(P,Q) \in \mathcal{B}$.

Strong bisimilarity is the union of all strong bisimulations. In words, the above definition says that two states in a transition system are bisimilar, if they can change to again bisimilar states, and interact on the same actions in doing so.

Example 2.2.2. Following Definition 2.2.2, the two processes E_6 and $\boxed{E_1 \;\; E_1}$ *depicted in Figure 2.5 are bisimilar. To facilitate the inspection of this claim* *we have shaded strongly bisimilar states with the same pattern. Note that the right process is obtained by composing in parallel two (one-place) buffers E_1 without any synchronisation. They behave like a two-place buffer E_6.*

Strong bisimilarity gives us appropriate means to compare interactive processes with respect to their branching structure. However, some subtle problems still have to be addressed that will become technically important in later chapters. For example, it is neither clear whether $\sim$ is itself a strong bisimulation, nor is it self-evident that $\sim$ is an equivalence relation, i.e., that it is reflexive, transitive and symmetric. This is basically due to the fact that $\sim$ is defined as a union of binary relations, that are not necessarily equivalence relations. In addition, even the union of equivalence relations is not necessarily transitive.

Lemma 2.2.1. *Strong bisimilarity*
- *is an equivalence relation on $\mathcal{S}^{\text{all}}$,*
- *is a strong bisimulation on $\mathcal{S}^{\text{all}}$, and*
- *is the largest strong bisimulation on $\mathcal{S}^{\text{all}}$.*

The style of the definition of bisimulation is sometimes called *coinductive*, since it borrows the concept of coinduction from category theory [123]. Roughly, a coinductive definition characterises the largest set (while *in*duction characterises the smallest set) satisfying an inductive definition. As we will see later, a coinductive definition is very beneficial for the development of efficient algorithms and simple proof strategies. Therefore, we shall evaluate other equivalences with respect to the question whether they are defined coinductively.

We will later also rely on an alternative characterisation of strong bisimilarity, borrowed from [76], that defines $\sim$ as the union of equivalence relations. It makes use of a (boolean) function $\gamma_{\text{o}} : S \times Act \times 2^S \mapsto \{\texttt{true}, \texttt{false}\}$. $\gamma_{\text{o}}(P, \mathsf{a}, C)$ is true iff P can evolve to a state contained in a set of states C by interaction on a.

Definition 2.2.3.

$$\gamma_{\text{o}}(P, \mathsf{a}, C) := \begin{cases} \texttt{true} & \textit{if there is } P' \in C \textit{ such that } P \xrightarrow{\mathsf{a}} P', \\ \texttt{false} & \textit{otherwise.} \end{cases}$$

With this definition, bisimilarity can be expressed as 'having the same interaction possibilities to evolve into the same set of behaviours', where the latter sets are classes of equivalent behaviour.

Lemma 2.2.2. *An equivalence relation $\mathcal{E}$ on S is a strong bisimulation if $(P, Q) \in \mathcal{E}$ implies for all $\mathsf{a} \in Act$ and all equivalence classes C of $\mathcal{E}$ that*

$$\gamma_{\text{o}}(P, \mathsf{a}, C) = \gamma_{\text{o}}(Q, \mathsf{a}, C).$$

Example 2.2.3. If we use  *to denote the set of states shaded like*  *in Figure 2.5, and similar with*  *and* *, then each of these sets is a class of an equivalence relation $\mathcal{E}$ satisfying Lemma 2.2.2. In particular, we compute the following values for each of the states in the respective class. All other combinations return* `false`.

$\gamma_O($  $, \mathtt{in},$  $) = \mathtt{true}$ $\qquad$ $\gamma_O($  $, \mathtt{out},$  $) = \mathtt{true}$

$\gamma_O($  $, \mathtt{in},$ 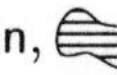 $) = \mathtt{true}$ $\qquad$ $\gamma_O($  $, \mathtt{out},$  $) = \mathtt{true}$

Let us now turn our attention to another important property of bisimilarity. Since we have introduced two composition operators, it is interesting to investigate whether $\sim$ gives rise to a proper notion of equality. A proper notion of equality should be preserved in the context of composition. In the above example, we have seen that E_6 and $\boxed{E_1 \;\; E_1}$ are bisimilar. They both describe the behaviour of a buffer with two places. However, it is not clear whether we can use either of them in a larger composition context and obtain again equivalent overall behaviours. As a general property, this is highly desirable, because it allows to replace a component of a parallel composition by an equivalent one, without affecting the behaviour of the complete parallel composition. The general requirement is called *substitutivity* of an equivalence relation. Algebraically it boils down to show that $\sim$ is a *congruence* relation with respect to the operators.

Theorem 2.2.1. *Strong bisimilarity is substitutive with respect to parallel composition and abstraction, i.e.,*

$$
\begin{array}{lll}
P_1 \sim P_2 & \textit{implies} & P_1 \;\underline{\overline{\mathtt{a}_1\ldots\mathtt{a}_n}}\; P_3 \;\sim\; P_2 \;\underline{\overline{\mathtt{a}_1\ldots\mathtt{a}_n}}\; P_3, \\
P_1 \sim P_2 & \textit{implies} & P_3 \;\underline{\overline{\mathtt{a}_1\ldots\mathtt{a}_n}}\; P_1 \;\sim\; P_3 \;\underline{\overline{\mathtt{a}_1\ldots\mathtt{a}_n}}\; P_2, \textit{ and} \\
P_1 \sim P_2 & \textit{implies} & \boxed{P_1 \;\; \mathtt{a}_1\ldots\mathtt{a}_n} \;\sim\; \boxed{P_2 \;\; \mathtt{a}_1\ldots\mathtt{a}_n}\,.
\end{array}
$$

Strong bisimilarity shares this substitutivity property with trace equivalence, but, in addition, it respects the branching structure of a process and therefore it does preserve deadlocks. Furthermore, it is defined coinductively. These properties are the main reasons why bisimilarity is central in the area of equivalences for process algebras. It is easy to define, easy to prove and is mathematically elegant.

2.2.2 Weak Equivalences

Hitherto we have only discussed equivalences that treat internal actions in the same way as external actions. In particular, internal actions have to be simulated stepwise in order to be able to establish strong bisimilarity between two processes. This is surely counterintuitive, because strong bisimilarity is meant to characterise the behaviour of two processes by means of their potential of interaction. But since internal actions are precluded from possible

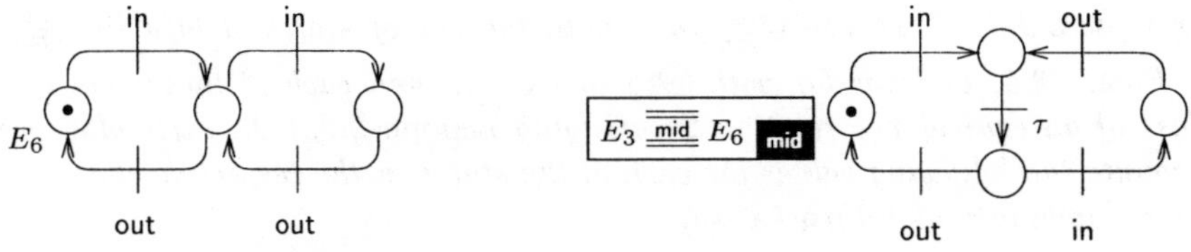

Figure 2.6. Not strongly bisimilar interactive processes

interactions, there seems to be no specific need to distinguish between two processes only because one of them may at some state do an internal state change while the other does not.

Example 2.2.4. We have discussed before that a serial connection of two one-place buffers as in $E_3 \,\|\text{mid}\|\, E_5$ mid *behaves like a two-place buffer. However, it is not possible to construct a strong bisimulation between this process and* E_6 *even though* E_6 *appears as a canonical representation of a two place buffer. The reason is that we have to bisimulate an internal* $\xrightarrow{\tau}$ *transition of* $E_3 \,\|\text{mid}\|\, E_5$ mid*, but* E_6 *does not possess such a transition at all (Figure 2.6).*

In order to abstract from internal moves, it is natural to neglect them as far as they do not influence the future behaviour of a process. It seems wise to introduce a notion of observable steps of a process, that consist of a single *external* action preceded and followed by an arbitrary number (including zero) of internal steps [145]. Technically this is achieved by deriving a 'weak' transition relation $\Longrightarrow$ from the 'strong' transition relation $\longrightarrow$.

Definition 2.2.4. *For internal actions,* $\overset{\tau}{\Longrightarrow}$ *is defined as the reflexive and transitive closure* $\xrightarrow{\tau}^{*}$ *of the relation* $\xrightarrow{\tau}$ *of internal transitions. External weak transitions are then obtained by defining* $\overset{a}{\Longrightarrow}$ *to denote* $\overset{\tau}{\Longrightarrow}\xrightarrow{a}\overset{\tau}{\Longrightarrow}$.

Note that a weak internal transition $\overset{\tau}{\Longrightarrow}$ is possible without actually performing an internal action, because $\xrightarrow{\tau}^{*}$ contains the reflexive closure, i.e., the possibility not to move at all. In contrast, a weak external transition $\overset{a}{\Longrightarrow}$ must contain exactly one transition $\xrightarrow{a}$ preceded and followed by arbitrary (possibly empty) sequences of internal moves.

Example 2.2.5. For the processes E_6 *and* $E_3 \,\|\text{mid}\|\, E_5$ mid *the weak transition relation is depicted in Figure 2.7.*

With this relation, weak trace equivalence and weak bisimilarity arise by simply replacing strong by weak transitions in Definition 2.2.1, respectively Definition 2.2.2. Since weak trace equivalence inherits the problems of its strong counterpart, we are not interested in this relation.

Definition 2.2.5. *A binary relation* $\mathcal{B}$ *on* $\mathcal{S}^{\text{all}}$ *is a weak bisimulation if* $(P, Q) \in \mathcal{B}$ *implies for all* $a \in Act$:

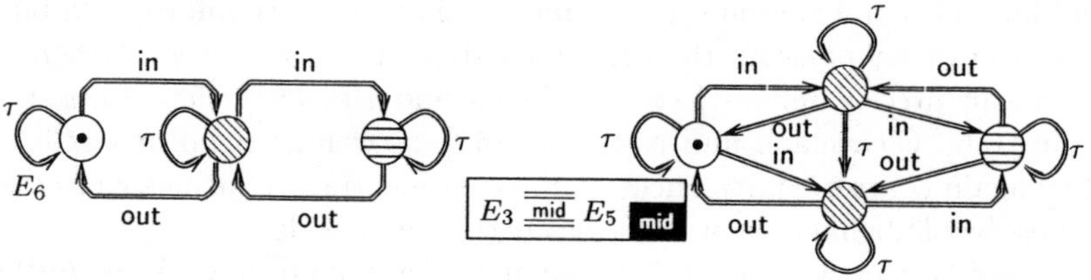

Figure 2.7. Two weakly bisimilar interactive processes

– $P \overset{\mathsf{a}}{\Longrightarrow} P'$ *implies* $Q \overset{\mathsf{a}}{\Longrightarrow} Q'$ *for some* Q' *such that* $(P', Q') \in \mathcal{B}$,
– $Q \overset{\mathsf{a}}{\Longrightarrow} Q'$ *implies* $P \overset{\mathsf{a}}{\Longrightarrow} P'$ *for some* P' *such that* $(P', Q') \in \mathcal{B}$.

Two interactive processes, P and Q, are weakly bisimilar, written $P \approx Q$, if they are contained in some weak bisimulation $\mathcal{B}$.

Weak bisimilarity has the same basic properties as strong bisimilarity (cf. Lemma 2.2.1), and it also is a proper notion of equivalence, because it is a substitutive relation.

Lemma 2.2.3. *Weak bisimilarity*

– *is an equivalence relation on $\mathcal{S}^{\mathrm{all}}$,*
– *is a weak bisimulation on $\mathcal{S}^{\mathrm{all}}$, and*
– *is the largest weak bisimulation on $\mathcal{S}^{\mathrm{all}}$.*

Theorem 2.2.2. *Weak bisimilarity is substitutive with respect to parallel composition and abstraction.*

Example 2.2.6. We have pointed out that the processes E_6 and $E_3 \overset{=}{\underline{\text{mid}}} E_5$ mid are not strongly bisimilar. But they are weakly bisimilar according to Definition 2.2.5. To illustrate this we have, again, shaded bisimilar states with the same pattern. A crucial aspect is that the weak internal transition $\bigcirc \overset{\tau}{\Longrightarrow} \bigcirc$ of the right process can be simulated by the left process because $\overset{\tau}{\Longrightarrow}$ contains the reflexive closure.

This example shows that weak bisimilarity is quite an appropriate notion to compare the behaviour of components, where some actions have been internalised. Furthermore, substitutivity ensures that equivalent components can be exchanged by each other inside a larger composition context without affecting the behaviour of the composite process. In addition, weak bisimilarity is defined coinductively and it inherits deadlock preservation from strong bisimilarity.

Nevertheless, weak bisimilarity is not as outstanding among the vast number of weak relations as strong bisimilarity is among the strong relations. It has been argued in both directions. Some, e.g. van Glabbeek and Weijland [73] and Montanari and Sassone [147] point out that weak bisimilarity is a bit too coarse to exactly preserve the branching structure of a process. Others, like Darondeau [57], Valmari [182], de Nicola and Hennessy [58], Parrow and Sjödin [156], Cleaveland and Natarjan [149], as well as Brinksma *et al.* [39] define again coarser equivalences and argue that these relations characterise the observable behaviour of processes better than $\approx$ does.

It may be worth to point out that it is desirable to have an equivalence notion that is just as coarse as possible, i.e., that identifies as many as possible of those processes that exhibit the same behaviour (where the word 'same' is still a matter of perspective). It is helpful to think of an equivalence as a netting, where each mesh of the netting surrounds a certain equivalence class of behaviour. If the netting is too fine, some processes may lie in different meshes, even though their behaviour is indistinguishable. But if the netting is coarser than required, two processes may be in the same mesh, even though they behave differently. This is indeed the case for (strong and weak) trace equivalence, where processes with different deadlock behaviour fall into the same class.

From this point of view, (fair) testing equivalences seem to be the right choice [58, 149, 39], because they are indeed as coarse as possible with respect to a rather natural scenario of what an observer is able to test. However, we omit the details of their definition, basically because testing equivalences do not possess coinductive definitions. As mentioned before, this style of definition is favourable because it allows mathematically elegant proof techniques.

For mechanised verification purposes, the time and space complexity of deciding equivalences for finite state systems is a crucial aspect. Efficient algorithms are known for bisimilarity, answering the question whether two processes do or do not behave equivalently. Remarkably, from an algorithmic viewpoint, the preference between fine and coarse relations appears reversed. Roughly speaking, the finer an equivalence is, the better is its computational complexity. This will be made more precise in the next section. The practically most efficient algorithms are known for branching bisimilarity [86], a relation introduced by van Glabbeek and Weijland, that will also be of some importance in later chapters. Therefore we give its definition and motivation here.

Example 2.2.7. Figure 2.8 contains four slightly different interactive processes. All the four processes are weakly bisimilar. This can be checked by means of the weak transition relation, depicted in the second row of this figure (for the sake of clarity, we have omitted loops of weak internal transitions that each state possesses). However, there is a subtle difference among them that is not *visible in the weak transitions. In particular,* E_{18} *and* E_{19} *possess the possibility to move with* b *into the class without passing through the class. On the contrary every computation in* E_{16} *and* E_{17} *leading from to is forced to pass through. Therefore, in* E_{16} *and* E_{17} *the decision to discard* a *has to be taken* after b *has been executed while* E_{18} *and*

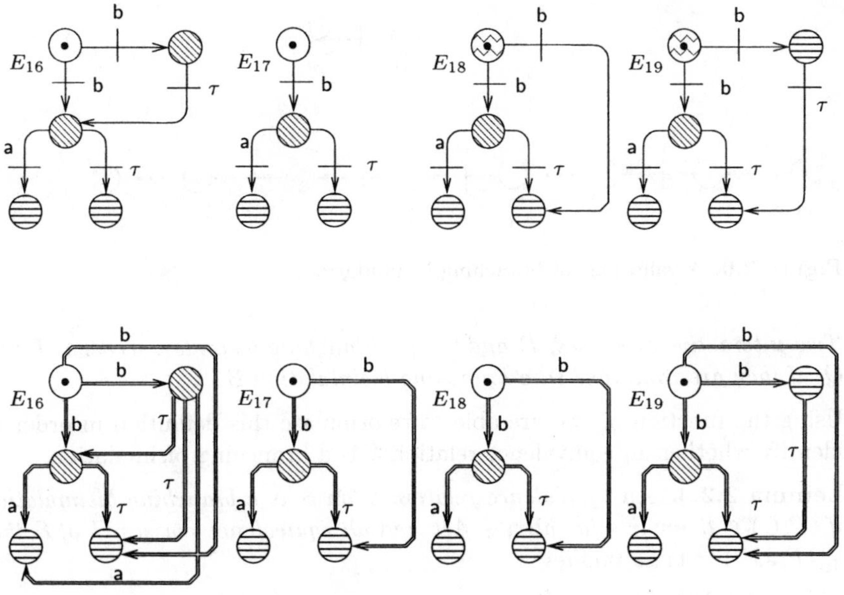

Figure 2.8. The difference between weak and branching bisimilarity

E_{19} may decide to discard a *already* before *interacting on* b. *This difference is not evident in the weak transition relation. Especially, E_{17} and E_{18} are isomorphic with respect to their weak transitions.*

This example reflects the main criticism why weak bisimilarity may appear too coarse. Strong bisimilarity has the property that any computation in one process corresponds to a computation in the other in such a way that all intermediate steps correspond as well. This is not true for weak bisimilarity, as exemplified in the above example.

In order to overcome this lack, van Glabbeek and Weijland postulate a branching condition. The basic idea is that in order to simulate a step $P \xrightarrow{\mathsf{a}} P'$ by $Q \stackrel{\mathsf{a}}{\Longrightarrow} Q'$, i.e., a sequence $Q \stackrel{\tau}{\Longrightarrow}\xrightarrow{\mathsf{a}}\stackrel{\tau}{\Longrightarrow} Q'$, all steps preceding the transition $\xrightarrow{\mathsf{a}}$ have to remain inside the class of P (and Q) and all steps following this transition have to stay inside the class of P' (and hence of Q'). This is schematically represented in Figure 2.9. To achieve this property it is sufficient to demand that Q performs arbitrarily many internal actions leading to a state Q'' which is still equivalent to P and then to directly reach the equivalence class of P' by performing action a [73].

Definition 2.2.6. *A symmetric relation $\mathcal{B}$ on $\mathcal{S}^{\text{all}}$ is a branching bisimulation if $(P, Q) \in \mathcal{B}$ implies for all* $\mathsf{a} \in$ *Act that $P \xrightarrow{\mathsf{a}} P'$ implies*

- $\mathsf{a} = \tau$ *and $P' \mathcal{B}\, Q$, or*
- $Q'' \xrightarrow{\mathsf{a}} Q'$ *for some Q'', Q' such that $Q \stackrel{\tau}{\Longrightarrow} Q''$ and $P \mathcal{B} Q''$ and $P' \mathcal{B} Q'$.*

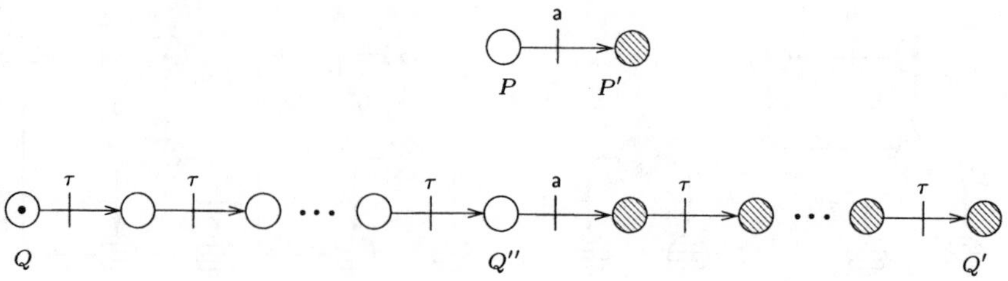

Figure 2.9. A valid pair of branching bisimulation

Two interactive processes, P and Q, are branching bisimilar, written $P \approx_{br} Q$, if they are contained in a branching bisimulation $\mathcal{B}$.

Using the function γ_O we are able to reformulate this definition in order to identify whether an equivalence relation $\mathcal{E}$ is a branching bisimulation.

Lemma 2.2.4. *An equivalence relation $\mathcal{E}$ on S is a branching bisimulation if $(P, Q) \in \mathcal{E}$ implies for all $\mathsf{a} \in Act$ and all equivalence classes C of $\mathcal{E}$ that $\gamma_O(P, \mathsf{a}, C) = \texttt{true}$ implies*

– $\mathsf{a} = \tau$ *and* $Q \in C$, *or*
– $\gamma_O(P, \mathsf{a}, C) = \gamma_O(Q'', \mathsf{a}, C)$ *for some* Q'' *such that* $Q \stackrel{\tau}{\Longrightarrow} Q''$ *and* $P\mathcal{E}Q''$.

Example 2.2.8. *E_{16} and E_{17} are branching bisimilar. The equivalence classes of a relation $\mathcal{E}$ satisfying Lemma 2.2.4 are depicted in the first row of Figure 2.8. One interesting case is that we have to simulate the transition* $\gamma_O($$, \mathsf{a},$ $) = \texttt{true}$ *of E_{17} in E_{16}. Indeed, each of the states shaded*  *of E_{16} may internally evolve to a state that still belongs to* , *and that satisfies* $\gamma_O($$, \mathsf{a},$ $) = \texttt{true}$.

E_{17} and E_{18} are not branching bisimilar because the classes  *and*  *can be distinguished. In particular,* $\gamma_O($$, \mathsf{b},$ $) = \texttt{true}$, *but E_{17} is not able to internally evolve to a state belonging to the same class (i.e.,* *) such that* $\gamma_O($$, \mathsf{b},$ $) = \texttt{true}$.

Branching bisimilarity fulfils the essential requirements for a proper notion of equality. It preserves deadlocks, it is an equivalence relation [22], is the largest branching bisimulation on a given transition system, and is a congruence with respect to parallel composition and abstraction. In addition, it is finer than weak bisimilarity, i.e., $\approx_{br} \subseteq \approx$. This inclusion is strict, as exemplified by the above example, where $E_{17} \not\approx_{br} E_{18}$, but $E_{17} \approx E_{18}$.

2.3 Algorithmic Computation of Equivalences

In the previous section, strong and weak bisimilarity arose as natural notions of equivalences on interactive processes. This section deals with algorithms to decide whether two interactive processes P and Q are equivalent with respect to either notion of bisimilarity.

2.3.1 Strong Bisimilarity

We begin with the basic algorithm for strong bisimilarity. The algorithm computes strong bisimilarity on a given *finite* transition system. 'Computing a bisimilarity' addresses the question, which states of a given transition system are bisimilar. To be precise, the algorithm factors the whole transition system into equivalence classes of bisimilar states. If this transition system is the union of two state spaces of, say P and Q, then this algorithm indirectly answers the question whether P and Q are bisimilar. They are bisimilar if their initial states fall into the same equivalence class of behaviours, otherwise they are not. This may seem to be a bit too much of work, because a whole transition system is factored just to decide bisimilarity of a single pair of states. Of course, this evaluation is wrong, because bisimilarity manifests itself as 'being able to stepwise simulate each other' which also involves all subsequent steps either system may undertake.

The global tactic of bisimilarity computation by factoring the state space is known as *partition refinement.* A partitioning of a state space is, loosely speaking, a netting where each mesh of the netting corresponds to a partition, containing some states. The algorithm should obviously terminate once the meshes correspond to the equivalence classes of bisimilarity. This is the case indeed, and it is achieved by successive refinement of the netting. Starting with a netting with a single (rather large) mesh, the meshes become finer and finer until no further refinement is needed, or, in algebraic terms, a fixed-point is reached. This fixed-point is the desired result.

Before we address the question how new strings are successively woven into the netting in order to refine the current partitioning we characterise strong bisimilarity as a fixed-point.

Theorem 2.3.1. *Let P be an interactive process with finite (reachable) state space S.*
Strong bisimilarity on S is the unique fixed-point of[1]

- $\smile_0 = S \times S,$
- $\smile_{k+1} =$
$$\{(P, Q) \in \smile_k \mid (\forall \mathsf{a} \in Act)(\forall C \in S/\smile_k)\ \gamma_{\mathrm{O}}(P, \mathsf{a}, C) = \gamma_{\mathrm{O}}(Q, \mathsf{a}, C)\}.$$

[1] We are a bit sloppy here (and in the sequel). Definition 2.2.2 defines $\sim$ as a relation on the global state space $\mathcal{S}^{\mathrm{all}}$ while here we compute only a particular subset of $\sim$, restricted to a subset S of this global state space, specifically the state space of some process(es). The explicit restriction to a subset is needed since the algorithm requires a finite state space. It computes bisimilarity on a given state space, but not on $\mathcal{S}^{\mathrm{all}}$ which, as we will learn in Chapter 5, is indeed infinite. So, 'strong bisimilarity on S' refers to $\sim \cap (S \times S)$.

Table 2.2. Algorithm for computing strong bisimilarity classes

Input:	labelled transition system $(S, Act, \longrightarrow)$
Output:	$S/\sim$
Method:	$Part := \{S\}$;
	$Spl := Act \times \{S\}$;
	While *Spl* not empty **do**
	Choose (a, C) **in** *Spl*;
	$Old := Part$;
	$Part := Refine(Part, \mathsf{a}, C)$;
	$New := Part - Old$;
	$Spl := (Spl - \{(\mathsf{a}, C)\}) \cup (Act \times New)$;
	od
	Return *Part*.

Milner [145] has shown that this property holds even for infinite state spaces. However, some notation has to be clarified. $S/\smile_k$ denotes the quotient of S under $\smile_k$, i.e., the set of equivalence classes of S with respect to the equivalence relation $\smile_k$. This raises the question whether each $\smile_k$ is an equivalence relation indeed, because otherwise $S/\smile_k$ would not be well-defined. A simple induction may be of help: First, $\smile_0$ is a trivial equivalence relation. Second, assuming that $\smile_k$ is an equivalence relation, $\smile_{k+1}$ also is, essentially because '=', equality on boolean values, is an equivalence relation. Be reminded that $\gamma_O(P, \mathsf{a}, C)$ is a boolean function that returns `true` iff P can move into the class C by performing an action a (Definition 2.2.3).

The refinement step from $\smile_k$ to $\smile_{k+1}$ rules out those pairs (P', Q') from $\smile_k$ for which γ_O produces (at least one) different truth value. In other words P' and Q' are split into different classes (of $\smile_{k+1}$), if one of them is able to move into a class C (of $\smile_k$) by performing an action a while the other one is not. So, (a, C) is a specific reason why (P', Q') has to be split. A pair (a, C) is therefore called a *splitter*.

Returning to the netting illustration, a splitter (a, C) is a means to weave a new string into the netting. Such a new string splits all the meshes for which $\gamma_O(\_, \mathsf{a}, C)$ returns different truth values into two finer meshes such that all states with the same truth value are contained in the same finer mesh. Since each mesh is basically a set of states, the netting is a *set of sets* of states corresponding to a partitioning of the state space. Formally the procedure of refining a partitioning *Part* by means of a splitter (a, C) can be defined by

$$Refine(Part, \mathsf{a}, C) :=$$
$$\left(\bigcup_{X \in Part} \left(\bigcup_{v \in \{\texttt{true},\texttt{false}\}} \left\{\{P \in X \mid \gamma_O(P, \mathsf{a}, C) = v\}\right\}\right)\right) - \{\emptyset\}.$$

Each mesh X is refined into two meshes, one for each possible value of γ_O. Since some of these meshes may be empty, the empty set is eventually extracted. This refinement step is the core of our global algorithm depicted in Table 2.2. We initialise this algorithm with a partitioning $\{S\}$ (a single mesh) and a set of possible splitters Spl, containing (a, S) for each action $a \in Act$. In the body of the algorithm each of the splitters is iteratively applied to refine the current partitioning. After each refinement, New contains all new (and thus finer) meshes. Each of them gives rise to new splitters, one for each action $a \in Act$. All these splitters are added to Spl while the currently processed splitter is removed.

The correctness of this partition refinement algorithm follows from a few observations [125]. First, it is easy to see that for each partitioning $Part$, $Refine(Part, a, C)$ is finer than $Part$ (but not necessarily *strictly* finer)[2]. In addition, $Refine(Part, a, C)$ is coarser than $S/\sim$ if $Part$ is coarser than $S/\sim$. If the set Spl is empty and $Part$ is coarser than $S/\sim$, then $Part$ coincides with $S/\sim$. The algorithm is correct because the initial partition is coarser than $S/\sim$. It terminates because no splitter is processed twice and the number of possible splitters is bounded by $|Act|2^{|S|}$.

Example 2.3.1. We illustrate the basic algorithm by means of an example. We aim to verify the equivalence classes of $\boxed{E_1 \;\; E_1}$ depicted in Figure 2.5. As before, we use shading of states to distinguish states with different properties, represented in Figure 2.10. We initialise the algorithm with $Part := \{\,$⬤$\,\}$, a single partition containing all states, and with two splitters, $(\text{in},$⬤$)$ and $(\text{out},$⬤$)$. This situation is represented in the first column of Figure 2.10.

We start partition refinement by choosing the splitter $(\text{in},$⬤$)$ and computing the values $\gamma_0(\_, \text{in},$⬤$)$ for all the states in ⬤. One of the states returns false *while the others return* true. *As a result, we have to refine partition ⬤ into two partitions, thus $Part := \{\,$◇$\,,$◈$\}$. Adding new splitters to Spl (and removing splitter $(\text{in},$⬤$))$ leads to the situation depicted in the second column of this figure.*

Continuing with splitter $(\text{out},$⬤$)$ returns different truth values of γ_0 in partition ◇ . Therefore, we split this partition into ◯ and ◈. Part becomes $\{$◯$,$◈$,$◈$\}$ and New is $\{$◯$,$◈$\}$. We update Spl accordingly. This leads to the situation depicted in the rightmost column of Figure 2.10.

Subsequent refinement steps do not reveal any distinction inside each of the classes ◯, ◈ and ◈. The algorithm terminates when Spl is empty. It has produced the equivalence classes already highlighted in Figure 2.5.

As illustrated by the above example, this algorithm works fine, but its complexity is relatively bad. In particular, the administration of splitters can be improved a lot. For instance, when updating Spl it is plausible to remove unprocessed splitters (a, C) from Spl, if C has been refined during the last step (i.e., if $C \in (Old - Part)$) because finer splitters will be processed anyway.

---

[2] To order partitionings in this sense, we use the equivalence $(\bigcup_{X \in Part'} X \times X)$ induced by $Part'$.

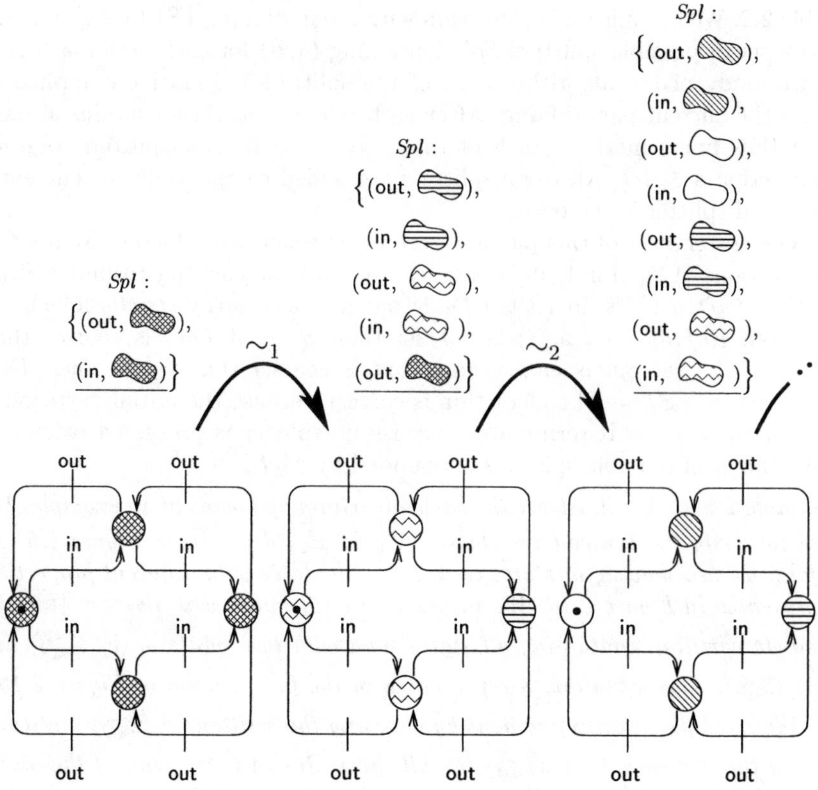

Figure 2.10. Stepwise partition refinement

Further improvements are possible. We refer to [63] for the details of an implementation that requires $\mathcal{O}(m_I)$ space and $\mathcal{O}(m_I \log n)$ time. Here n is the cardinality of the state space $|S|$ and m_I the number of transitions, i.e., $|\longrightarrow\!\!\!\!\!\!\rightarrow|$. This algorithm was first proposed by Paige and Tarjan [154]. It is worth to point out that most of the coarser equivalences, including testing equivalence and trace equivalence, are PSPACE-complete [125].

2.3.2 Weak Bisimilarity

Weak bisimilarity can be computed in the same way, only replacing the transition relation $\longrightarrow\!\!\!\!\!\!\rightarrow$ by $\Longrightarrow$. In other words, the algorithm is as before, but partitions the state space with respect to $\Longrightarrow$ instead of $\longrightarrow\!\!\!\!\!\!\rightarrow$. For this purpose we define a function $\boldsymbol{\gamma}_{\mathrm{O}}$ as the straightforward adaption of γ_{O} by

$$\boldsymbol{\gamma}_{\mathrm{O}}(P, \mathsf{a}, C) := \begin{cases} \texttt{true} & \text{if there is } P' \in C \text{ such that } P \stackrel{\mathsf{a}}{\Longrightarrow} P', \\ \texttt{false} & \text{otherwise.} \end{cases}$$

Table 2.3. Algorithm for computing weak bisimilarity classes

Input:	labelled transition system $(S, Act, \longrightarrow)$
Output:	$S/\approx$
Method:	Compute $\Longrightarrow$ from $\longrightarrow$;
	$Part := \{S\}$;
	$Spl := Act \times \{S\}$;
	While *Spl* not empty **do**
	Choose (a, C) **in** *Spl*;
	$Old := Part$;
	$Part := \mathit{Refine}(Part, \mathsf{a}, C)$;
	$New := Part - Old$;
	$Spl := (Spl - \{(\mathsf{a}, C)\}) \cup (Act \times New)$;
	od
	Return *Part*.

The global strategy to compute weak bisimilarity remains as in Table 2.2. The only change concerns the refinement step where $\boldsymbol{\gamma}_{\mathrm{O}}$ replaces γ_{O},

$$\mathit{Refine}(Part, \mathsf{a}, C) :=$$

$$\left(\bigcup_{X \in Part}\left(\bigcup_{v \in \{\mathtt{true},\mathtt{false}\}}\Big\{\{P \in X \mid \boldsymbol{\gamma}_{\mathrm{O}}(P, \mathsf{a}, C) = v\}\Big\}\right)\right) - \{\emptyset\}.$$

However, to make use of the function $\boldsymbol{\gamma}_{\mathrm{O}}$, the algorithm has to compute the weak transition relation $\Longrightarrow$ from $\longrightarrow$ *a priori*. This leads to an algorithm to compute weak bisimilarity depicted in Table 2.3.

As a matter of fact, the computation of $\Longrightarrow$ dominates the complexity of partition refinement, basically because the reflexive and transitive closure $\xrightarrow{\tau}{}^{*}$ of internal moves has to be computed in order to build the weak transition relation. The usual way of computing a transitive closure has cubic complexity. Some improvements are known for this task, see, for instance, [52] for an algorithm with $\mathcal{O}(n^{2.376})$ time requirements. In any case, this is the computationally expensive part, since the time complexity of partition refinement is $\mathcal{O}(m'_{I} \log n)$ (and space complexity is $\mathcal{O}(m'_{I})$), where m'_{I} is the number of weak transitions, i.e., $|\Longrightarrow|$. Note that m'_{I} is bounded by $n^2|Act|$ which is of order $\mathcal{O}(n^2)$.

Computing branching bisimulation equivalence classes [86] is based on the same partition refinement strategy, but its implementation proceeds in a different way, compared to Paige and Tarjan [154]. It's space complexity is better, it requires $\mathcal{O}(m''_{I})$ space where m''_{I} is m_{I} plus the number of $\xrightarrow{\tau}{}^{+}$ transitions. In fact, during initialisation only the computation of $\xrightarrow{\tau}{}^{+}$ is necessary instead of the full closure $\overset{\mathsf{a}}{\Longrightarrow}$ for each $\mathsf{a} \in Act$. The time complexity to do so is, of course, still $\mathcal{O}(n^{2.376})$. In the style of [86] the partition refinement part of the algorithm requires $\mathcal{O}(n\, m''_{I})$ time. Bouali has adopted this

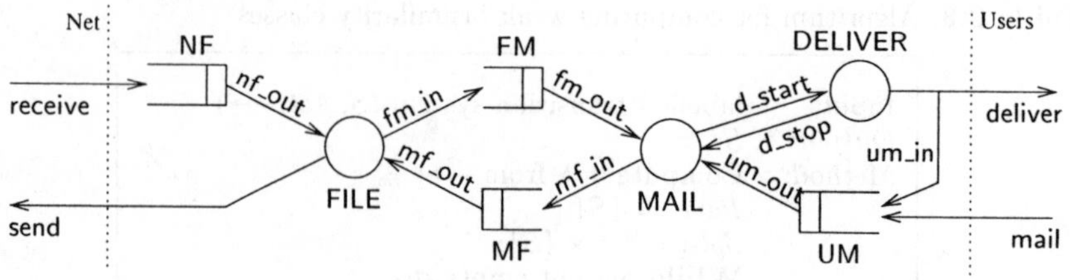

Figure 2.11. The mail system example

branching bisimulation machinery to compute also weak bisimulation [30]. In this way weak bisimulation requires the same order of space and time.

From a practical point of view the preference between these two implementations, the one of Paige and Tarjan and the one of Bouali, is not clear. The full closure combined with Paige and Tarjan's technique is inferior with respect to space complexity while it is, in the worst case, superior with respect to time complexity. The worst case arises for a fully connected transition system, i.e., $\dashrightarrow = S \times Act \times S$. In this case, Bouali requires $\mathcal{O}(n^3)$ time while the Paige and Tarjan implementation (plus full closure) requires $\mathcal{O}(n^{2.376})$ time.

However, the complexity of the necessary closure operations is always better in Bouali's implementation. In addition, also the overall time demands are often superior: In many cases the density of transitions is far lower than the worst case. Usually each $\xrightarrow{a}$ is only connecting few pairs of states, compared to $S \times S$. Thus m_I tends to be drastically smaller than the possible worst case $|Act|n^2$. Then it depends mainly on the number of internal transitions whether the time $\mathcal{O}(n\, m_I'')$ is better than $\mathcal{O}(m_I' \log n)$, because this governs the size of m_I' compared to m_I''.

2.4 An Application Example: Brebner's Distributed Mail System

In this section we illustrate the main ingredients we have presented in the previous sections and their benefits. We build upon a case study of G. Brebner [37], who described the use of CCS [145] to investigate a problem which arose in a distributed electronic mail system. Practical experience showed that the system contained a deadlock and this was confirmed by an analysis of the CCS model. Here, we use interactive processes to represent a (slightly simplified) system.

We first describe the global structure of the specification and then turn our attention to the behaviour of each component. Afterwards we apply the operational semantics in order to generate the underlying transition system, and subsequently aggregate it by means of the algorithms of Section 2.3.

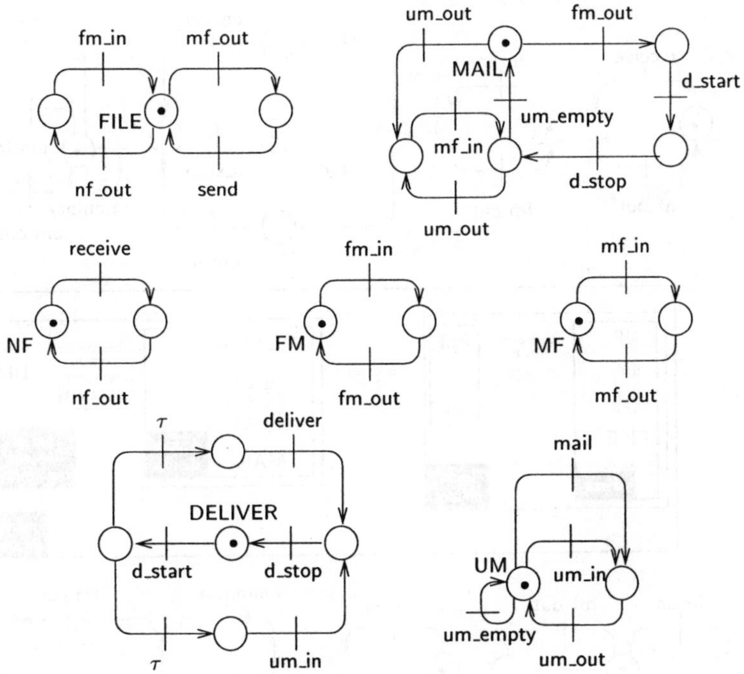

Figure 2.12. Interactive processes of the mail system

Finally, we show that the compositional structure of the specification can be exploited by compositional aggregation circumventing the generation of large portions of the state space.

Specification. The system provides a bidirectional electronic mail connection between local users and the rest of the world, the net. It consists of three processes and four buffers. We explain the structure, depicted in Figure 2.11, from left to right. The process FILE manages the file transfer, in particular it collects incoming mail from a buffer NF connected to the net and delivers it to the MAIL process via an internal buffer FM, or takes outgoing mail from an internal MF buffer to deliver it to the net. A rather abstract view on the interactions of the process FILE is depicted in Figure 2.12 (upper left).

The MAIL process handles incoming mail or forwards outgoing mail, contained in the buffer UM to the FILE process via buffer MF. It will process all outgoing user mail, until the UM buffer is empty (um_empty), before handling incoming mail. By means of action **d_start**, incoming mail is passed to the process DELIVER which evaluates the header of the mail. The MAIL process waits for a signal of the DELIVER process (d_stop) before processing further mail. This behaviour is represented in Figure 2.12 (upper right).

The process DELIVER typically delivers mail to the user; alternatively it may forward the mail to another site by putting it into the buffer UM

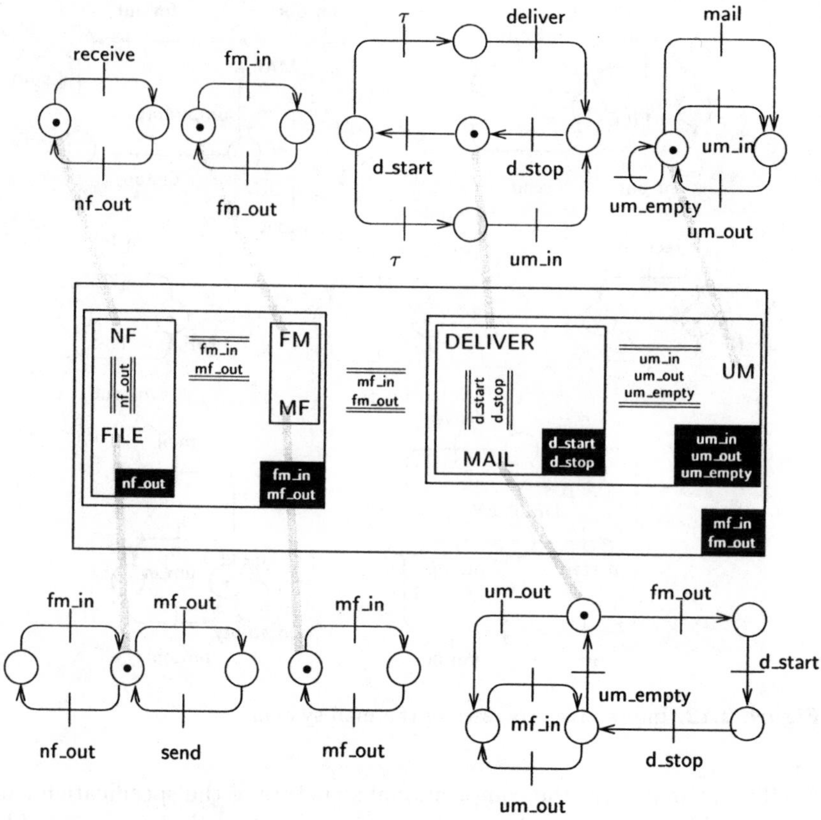

Figure 2.13. Hierarchical structure of the mail system specification

responsible for outgoing mail. This is the case if for instance the recipient of the mail has set up a forwarding procedure to another site, as this user has temporarily moved there. The decision between forwarding and delivering is internal, since it only depends on the information contained in the header of the mail and local information about which user requires forwarding. The decision is modelled as internal nondeterminism in Figure 2.12 (lower left) by means of two $\xrightarrow{\tau}$ transitions emanating the state reached after d_start.

In this introductory example we model all the buffers with one place only. This is sufficient to analyse the functionality of the system. All buffers have the structure of a one-place buffer (cf. E_1 in Figure2.1), except buffer UM. This buffer provides the possibility to test for emptiness (um_empty). In addition it is able to accept mail from different origins, namely either from the users (mail) or from the process DELIVER (um_in).

The overall system is modelled by the following hierarchy of parallel composition and abstraction of components that nicely reflects the structure,

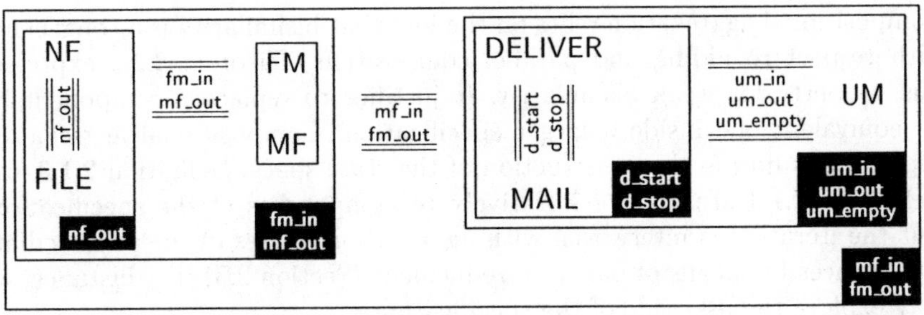

.

The detailed behaviour of each component is depicted in Figure 2.13. This system is again an interactive process, it interacts with the environment on receive and send on the net side and with deliver and mail on the user side. All other activities of this process are internalised by means of abstraction.

Analysis. In order to explore the behaviour of this specification, we iteratively apply Definition 2.1.3 and Definition 2.1.4. This leads to a transition system with 413 states and 1140 transitions. By applying the algorithms sketched in Section 2.3 this transition system can be aggregated to 356 states, in the case of strong bisimilarity, and 160 states in the case of weak bisimilarity. Indeed, the inspection of either of these transition systems reveals that the system contains a deadlock.

G. Brebner [37] has given a detailed explanation of the deadlock situation[3]. It arises when buffer UM is full while DELIVER is trying to forward a mail back to the net by means of um_in. In this case the system is stuck, since MAIL is waiting for DELIVER to finish delivery (d_stop) before handling further outgoing mail from UM. The existence of a deadlock is independent from the buffer size.

Compositional Analysis. We have used abstraction to hide actions that are not required for synchronisation outside a certain level of the hierarchical specification. This is not only preferable in order to enhance the understanding of the example. It is also valuable in order to apply *compositional aggregation techniques* to the generation of the state space. In this example, the original state space consists of 413 states even though the relevant information is contained in the weak bisimilar variant that only consists of 160 states. Compositional aggregation can avoid to construct the whole original state space in favour of a smaller one. For this small example this is surely not required, because the state space is manageable without problems. However, since compositional aggregation has turned out to be of crucial importance to cope with systems of billions of states (cf. [44], for instance), we explain it by means of our introductory example. The strategy of compositional aggregation will re-appear in later chapters.

[3] To be precise, the system contains three different deadlocking situations; the most interesting one is discussed here.

Compositional aggregation relies on the fact that bisimilarity is a congruence with respect to hiding and parallel composition. Theorem 2.2.2 expresses this property for weak bisimilarity. It justifies to replace a component by an equivalent one inside a larger specification. Compositional aggregation exploits this during the construction of the state space. Definition 2.1.3 and Definition 2.1.4 are applied iteratively to components of the specification. But the iteration is interwoven with aggregation phases of the intermediate state spaces by means of partition refinement (Section 2.3). For instance, we may explore the user side of the specification,

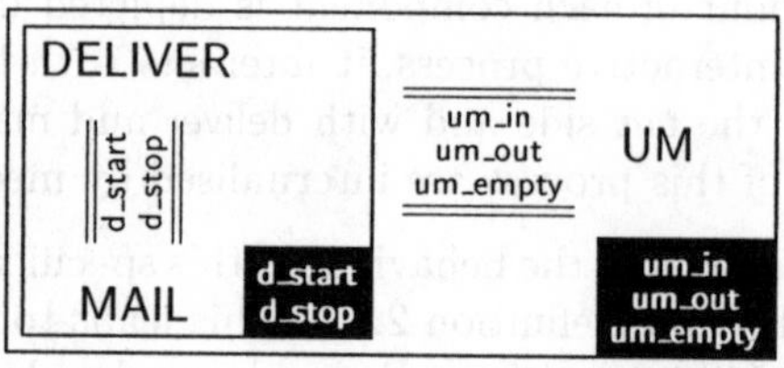

and obtain a process with 19 states and 31 transitions. Applying the weak bisimulation algorithm (Table 2.3) factors the state space into equivalently behaving classes of states. By representing each class by a single state we obtain an aggregated process, say U, which consists of 10 states and 17 transitions. Similarly the net side,

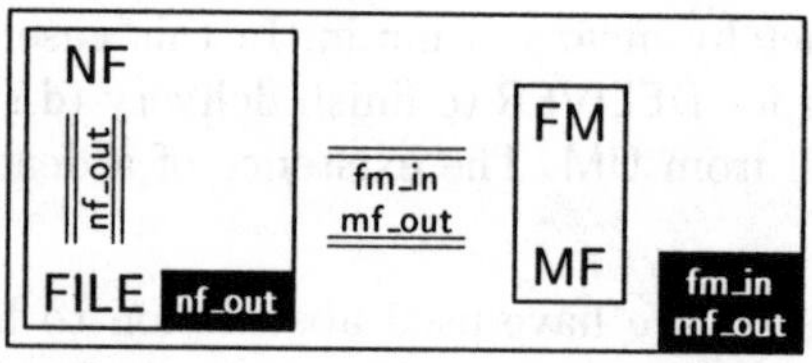

consists of 25 states and 58 transitions, but can be aggregated to a weakly bisimilar process N with 20 states and 48 transitions. Due to the congruence result (Theorem 2.2.2) we are free to use these smaller interactive processes, U and N, instead of the original user, respectively net side specifications.

Thus it is sufficient to investigate $N \overline{\overline{\text{mf\_in,fm\_out}}} \ U \underline{\underline{\text{mf\_in.fm\_out}}}$. This aggregated specification turns out to have 189 states and 528 transitions. This is rather close to the minimally required state space, 160 states (though not overwhelming compared to 413 states). For larger examples the gain in state space aggregation can amount to several orders of magnitudes, especially if this strategy is applied iteratively, along the hierarchy of parallel composition and abstraction, as in [44].

2.5 Discussion

In this chapter we have introduced a rudimentary calculus of interactive processes. It consists of labelled transition systems with initial states, together with two composition operators, parallel composition and hiding. The notation we are using for interactive processes as well as our composition operators are a bit unusual.

Interactive processes have an intuitive correspondence to Petri nets. Indeed, they correspond to (transition labelled) state machines, the subclass of concurrency free Petri nets. Note that parallel composition does not introduce *true* concurrency into the framework of interactive processes, since parallelism is modelled by interleaving the different possibilities of interaction nondeterministically. Therefore the result of parallel composition is again an interactive process that does not exhibit explicit concurrency.

On this model we have discussed the question what requirements should be fulfilled by a reasonable equivalence notion. Essentially recalling the motivations for strong and weak bisimilarity, we have highlighted that a reasonable equivalence

- should only equate processes with the same behaviour,
- should be substitutive with respect to the operators,
- should be decidable in an efficient way, and
- should allow one to abstract from internal computation.

The mathematical elegance of bisimulation (due to its coinductive definition) turned out to be also valuable for the computational complexity of deciding these relations. We have sketched algorithms to compute strong, weak and branching bisimilarity by means of partition refinement. In Section 2.4 we have exemplified the benefits of (weak) bisimulation for compositional aggregation by means of a mail system example. These relations and algorithms form the basis of the results and techniques developed in the next chapters where we will incorporate stochastic time into this framework.

3. Markov Chains

This chapter deals with a particular class of stochastic models that form a cornerstone of this book. Stochastic models are widely used to describe phenomena that change randomly as time progresses. We focus on Markov chains, as simple and adequate models for many such phenomena. More precise, we cover discrete- as well as continuous-time Markov chains. We discuss details of their analysis to set the ground for the requirements of later chapters, and introduce useful equivalence relations for both types of models. These relations are defined in the style of bisimilarity and are akin to the notion *lumpability* on Markov chains. Furthermore, we present efficient algorithms to compute these relations, which, as a side result, can be used to compute the 'best possible' lumping of a given Markov chain.

3.1 Stochastic Processes

A *stochastic process* $\{X_t \mid t \in T\}$ is a family of random variables X_t defined over the same probability space and taking values in a set S, usually referred to as the *state space* of a process. The parameter set T is often interpreted as time, and is sometimes called the *time range*. Each random variable X_t describes a snapshot random distribution on the state space S of the process at time t. The time range can be either discrete or continuous. This distinction separates two classes of stochastic processes, *discrete-time stochastic processes* and *continuous-time stochastic processes*. For simplicity, we assume that T is a subset of the nonnegative real numbers $\mathbb{R}^+$ in the continuous-time case, while in the discrete-time case we identify T with the set of natural numbers $\mathbb{N}$ (including 0).

Example 3.1.1. An example of a stochastic process $\{X_t \mid t \in T\}$ may be the evolution of temperature at a specific place, say in Papenburg, Germany. The state space S of such a process will be a reasonable temperature range. Each random variable X_t will then give a probability for any possible temperature at time t. If we are, for instance, considering the maximal or average temperature per day, *the time range can be discrete. This leads to a discrete-time stochastic process where X_0 describes a probability distribution on the* *temperature range of the very first day, X_1 of the day after, and so on. Alternatively we may describe the continuous evolution of temperature by means of a continuous-time stochastic process (i.e., choosing a continuous time range).*

A *Markov process* is a stochastic process that satisfies an additional requirement. This *Markov property* requires that, for any given time instant (say t_n) the future behaviour, for instance the value of $X_{t_{n+1}}$, is totally independent of its history, i.e., the values of $X_{t_{n-1}}$, $X_{t_{n-2}}$, and so on. It only depends on the state occupied at the current time instant t_n, given by the value of X_{t_n}.

In mathematical terms the Markov property requires that, for each sequence of time instances $t_{n+1} > t_n > t_{n-1} > t_{n-2} > \ldots > t_0$ (of arbitrary length n), we have that for each (measurable) subset A of states,

$$\begin{aligned} &Prob\{X_{t_{n+1}} \in A \mid X_{t_n} = P_n, X_{t_{n-1}} = P_{n-1}, X_{t_{n-2}} = P_{n-2}, \ldots, X_{t_0} = P_0\} \\ &\quad = \; Prob\{X_{t_{n+1}} \in A \mid X_{t_n} = P_n\}. \end{aligned} \tag{3.1}$$

Thus, the fact that the process was in state P_{n-1} at time t_{n-1}, in state P_{n-2} at time t_{n-2}, and so on, up to the fact that it was in state P_0 at time t_0 is entirely irrelevant. The state X_{t_n} contains all relevant history information to determine the random distribution on S at time t_{n+1}.

This property owes its name to A.A. Markov, who studied processes with this property at the beginning of the last century [137]. To be precise, the above definition is tailored for continuous-time Markov processes. In the discrete-time case, the Markov property becomes somewhat simpler, since we do not have to bother about arbitrary sequences of time instances. Instead, we consider the (unique) sequence that contains all former time instances. Since we identified T with $\mathbb{N}$ we simply require for arbitrary $t \in \mathbb{N}$,

$$\begin{aligned} &Prob\{X_{t+1} \in A \mid X_t = P_t, X_{t-1} = P_{t-1}, X_{t-2} = P_{t-2}, \ldots, X_0 = P_0\} \\ &\quad = \; Prob\{X_{t+1} \in A \mid X_t = P_t\}. \end{aligned} \tag{3.2}$$

Example 3.1.2. If the probability to reach a certain maximal temperature on a specific day does not depend on the temperature reached the days before, the temperature evolution could describe a discrete-time Markov process. Of course, even in Papenburg, this is rarely the case.

It is worth to point out that the Markov property does not imply that the future behaviour is independent of the current time instant t. If the value X_t does depend on t, the process is said to be *inhomogeneous*. However, throughout our discussion in the remainder of this book we shall assume that Markov processes are independent of the time instant of observation. In this case, a Markov process is said to be *homogeneous*; we gain the freedom to arbitrarily choose the origin of the time axis. In technical terms, homogeneity requires that we have (for $t' \geq t$, and measurable $A \subseteq S$),

$$Prob\{X_{t'} \in A \mid X_t = P\} \;=\; Prob\{X_{t'-t} \leq P' \mid X_0 = P\}. \tag{3.3}$$

Example 3.1.3. If the probability to reach a certain maximal temperature on a day does not depend on the date of this day (relative to the first day of observation) the process is homogeneous.

The last simplification that we will impose concerns the state space S of a homogeneous (discrete- or continuous-time) Markov processes. Similar to the time range, the state space can be either discrete or continuous. We only consider discrete state spaces. This class of Markov processes is widely known as *Markov chains*.

Example 3.1.4. If the temperature is measured in real numbers, the state space would be continuous. By dividing the temperature range into intervals, we obtain a discrete state space. In particular we may choose, say, three intervals of high, medium *and* low *temperature, then the state space is discrete, containing three states.*

In summary, we have made three major restrictions starting from the very general model of stochastic processes. Apart from the Markov property we have required homogeneity as well as discrete state spaces. The resulting class of homogeneous discrete-time and continuous-time Markov chains is admittedly one of the simplest classes of stochastic processes at all. Nevertheless Markov chains are used to model a large variety of real world applications, partly because they closely model some real situations, partly because of their simplicity when it comes to numerical analysis by means of efficient algorithms [181]. The enormous amount of literature that exists on this subject testifies this.

3.2 Discrete-Time Markov Chains

A discrete-time Markov chain (DTMC) is a Markov process with discrete time range and discrete state space. The discreteness of the state space ensures, together with our assumption of time homogeneity (3.3), that the Markov property (3.2) can be reformulated as

$$\begin{aligned} &Prob\{X_{t+1} = P' \mid X_t = P, X_{t-1} = P_{t-1}, X_{t-2} = P_{t-2}, \ldots, X_0 = P_0\} \\ &\quad = \quad Prob\{X_{t+1} = P' \mid X_t = P\} \\ &\quad = \quad Prob\{X_1 = P' \mid X_0 = P\}. \end{aligned}$$

It is important to note that this expression denotes the probability to reach state P' from state P in a single time step and that this probability is independent of the actual time instant of observation. In other words, it is the one-step transition probability between two states. This remark builds a bridge to our notion of transition systems introduced in Chapter 2. If we let p denote $Prob\{X_1 = P \mid X_0 = P'\}$ (or $Prob\{X_2 = P \mid X_1 = P'\}$, and so on), we can neatly represent this one-step transition probability from P to P' by

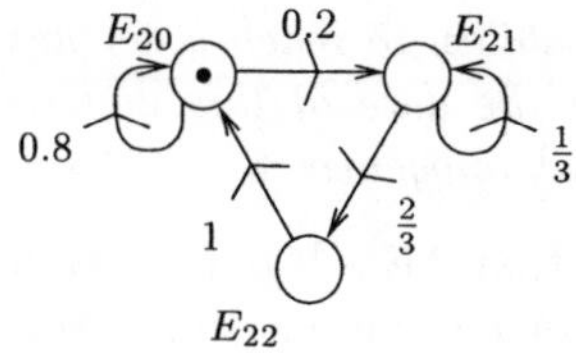

Figure 3.1. Probabilistic chain

means of a *probabilistic transition* $P \overset{p}{\twoheadrightarrow} P'$ where a transition is labelled with a probability instead of an action, as it has been the case in Chapter 2.

Definition 3.2.1. *A probabilistic transition system (PTS) is a tuple* $(S, \twoheadrightarrow)$, *where*

- S *is a nonempty set of states, and*
- $\twoheadrightarrow$ *is a* probabilistic transition relation, *a subset of* $S \times \mathbb{R}^+ \times S$ *such that, for each state, probabilities of outgoing probabilistic transitions cumulate to 1.*

In this definition we have added a restriction essentially saying that for all states the cumulative probability to move somewhere in a single step is 1. This should be an obvious requirement to obtain a proper probability distribution among successor states.

Now the question arises in which way DTMC and PTS are related. Indeed, a probabilistic transition system gives us almost all the necessary information to completely determine a specific DTMC. It defines all (nonzero) one-step transition probabilities on the state space. The only information that is missing to determine the dynamics of this chain is the shape at the beginning of its observation, i.e., at time instant 0. This information might be a particular *initial state* or, more generally, assign probabilities to different initial states, by means of an *initial probability distribution*. Since the latter can be encoded (by adding probabilistic transitions out of a single, auxiliary initial state) into the former, we restrict ourselves to those DTMC that possess a single initial state P. We call them *probabilistic chains*.

Definition 3.2.2. *A probabilistic chain is a triple* $(S, \twoheadrightarrow, P)$, *where*

- $(S, \twoheadrightarrow)$ *is a probabilistic transition system, and*
- $P \in S$ *is the initial state.*

Example 3.2.1. Figure 3.1 contains an example of a probabilistic chain. As before, the initial state is marked by $\odot$. *A possible interpretation of this chain is that it describes the evolution of temperature between the three states* E_{20}, E_{21} *and* E_{22}, *representing high, medium, respectively low temperature. So, if the temperature is high it will stay high with a probability of* 0.8 *while it will become medium with probability* 0.2 *on the day after. (We will not bother about the plausibility of this specific example).*

In this example we have not used the full expressive power of a transition *relation* $S \times \mathbb{R}^+ \times S$. As opposed to a transition *function* $S \times S \mapsto \mathbb{R}^+$ there can be multiple probabilistic transitions between two states, each labelled with a different probability. However, this degree of detail is not representable in a DTMC. Therefore, probabilities of such 'parallel' transitions have to be cumulated in order to obtain the proper one-step transition probability. We have chosen a relation instead of a function for technical reasons, essentially to ensure compatibility among the different parts of our dynamic process model.[1] For the moment we exclude probabilistic chains with 'parallel' transitions by requiring that for each pair (P, Q) of states it holds that $|(\twoheadrightarrow \cap(\{P\} \times \mathbb{R}^+ \times \{Q\}))| \leq 1$.

Let us highlight another important property of DTMC and thus of probabilistic chains by means of the above example. The chain E_{20} stays in this (initial) state with probability 0.8 during the first step. Similarly, the probability to (still) stay in this state after two steps amounts to $0.8 \cdot 0.8$. It is easy to calculate that the *sojourn time* SJ_P, i.e., the number of consecutive time steps the process remains in a given state P before exiting, is geometrically distributed, i.e.,

$$Prob\{SJ_P = i\} = p^i(1-p).$$

For the initial state of our example, $p = 0.8$ and hence we have, for instance, $Prob\{SJ_{E_{20}} = 2\} = 0.8^2 \cdot 0.2 = 0.0128$. It is worth to point out that this discrete distribution is *memoryless*. This refers to the fact that the information that we have been in a state for a certain amount of time is irrelevant for the distribution of the residual sojourn time. It remains geometrically distributed with the same parameters. This *memoryless property* is a natural consequence of the Markov property. Since the future of a Markov chain depends on the present state only, but not on the states observed at preceding time instances, the sojourn time cannot depend on the time already spent in the present state. Returning to our example, the probability to stay in state E_{20} for precisely 1002 consecutive time steps is also 0.0128, under the assumption that we get interested in this probability after 1000 time steps. In technical terms we have

$$Prob\{SJ_{E_{20}} = 1002 \mid SJ_{E_{20}} > 1000\} = Prob\{SJ_{E_{20}} = 2\} = 0.0128.$$

The class of geometric distributions is the only class of memoryless discrete probability distributions.

3.3 Continuous-Time Markov Chains

The definition of continuous-time Markov chains (CTMC) is slightly more involved compared to DTMC. However it is worth to thoroughly introduce

[1] Indeed, we will later allow multiple transitions labelled with the *same* probability, by using a *multi*-relation instead of a relation.

them here, since CTMC form the base of contemporary performance evaluation methodology. A CTMC is a Markov process with discrete state space but continuous time range. As with DTMC, we reformulate the Markov property (3.1), adjusting some notation, for $t_n + \Delta t > t_n > t_{n-1} > t_{n-2} > \ldots > t_0$, as follows:

$$\begin{aligned} & Prob\{X_{t_n+\Delta t} = P' \mid X_{t_n} = P, X_{t_{n-1}} = P_{t_{n-1}}, X_{t_{n-2}} = P_{t_{n-2}}, \ldots, X_{t_0} = P_{t_0}\} \\ &\quad = \; Prob\{X_{t_n+\Delta t} = P' \mid X_{t_n} = P\} \\ &\quad = \; Prob\{X_{\Delta t} = P' \mid X_0 = P\}. \end{aligned}$$

As in the discrete-time case, this probability is (due to time homogeneity (3.3)) independent of the actual time instant t_n (or t' or 0) of observation. Nevertheless it *does* depend on the length of the time *interval* Δt. It requires some limit calculation to deduce that we are confronted with a *linear dependence* [128]. More precise, for every pair of states P and P', there is some parameter λ such that (for sufficiently small Δt)

$$Prob\{X_{\Delta t} = P' \mid X_0 = P\} = \lambda \Delta t + o(\Delta t) \,,$$

where $o(\Delta t)$ subsumes the probabilities to pass through intermediate states between P and P' during the interval Δt. The quantity λ is thus a transition *rate*, a nonnegative real value that scales how the (one step) transition probability between P and P' increases with time. Here, we have implicitly assumed that state P is different from P'. If, otherwise, state P and P' coincide, the probability to stay in state P during an interval Δt (and hence $Prob\{X_{\Delta t} = P \mid X_0 = P\}$) decreases with time, starting from 1 if $\Delta t = 0$. The corresponding transition rate is thus a negative real value. It is implicitly determined by the increasing probability to leave state P; that is, it is the negative sum of the respective transition rates.

Unlike transition probabilities (in the CTMC setting), transition rates do not depend on the length of time intervals. In addition, the probabilistic behaviour of a CTMC is completely described by the initial state (or distribution) and the transition rates between distinct states. We therefore proceed as in the discrete-time case and fix a CTMC by means of a specific transition relation, $P \xrightarrow{\lambda} P'$, defined on a state space S, together with an initial state P. We call them *Markovian chains.*[2]

Definition 3.3.1. *A Markovian transition system is a tuple* $(S, \longrightarrow)$, *where*

- S *is a nonempty set of states, and*
- $\longrightarrow$ *is a* Markovian transition relation, *a subset of* $S \times \mathbb{R}^+ \times S$.

A Markovian chain is a triple $(S, \longrightarrow, P)$, *where*

- $(S, \longrightarrow)$ *is a Markovian transition system, and*
- $P \in S$ *is the initial state.*

[2] Note that we make a difference between Markov chains and Markov*ian* chains, the latter being a special case of the former.

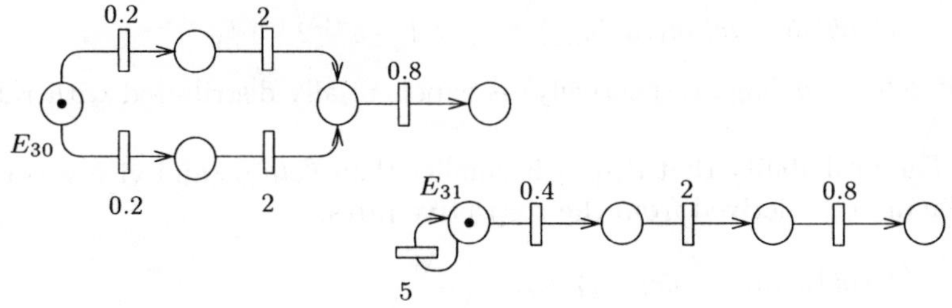

Figure 3.2. Two Markovian chains

Example 3.3.1. Figure 3.2 contains two examples of Markovian chains, E_{30} and E_{31}.

Markovian chains are a bit more expressive than CTMC. Similar to PTS compared to DTMC, 'parallel' Markovian transitions are not representable directly in a CTMC. We therefore restrict ourselves to those Markovian chains satisfying that for each pair (P, Q) of states it holds that $|(\dashrightarrow \cap (\{P\} \times \mathbb{R}^+ \times \{Q\}))| \leq 1$.

Furthermore, loops of Markovian transitions (for instance $E_{31} \overset{5}{\dashrightarrow} E_{31}$) are irrelevant for the probabilistic behaviour of the associated CTMC. Loops can be ignored because the probability to stay in a state decreases with a rate cumulated from all the rates leading away from this state (0.4, in this example). We could easily avoid such unnecessary loops, by requiring that the transition relation is irreflexive.

Complementary to the role of geometric distributions in the discrete-time case, the sojourn time distribution SJ_P for any state of a CTMC is *exponentially distributed.* To illustrate why the sojourn time is given by an exponential distribution, we first require to highlight some important properties enjoyed by exponential distributions. They will also be crucial for many explanations in later chapters.

(A) An exponential distribution $Prob\{delay \leq t\} = 1 - e^{-\lambda t}$ is characterised by a single parameter λ, a positive real value, usually referred to as the *rate* of the distribution. The mean duration of this delay amounts to $1/\lambda$ time units.

(B) In correspondence to geometric distributions in the discrete-time setting, the class of exponential distribution is the only class of *memoryless* continuous probability distribution. The remaining delay after some time t_0 has elapsed is a random variable with the same distribution as the whole delay:

$$Prob\{delay \leq t + t_0 \mid delay > t_0\} = Prob\{delay \leq t\}. \tag{3.4}$$

(C) The class of exponential distributions is closed under minimum, which is exponentially distributed with the sum of the rates:

$$Prob\{\min(delay_1, delay_2) \leq t\} = 1 - e^{-(\lambda_1+\lambda_2)t}, \tag{3.5}$$

if $delay_1$ ($delay_2$, respectively) is exponentially distributed with rate λ_1 (λ_2).

(D) The probability that $delay_1$ is smaller than $delay_2$ (and vice versa) can be directly derived from the respective rates:

$$Prob\{delay_1 < delay_2\} = \frac{\lambda_1}{\lambda_1 + \lambda_2}, \tag{3.6}$$

$$Prob\{delay_2 < delay_1\} = \frac{\lambda_2}{\lambda_1 + \lambda_2}. \tag{3.7}$$

(E) The continuous nature of exponential distributions ensures that the probability that both delays elapse at the same time instant is zero.

With these properties, we can turn our attention to the sojourn time distributions in a Markovian chain. For each state P, there is some parameter λ such that

$$Prob\{SJ_P \leq t\} = 1 - e^{-\lambda t}.$$

In the case of E_{30}, for instance, $\lambda = 0.4$ and therefore $Prob(SJ_{E_{30}} \leq 2) = 1 - e^{-0.8} = 0.55$. The quantity 0.4 is the rate of leaving state E_{30} and appears as the parameter of this distribution. If multiple Markovian transitions emanate a state P, the parameter of the sojourn time distribution SJ_P is obtained by cumulating all the transition rates (except for loops, of course).

In a sense, each Markovian transition $P \overset{\lambda_i}{\longrightarrow} P_i$ (with $P \neq P_i$) contributes a clock C_i with an exponentially distributed expiration time to the sojourn time of P, determined by $\lambda = \sum \lambda_i$. The interpretation is as follows. All these clocks are (re)initialised whenever the state is entered. The sojourn time elapses as soon as either of the clocks, say C_j, has elapsed, and the system jumps to the respective successor state P_j. The sojourn time is hence given by a *minimum* of exponential distributions. Since according to property (C) the minimum of such distributions is exponentially distributed with the sum of the rates, this interpretation explains why the parameter λ of SJ_P is cumulated from the individual transition rates λ_i. Furthermore, the probability to reach state P_j is given by the fraction $\lambda_j / \sum \lambda_i$, the probability that clock C_j is first to elapse.

Since the sojourn time distribution is exponential, it is memoryless, as in the discrete-time case, and we get[3]

$$Prob\{SJ_P \leq t + \Delta t \mid SJ_P > t\} = Prob\{SJ_P \leq \Delta t\} = 1 - e^{-\lambda \Delta t}.$$

Thus, we obtain, for instance, $Prob\{SJ_{E_{30}} \leq 1002 \mid SJ_{E_{30}} > 1000\} = Prob\{SJ_{E_{30}} \leq 2\} = 0.55$, in analogy to the discrete-time case.

[3] As a consequences of the memoryless property the re-initialisation of those clocks that do not have expired is irrelevant when reentering a state.

3.4 Analysing Markov Chains

The study of Markov chains is usually aiming to condense relevant time-dependent or long-term information out of the detailed description of the chain. Typical measures of interest include

- mean time MTA until an absorbing state is reached,
- mean time $MTFP(P)$ until the first passage of a specific state P,
- probability $Prob\{X_t = P\}$ to be in a specific state at a specific point in time t,
- limiting probability $\lim_{t\to\infty} Prob\{X_t = P\}$ to be in a specific state provided that the chain has reached an equilibrium.

More advanced measures can be defined using a temporal logic, as e.g. in [17], but of course, not every measure is meaningful for every Markov chain. For instance, it is only reasonable to ask for the mean time until absorption if the chain does indeed contain absorbing states, i.e., states that do not possess successor states, and if at least one of these states is reachable from the initial state. Similarly, the existence of an equilibrium can only be guaranteed under some conditions.

For a homogeneous (discrete- or continuous-time) Markov chain with finite state space, an equilibrium is known to exist if the Markov chain is *ergodic*. Ergodicity, in turn requires ***irreducibility*** and, in the discrete-time case, ***aperiodicity***. A chain is irreducible if any state is reachable from any other state by means of a sequence of (probabilistic, respectively Markovian) transitions. (Note that irreducibility implies the absence of absorbing states.) Aperiodicity refers to the number of steps needed from state P to return to state P. If the period of P, the greatest common divisor of all such numbers, amounts to 1, then the state is said to be aperiodic. An irreducible chain is aperiodic if one of its states is, since all states in such a chain have the same period.

Example 3.4.1. The probabilistic chain E_{20} of Figure 3.1 describes an irreducible, aperiodic DTMC. It is thus ergodic, whence we know that an equilibrium exists. On the contrary, the Markovian chain E_{30} (Figure 3.2) is not irreducible.

Supposing that we are studying a probabilistic chain $(S, \twoheadrightarrow, P)$ with an ergodic DTMC X, we can now turn our attention towards the computation of the limiting state probabilities in the equilibrium, $\lim_{t\to\infty} Prob\{X_t = P\}$, denoted π_P in the sequel. The computation is usually called ***steady state analysis***, as opposed to ***transient analysis*** that evaluates the time dependent state probabilities $Prob\{X_t = P\}$. We refer to [181, 91] for a discussion of numerical algorithms for transient analysis.

State probabilities give very fine grain information about the Markov chain. This is often required to get some particular insight into the behaviour. If we define $lab(P, Q) = p$ in case there is a transition $P \overset{p}{\twoheadrightarrow} Q$

and $lab(P,Q) = 0$ otherwise, then we can obtain the probabilities π_P by solving the set of linear equations

$$\pi_P = \sum_{P' \in S} \pi_{P'} \; lab(P', P) \qquad \text{for each } P \in S. \tag{3.8}$$

Any solution of these equations $\tilde{\pi} = (\tilde{\pi}_P, P \in S)$ will be called an *unnormalised steady state distribution.* Due to the condition of irreducibility, several of these solutions exist. In order to obtain a proper probability distribution on S, the solution has to be normalised by taking into account that the probability to be in any of the states of the chain always cumulates to 1, i.e., $\sum_{P \in S} \pi_P = 1$.

Example 3.4.2. The steady state distribution of the probabilistic chain of Figure 3.1 satisfies the equations

$$\begin{aligned} \pi_{E_{20}} &= 4/5\, \pi_{E_{20}} + \pi_{E_{22}} \\ \pi_{E_{21}} &= 1/3\, \pi_{E_{21}} + 1/5\, \pi_{E_{20}} \\ \pi_{E_{22}} &= 2/3\, \pi_{E_{21}}. \end{aligned}$$

As a result of normalisation with $\pi_{E_{20}} + \pi_{E_{21}} + \pi_{E_{22}} = 1$ *we obtain* $\pi_{E_{20}} = 2/3$, $\pi_{E_{21}} = 1/5$, *and* $\pi_{E_{22}} = 2/15$.

Analysing steady state probabilities of a Markovian chain describing an ergodic CTMC proceeds in precisely the same way, apart from the fact that transition *rates* replace transition *probabilities* in function *lab* and that $lab(P,P) = -\lambda$ where λ is the rate of the sojourn time SJ_P. Solving a set of linear equations is usually performed by means of matrix calculations and efficient methods exist, tailored to the setting of Markov chains [181].

State probabilities are often the starting point to obtain more general insight into the behaviour of a chain. Indeed, they form the basis of a wide class of measures that appear (in the simplest cases) as weighted sums of such probabilities. To obtain a particular measure each state P is equipped with a *reward* $\Re(P)$ provided by a real valued *reward function* $\Re : S \mapsto \mathbb{R}$.

For instance if the chain models an unreliable system, we can isolate those states where the system is down. They obtain a reward $\Re(s) = 1$ while all others obtain reward 0. A measure of interest, the probability that the system is down (in the equilibrium), then arises as the weighted sum

$$\sum_{P \in S} \pi_P \, \Re(P). \tag{3.9}$$

Example 3.4.3. We have earlier provided an interpretation of chain E_{20} *as a model of temperature evolution. If we are interested in the probability for having a modest or hot climate, we may define that* $\Re(E_{20}) = \Re(E_{21}) = 1$ *and* $\Re(E_{22}) = 0$. *We thus calculate the weighted sum* $\sum_{P \in \{E_{20}, E_{21}, E_{22}\}} \pi_P \; \Re(P) = 2/3 + 1/5 = 13/15$ *as the steady state probability for this situation.*

The same formula (3.9) can be used to compute various other measures. If, for instance, the system under investigation contains a queue we can compute the mean queue length by assigning $\Re(P) = n$ iff precisely n places are occupied in the queue. More complicated reward structures are known, that cover performability and dependability aspects, see e.g. [92].

3.5 Equivalences on Markov Chains

Strong and weak bisimilarities, as introduced in Chapter 2, are central in the theory of process algebraic equivalences. Apart from their theoretical importance, a practical merit is the possibility of behaviour preserving state space aggregation, as exemplified in Section 2.4. This is achieved by neglecting the identity of states in favour of equivalence classes of states exhibiting identical behaviours. We follow the same spirit in the context of Markovian (and later also of probabilistic) chains. We achieve a state space aggregation method that is profitable in order to allow one to analyse complex Markov chains.

3.5.1 Bisimilarity on Markovian Chains

We begin our efforts to define a bisimulation style equivalence in the context of Markovian chains. For a given chain, assume that we are only interested in probabilities of equivalence classes of states with respect to some equivalence $\sim$ (that we are aiming to define) instead of probabilities of states. Any such equivalence preserving view on a Markovian chain gives rise to an aggregated stochastic process $\widetilde{X} = \{\widetilde{X}_t | t \in T\}$. It can be defined on the state space $S/\sim$, the set of the equivalence classes with respect to $\sim$, by

$$Prob\{\widetilde{X}_t = C\} := Prob\{X_t \in C\} \qquad \text{for each } C \in S/\sim. \tag{3.10}$$

$\widetilde{X}$ is a discrete state space stochastic process, but it is not necessarily a CTMC, let alone a time homogeneous one. However sufficient conditions exist such that $\widetilde{X}$ is again a time homogeneous CTMC. They impose restrictions on the shape of the sets C and are known as lumping conditions [128]. We approach them from a different perspective, namely by constraints on the equivalence $\sim$, similar to [42, 114]. Anticipating the technical details, we achieve that $\widetilde{X}$ is a homogeneous CTMC, if $\sim$ is a variant of bisimulation. The difficulty is that we have to equate not only qualities but also *quantities*, for example transition rates of moving from one state to an equivalence class. In contrast, bisimilarity only talks about a (logical) quality: Either there is a move from a state into a class possible or it is impossible, but *tertium non datur*.

The bridge to *quantify* strong bisimilarity is an alternative characterisation of (ordinary) strong bisimilarity that we have mentioned as Lemma 2.2.2.

To recall its essentials, note that it uses a predicate $\gamma_{\text{O}} : S \times Act \times 2^S \mapsto \{\texttt{true}, \texttt{false}\}$ that is true iff P can evolve to a state contained in a set of states C (by interaction on action a). Bisimilarity then occurs as the union of all equivalence relations that equate two states if they posses the same γ_{O} values (for each possible combination of action a and equivalence class C).

We follow this style of definition but replace the predicate γ_{O} by a (non-negative) real-valued function $\gamma_{\text{M}} : S \times 2^S \mapsto \mathbb{R}^+$, that calculates the cumulative rate to reach a set of states C from a single state R by

$$\gamma_{\text{M}}(R, C) = \sum \{\!|\lambda | R \xrightarrow{\lambda} R' \text{ and } R' \in C|\!\}.$$

In this definition we let $\sum \{\!|\ldots|\!\}$ denote the sum of all elements in a multiset (of transition rates), where $\{\!|\ldots|\!\}$ delimits this multiset. The need for this notational burden is best explained by means of an example.

Example 3.5.1. Considering Figure 3.2, the cumulative rate to reach any state in S from state E_{30} is $\gamma_{\text{M}}(E_{30}, S) = \sum \{\!|0.2, 0.2|\!\}$ which amounts to 0.4 due to our definition. Note that by the additional restriction imposed below Definition 3.3.1 multiple transitions labelled with the same rate are prohibited between pairs of states, but they may anyway occur from a state to multiple successor states, as in this example.

We are now ready to lift bisimilarity to the setting of Markovian chains, along the lines of Definition 2.2.2 and Lemma 2.2.2.

Definition 3.5.1. *For a given Markovian chain $(S, \longrightarrow, P)$, an equivalence relation $\mathcal{E}$ on S is a Markovian bisimulation iff $P\mathcal{E}Q$ implies that for all equivalence classes C of $\mathcal{E}$ it holds that*

$$\gamma_{\text{M}}(P, C) = \gamma_{\text{M}}(Q, C).$$

Two states P and Q are Markovian bisimilar, written $P \sim_{\text{M}} Q$, if (P, Q) is contained in some Markovian bisimulation $\mathcal{E}$.

Thus $\sim_{\text{M}}$ is the union of all such equivalences. Indeed, it is itself a Markovian bisimulation and therefore the largest such relation.

Definition 3.5.2. *For a given Markovian chain $(S, \longrightarrow, P)$ and a Markovian bisimulation $\mathcal{E}$ on S, define an aggregated chain $(S/\mathcal{E}, \longrightarrow_{\mathcal{E}}, [P]_{\mathcal{E}})$ where the Markovian transition relation $\longrightarrow_{\mathcal{E}}$ is given by*

$$[P']_{\mathcal{E}} \xrightarrow{\lambda}_{\mathcal{E}} [Q']_{\mathcal{E}} \quad \textit{iff} \quad \gamma_{\text{M}}(P', [Q']_{\mathcal{E}}) = \lambda.$$

Example 3.5.2. With the notation introduced in Chapter 2 each of the sets , ,  , *and* 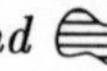 *appearing in Figure 3.3 is a class of an equivalence relation $\mathcal{E}$ on the state space of E_{30} satisfying Definition 3.5.1. In particular, we compute the values*

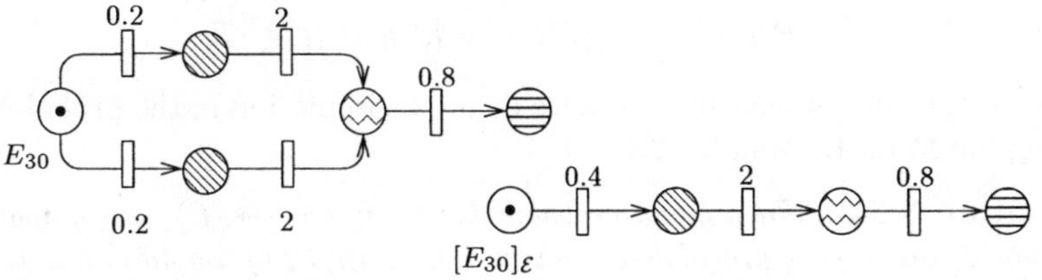

Figure 3.3. Markovian chain and its aggregated representative

$$\gamma_{\mathrm{M}}(\bigcirc\ ,\ \text{⬤}) = 0.4 \qquad \gamma_{\mathrm{M}}(\text{⬤}\ ,\ \text{⬤}\ ) = 2 \qquad \gamma_{\mathrm{M}}(\text{⬤}\ ,\text{⬤}) = 0.8$$

for the states in the respective classes, all other values of γ_{M} are zero. The aggregated Markov chain $[E_{30}]_{\mathcal{E}}$ obtained by applying Definition 3.5.2 is depicted on the right.

Theorem 3.5.1. *Let P be a Markovian chain, describing the CTMC X and let $\mathcal{E}$ be a Markovian bisimulation on the state space of P. The aggregated chain $P_{\mathcal{E}}$ describes a homogeneous CTMC $\widetilde{X}$ such that for all equivalence classes C of $\mathcal{E}$,*

$$Prob\{\widetilde{X}_t = C\} = Prob\{X_t \in C\}.$$

Proof. The conditions imposed on a Markovian bisimulation can be matched with the definition of lumpability [128, 42, 114].

As a particular consequence, the stochastic process induced by factoring with respect to a Markovian bisimulation is again a homogeneous CTMC.

As mentioned above this kind of aggregation is known as *lumping*. Lumping is usually formulated with respect to a *suitable* partitioning of the state space. Here, we have defined a *suitability criterion* in a coinductive way. Our partitioning is obtained via factoring with respect to a bisimulation.

The most profitable result obtained from this style of definition is that we are able to adopt partition refinement algorithms for computing bisimilarity to the setting of CTMC. As a result, we can efficiently compute suitable partitionings. The original definition of lumpability in [128] is not constructive, i.e., it does not give rise to an algorithmic way to compute suitable partitions. Therefore, the use of lumping had to rely on heuristics, e.g. by detecting symmetries in the model, as in [167, 46, 176].

3.5.2 Bisimilarity on Probabilistic Chains

With the notational and conceptual background of the previous section, it is rather straightforward to define a bisimulation style equivalence for probabilistic chains. In order to compare the probabilistic behaviour of two states we define a cumulative probability function $\gamma_{\mathrm{p}} : S \times 2^S \mapsto \mathbb{R}^+$ along the lines of γ_{M}.

$$\gamma_{\mathrm{P}}(R, C) = \sum \{\!|p | R \overset{p}{\twoheadrightarrow} R' \text{ and } R' \in C|\!\}.$$

As in the continuous-time case, we define bisimulation in the probabilistic setting based on Lemma 2.2.2.

Definition 3.5.3. *For a given probabilistic chain $(S, \twoheadrightarrow, P)$, an equivalence relation $\mathcal{E}$ on S is a probabilistic bisimulation iff $P\mathcal{E}Q$ implies that for all equivalence classes C of $\mathcal{E}$ it holds that*

$$\gamma_{\mathrm{P}}(P, C) = \gamma_{\mathrm{P}}(Q, C).$$

Two states P and Q are probabilistic bisimilar, written $P \sim_{\mathrm{P}} Q$, if (P, Q) is contained in some probabilistic bisimulation $\mathcal{E}$.

Thus $\sim_{\mathrm{P}}$ is the union of all such equivalences. It is itself a probabilistic bisimulation and therefore the largest such relation. However, it seems worth to remark a surprising difference to the continuous case. Probabilistic bisimilarity on a given probabilistic chain with state space S always leads to a trivial result: it equates all the states, because *by definition* $\gamma_{\mathrm{P}}(P, S) = 1$ holds for each of the states (Definition 3.2.1). Thus S is the only equivalence class of $\sim_{\mathrm{P}}$.

This observation does by far not imply that the concept of probabilistic bisimulation is useless. Indeed, any probabilistic bisimulation (including $\sim_{\mathrm{P}}$) can be used to aggregate the state space of a given probabilistic chain, by factoring the state space, i.e., considering classes of equivalent states instead of individual states. But not every bisimulation is appropriate to aggregate the state space, dependent on the information we would like to condense from the chain.

A bisimulation should respect the measures of interest we are aiming to obtain. In particular, if a reward function $\mathfrak{R}$ is provided, a bisimulation should not equate states with different reward values, because otherwise it would be impossible to associate rewards to classes of equivalent states. We therefore introduce the notion of $\mathfrak{R}$-preserving probabilistic bisimulation that is parametric in the reward function $\mathfrak{R}$.

Definition 3.5.4. *A probabilistic bisimulation $\mathcal{E}$ on S is an* $\mathfrak{R}$-preserving probabilistic bisimulation *iff $P\mathcal{E}Q$ implies $\mathfrak{R}(P) = \mathfrak{R}(Q)$. Two states P and Q are $\mathfrak{R}$-preserving probabilistic bisimilar, written $P \sim_{\mathrm{P}}^{\mathfrak{R}} Q$, if (P, Q) is contained in some $\mathfrak{R}$-preserving probabilistic bisimulation $\mathcal{E}$.*

Again, $\sim_{\mathrm{P}}^{\mathfrak{R}}$ is the union of all such equivalences and is the largest $\mathfrak{R}$-preserving probabilistic bisimulation. With this definition we may proceed as in the continuous-time setting and define an aggregated probabilistic chain that describes a homogeneous DTMC.

Definition 3.5.5. *For a given probabilistic chain $(S, \twoheadrightarrow, P)$ and a probabilistic bisimulation $\mathcal{E}$ on S, define an aggregated chain $(S/\mathcal{E}, \twoheadrightarrow_{\mathcal{E}}, [P]_{\mathcal{E}})$ where the probabilistic transition relation $\twoheadrightarrow_{\mathcal{E}}$ is given by*

$$[P']_{\mathcal{E}} \xrightarrow{p}_{\mathcal{E}} [Q']_{\mathcal{E}} \quad \textit{iff} \quad \gamma_{\text{P}}(P', [Q']_{\mathcal{E}}) = p.$$

Theorem 3.5.2. *Let P be a probabilistic chain, describing the DTMC X and let $\mathcal{E}$ be a ($\Re$-preserving) probabilistic bisimulation on the state space of P.*
The aggregated chain $P_{\mathcal{E}}$ describes a homogeneous DTMC $\widetilde{X}$ such that for all equivalence classes C of $\mathcal{E}$,

$$Prob\{\widetilde{X}_t = C\} = Prob\{X_t \in C\}.$$

The proof follows the continuous-time case (Theorem 3.5.1). The necessary lifting of the reward function from S to $S/\mathcal{E}$ is easy: for each class C define $\widetilde{\Re}(C) := \Re(P)$ (for some $P \in C$).

Thus we have established a valuable link between *lumping* and bisimulation. We have introduced this parametric refinement of $\sim_{\text{P}}$ mainly in order to avoid the somehow awkward phenomenon that probabilistic bisimilarity lumps all the states (which *is* indeed optimal if no reward function is provided). This effect was not observed in the context of Markovian chains.[4] But even in this context it may occur that, in the presence of a reward function, states with different rewards are equated by Markovian bisimilarity. In this situation, a refinement of $\sim_{\text{M}}$ that preserves rewards can be defined along the lines of Definition 3.5.4, see also [24]. Since reward functions will not be of particular importance in the remainder of this book we do not work out the details.

3.5.3 Weak Bisimulations

Hitherto we have studied only *strong* bisimilarity on Markovian and probabilistic chains. It seems to be equally worthwhile to investigate *weak* bisimilarity. For this purpose, several questions have to be addressed. We start with a discussion in the setting of Markovian chains.

Markovian Chains. First, what is the counterpart of a *weak transition* in terms of Markovian transitions? In the non-stochastic setting we have used a weak transition relation to successfully define weak bisimilarity. It was based on the distinction between internal actions (labelled τ) and external, observable actions. Such a distinction is not obvious for Markovian chains, because there is no notion of interaction with the external environment.

We may therefore refuse to think about weak relations on Markovian chains at all. Alternatively we may decide that either none, or all of the Markovian transitions are internal. In the former case, a weak Markovian bisimulation will not differ from its strong counterpart, because there is no

[4] The same phenomenon however arises if Definition 3.5.1 is slightly coarsened, such that loops are treated as being irrelevant (which they are indeed). This can be achieved by including a side condition saying that the cumulative rates of P and Q into class C only have to be compared if $P \notin C$.

internal transition that could be abstracted away. So, how about assuming that all Markovian transitions are internal? The corresponding weak transition relation would then combine sequences of Markovian transitions into a single 'weak' transition, in the same way as $\stackrel{\tau}{\Longrightarrow}$ combines sequences of $\stackrel{\tau}{\longrightarrow}$ transitions. For instance, a sequence $P \stackrel{\lambda}{\dashrightarrow} P' \stackrel{\mu}{\dashrightarrow} P''$ could be combined to a weak transition from P to P'' with a parameter ν. This parameter subsumes the exponentially distributed sojourn times in P and P', and it may, in general, be defined as a function $\phi(\lambda, \mu)$.

Unfortunately, the sequence of two (or more) exponentially distributed delays is no longer exponentially distributed. So, any particular choice of a function ϕ will introduce (a possibly severe) error in the model. In other words, replacing a sequence of Markovian transitions by a single *weak* Markovian transitions will lead to a CTMC where it is impossible to reconstruct the stochastic behaviour of the original chain. A result similar to Theorem 3.5.1 is thus not possible for any kind of weak Markovian bisimulation.

Probabilistic Chains. Turning our attention to the discrete-time case, we may again decide that either all probabilistic transitions are external or they are internal. The former case directly leads to the strong notion of bisimulation introduced earlier. In the latter case, we aim to combine sequences of probabilistic transition in a weak transition. So, we have to address the question, what the probability of a sequence of probabilistic transitions might be. Different to the setting of Markovian chains, there is a rather straightforward answer to this question: The probability of a sequence of probabilistic transitions is obviously the product of the individual one step transition probabilities.

We will follow this spirit by introducing a function γ_{P} that naturally extends the predicate γ_{O} to the probabilistic case. This predicate has been introduced in Section 2.3 (page 26) to describe the possibility to move into a set of states by means of a weak transition. Before introducing the function γ_{P} we point out the following property of $\gamma_{\mathrm{O}}(P, \tau, C)$. This predicate is `true` iff either $P \in C$ (the reflexive closure) or if there is some P' such that $P \stackrel{\tau}{\longrightarrow} P'$ and $\gamma_{\mathrm{O}}(P', \tau, C)$ (the transitive closure). In other words:

$$\gamma_{\mathrm{O}}(P, \tau, C) = \begin{cases} \texttt{true} & \text{if } P \in C, \\ \bigvee_{P' \in S} \left(\gamma_{\mathrm{O}}(P, \tau, \{P'\}) \wedge \gamma_{\mathrm{O}}(P', \tau, C)\right) & \text{else.} \end{cases}$$

This specific predicate indicates the *possibility* to move from state P into a set C by means of a (possibly empty) sequence of internal transitions. The function $\gamma_{\mathrm{P}}(P, C)$ we are aiming to define will indicate the *probability* to move from state P into set C by means of a (possibly empty) sequence of probabilistic transitions. With respect to an empty sequence (corresponding to the reflexive closure) this probability is obviously 1 if and only if P is already contained in C. If $P \notin C$ we cumulate the probabilities of all the different possibilities to leave P and to end in the set C. This can be elegantly defined as follows, with a clear correspondence to the predicate $\boldsymbol{\gamma}_{\mathrm{O}}$,

$$\boldsymbol{\gamma}_{\mathrm{P}}(P,C) = \begin{cases} 1 & \text{if } P \in C, \\ \sum\limits_{P' \in S} \left(\gamma_{\mathrm{P}}(P,\{P'\})\, \boldsymbol{\gamma}_{\mathrm{P}}(P',C)\right) & \text{else.} \end{cases}$$

With this function, weak probabilistic bisimulation is straightforward to define. Since we want to avoid the effect of equating all states (which will surely happen, since a weak relation has the potential of equating *even more* states than its strong counterpart) we parametrise the relation with a given reward function $\mathfrak{R}$.

Definition 3.5.6. *An equivalence relation $\mathcal{E}$ is a $\mathfrak{R}$-preserving weak probabilistic bisimulation iff $P\mathcal{E}Q$ implies $\mathfrak{R}(P) = \mathfrak{R}(Q)$ and that for all equivalence classes C of $\mathcal{E}$,*

$$\boldsymbol{\gamma}_{\mathrm{P}}(P,C) = \boldsymbol{\gamma}_{\mathrm{P}}(Q,C).$$

Two states P and Q are $\mathfrak{R}$-preserving weak probabilistic bisimilar, written $P \approx^{\mathfrak{R}}_{\mathrm{P}} Q$, if (P,Q) is contained in some $\mathfrak{R}$-preserving weak probabilistic bisimulation $\mathcal{E}$.

Again, it can be shown that $\approx^{\mathfrak{R}}_{\mathrm{P}}$ is an equivalence relation, a weak probabilistic bisimulation that preserves $\mathfrak{R}$ and therefore the largest such relation.

Despite these properties, it has to be addressed whether any such weak bisimulation has a plausible stochastic interpretation. For (strong) probabilistic bisimulations the correspondence to lumping (Theorem 3.5.2) provided an intuitive and valuable interpretation.

However, the weak probabilistic bisimulations do not provide a comparably valuable result. For ergodic chains we obtain that $P \approx^{\mathfrak{R}}_{\mathrm{P}} Q$ if $\mathfrak{R}(P) = \mathfrak{R}(Q)$ as well as $\pi_P = \pi_Q$ which is not truly helpful. For absorbing chains, the equivalence classes of $\approx^{\mathfrak{R}}_{\mathrm{P}}$ do not represent obviously valuable information. The reason is that $\boldsymbol{\gamma}_{\mathrm{P}}(P,C)$ does not allow comparing the number of (time) steps needed to move from state P into class C. In the discrete-time context however, this information is important but gets lost when building weak bisimulation classes. Thus a result similar to Theorem 3.5.2 is out of reach.

As a whole, strong bisimulations are very useful to aggregate discrete- or continuous-time Markov chains, while weak bisimulations are not in this context.

3.6 Algorithmic Computation of Equivalences

In this section we describe algorithms to compute Markovian bisimilarity, and later probabilistic bisimilarity. We build upon our experience gained in Section 2.3 and use a partition refinement approach. For this purpose

Table 3.1. Algorithm for computing Markovian bisimilarity classes

Input:	Markovian transition system $(S, \dashrightarrow)$
Output:	$S/\sim_{\mathrm{M}}$
Method:	$Part := \{S\}$;
	$Spl := \{S\}$;
	While Spl not empty **do**
	Choose C **in** Spl;
	$Old := Part$;
	$Part := M\_Refine(Part, C)$;
	$New := Part - Old$;
	$Spl := (Spl - \{C\}) \cup New$;
	od
	Return $Part$.

we reformulate Markovian bisimilarity as a fixed-point of successively finer relations.

Lemma 3.6.1. *Let P be an Markovian chain with finite (reachable) state space S.*
Markovian bisimilarity is the unique fixed-point of

- $\smile_0 = S \times S$,
- $\smile_{k+1} = \{(P, Q) \in \smile_k \mid (\forall C \in S/\smile_k)\ \gamma_{\mathrm{M}}(P, C) = \gamma_{\mathrm{M}}(Q, C)\}$.

The proof of this lemma is not very difficult. The refinement step from $\smile_k$ to $\smile_{k+1}$ rules out those pairs (P', Q') from $\smile_k$ for which γ_{P} produces different values. So, P' and Q' are split into different classes (of $\smile_{k+1}$), if the respective rates to move into class C (of $\smile_k$) differ. The class C is thus a *splitter*, a specific reason why (P', Q') have to be split.

The above lemma will be the basis of the algorithm for computing Markovian bisimilarity. Technically the procedure of refining a partitioning *Part* by means of a splitter C is performed by a function *M_Refine*, defined as follows,[5]

$$M\_Refine(Part, C) :=$$
$$\left(\bigcup_{X \in Part}\left(\bigcup_{v \in \mathbb{R}^+}\left\{\{P \in X \mid \gamma_{\mathrm{M}}(P, C) = v\}\right\}\right)\right) - \{\emptyset\}.$$

Each class X of *Part* is refined into finer classes, according to the different rates v of moving into class C. This function is an adaption of function

[5] Again we are a bit sloppy, because we use $\mathbb{R}^+$ as an index set, which, strictly speaking, leads to an uncountable union. However, only a finite union is needed, since the number of possibly different γ_{M} values is bounded by the number of states.

Refine, defined in Section 2.3. To point out the differences, we recapitulate its definition,

$$Refine(Part, a, C) := \left(\bigcup_{X \in Part} \left(\bigcup_{v \in \{\texttt{true},\texttt{false}\}} \left\{ \{P \in X \mid \gamma_{\mathrm{O}}(P, a, C) = v\} \right\} \right) \right) - \{\emptyset\}.$$

The notational adoptions are obvious. Remark that *M_Refine* possibly splits a class X with, say, m elements into m classes, specifically if all values of the real valued function γ_{M} are different. In contrast, function *Refine* splits each class X into at most two new parts, dependent on the truth values of predicate γ_{O}.

Function *M_Refine* is the heart of an algorithm to compute Markovian bisimilarity. This algorithm is depicted in Table 3.1. Though function *M_Refine* potentially splits more efficient than its non-stochastic counterpart, the worst case complexities of the two algorithms are the same.

Theorem 3.6.1. *The algorithm of Table 3.1 computes Markovian bisimilarity on S. It can be implemented with a time complexity of $\mathcal{O}(m_{\mathrm{M}} \log n)$ where m_{M} is the number of Markovian transitions and n is the number of states. The space complexity of this implementation is $\mathcal{O}(m_{\mathrm{M}})$.*

Proof. See Appendix A.1.

In the setting of probabilistic chains, the task of computing probabilistic bisimilarity (Definition 3.5.3) does not require any specific algorithm. We have already pointed out that S is the only equivalence class of $\sim_{\mathrm{P}}$, hence refinement is not required. The issue of interest in this setting is $\Re$-preserving probabilistic bisimilarity.

Again, we are able to characterise such a relation as a fixed point of successively finer relations. But this time we have to ensure that only states with the same reward are related. This requirement is incorporated into the initial relation of the refinement, $\smile_0$. Instead of $S \times S$ we start with an equivalence relation that equates precisely the states with the same reward. This equivalence $\smile_0$ is the coarsest reward preserving equivalence. Further refinement steps can then proceed as in the above cases.

Lemma 3.6.2. *Let P be an probabilistic chain with finite (reachable) state space S and let $\Re : S \mapsto \mathbb{R}$ be a reward function.*
$\Re$-preserving probabilistic bisimilarity is the unique fixed-point of

- $\smile_0 = \{(P', Q') \in S \times S \mid \Re(P') = \Re(Q')\}$
- $\smile_{k+1} = \{(P, Q) \in \smile_k \mid (\forall C \in S/\smile_k)\ \gamma_{\mathrm{P}}(P, C) = \gamma_{\mathrm{P}}(Q, C)\}$

This lemma can be turned into an algorithm, depicted in Table 3.2. The initialisation phase differs from the ones we have described before, in order to ensure preservation of rewards. All states with the same reward are grouped

Table 3.2. Algorithm for computing reward preserving probabilistic bisimilarity classes

Input:	Probabilistic transition system $(S, \twoheadrightarrow)$
	Reward function $\Re : S \mapsto \mathbb{R}$
Output:	$S/\sim_{\mathrm{P}}^{\Re}$
Method:	$Part := \bigcup_{v \in \mathbb{R}^+} \{\{P' \in S \mid \Re(P') = v\}\} - \{\emptyset\};$
	$Spl := Part;$
	While *Spl* not empty **do**
	Choose *C* **in** *Spl*;
	$Old := Part;$
	$Part := P\_Refine(Part, C);$
	$New := Part - Old;$
	$Spl := (Spl - \{C\}) \cup New;$
	od
	Return *Part*.

into the same initial partition. The body of this algorithm iteratively calls a function *P_Refine*, defined as follows (and as expected),

$$P\_Refine(Part, C) :=$$

$$\left(\bigcup_{X \in Part} \left(\bigcup_{v \in \mathbb{R}^+} \Big\{\{P \in X \mid \gamma_{\mathrm{P}}(P, C) = v\}\Big\}\right)\right) - \{\emptyset\}.$$

The computational effort to determine the initial partition is linear in the number of states. Since this is negligible compared to the complexity of partition refinement, the time and space complexity of the whole algorithm is the same as the one for Markovian bisimilarity.

3.7 Discussion

In this section we have introduced probabilistic and Markovian chains as operational models of discrete-time, respectively continuous-time Markov chains. We have discussed details of their analysis to set the ground for the requirements of later chapters.

The equivalence relations introduced in Section 3.5 are defined in a coinductive way based on the notion of bisimulation. Like any equivalence they partition the state space into equivalence classes of states. In this way, each equivalence induces an aggregated stochastic process with a discrete state space, where each state corresponds to an equivalence class.

The important result is that this aggregated process is again a Markov chain of the same type (discrete-time or continuous-time) if a bisimulation is used (Theorem 3.5.1 and 3.5.2). The Markov chain has been *lumped* by applying the respective bisimulation.

Probabilistic bisimulation on DTMC has first been introduced in [133], and has been transferred to CTMC under the name Markovian bisimulation in [94]. Hillston [114] has pointed out the relation between bisimulation and lumping (in the continuous-time case) but developed no algorithm. Buchholz [43] has proposed a partition refinement algorithm for this purpose. His algorithm has a time complexity of $\mathcal{O}(n\, m_M)$ (where n is the number of states and m_M the number of transitions) while the ones presented in Section 3.6 perform better; we have shown that they are of order $\mathcal{O}(m_M \log n)$. This complexity has also been mentioned without proof by Bernardo and Gorrieri [25], as well as Huynh and Tian who considered the discrete-time case [120]. The $\mathcal{O}(m_M \log n)$ complexity proof is joint work with Salem Derisavi and William H. Sanders [59], while the proof sketch in [96] is flawed. Since either of our algorithms computes the largest (probabilistic, respectively Markovian) bisimulation, they induce an aggregation which is optimal among the possible lumpings of a chain.

The original definition of lumping [128] does not give rise to an algorithmic procedure. Therefore, the use of lumping had to rely on heuristics, e.g. by detecting symmetries in the model, e.g. [46, 175]. Markovian and probabilistic bisimilarity improve this. However, in order to obtain a useful lumping, we have refined probabilistic bisimulation such that it preserves a given reward function. We will see later that lumping can aggregate the state space of a Markov chain by several orders of magnitudes, in particular as part of a compositional methodology.

Composition operators have not been an issue of this chapter. This is particularly reflected by the fact that our attempt to define a weak bisimulation on Markov chains was rather unsuccessful. In Chapter 2 the notion of weak bisimulation relied on the distinction between external and internal actions. While the former can be used in the context of synchronisation of components the latter can not. Since we have made no such distinction in the Markov chain models of this chapter, it is not surprising that the notion of weak bisimulation is not reasonable here.

For the sake of completeness we mention that by introducing a distinction between external and internal *probabilistic* transitions, the notion of weak probabilistic bisimilarity in the sense of Definition 3.5.6 turns out to quite valuable, it naturally extends non-stochastic weak (and also branching) bisimilarity. Together with an algorithm of cubic time complexity to compute this relation, this result is joint work with Christel Baier [16]. In contrast, a weak *Markovian* bisimulation suffers from the fact that sequences of exponential distributions are not exponentially distributed. Thus even a distinction between external and internal Markovian transitions will not lead to a reasonable definition.

4. Interactive Markov Chains

This chapter introduces the central formalism of this book, Interactive Markov Chains[1] (IMC). It arises as an integration of interactive processes and *continuous-time* Markov chains. There are different ways to combine both formalisms, and some of them have appeared in the literature. We therefore begin with a detailed discussion of the different integration possibilities and argue why we take which decision. As a result IMC combine the different ingredients as orthogonal to each other as possible. We proceed by defining composition operators for IMC. We then focus our attention on the discussion of strong and weak bisimilarity, incorporating the notion of *maximal progress* into the definitions. In order to efficiently compute these relations we develop algorithms that are more involved than the ones presented in earlier chapters. Anyhow, we prove that their computational complexity is not increased. A small example of using IMC to compositionally specify and aggregate the leaky bucket principle concludes this chapter.

4.1 Design Decisions

The operational representation of interactive processes and Markov Chains in terms of transition systems suggests that they are rather easy to combine. However, there is a crucial difference between both models, the presence, respectively absence of *nondeterminism.*

The notion of nondeterminism is essential for interactive processes as it is essential for process algebra in general. Nondeterminism is useful to model different important concepts that otherwise could not be expressed. They can be summarised as follows [171]:

Implementation freedom: An interactive process can be viewed as an abstract specification, and nondeterminism represents implementation freedom. That is, if for some state there are two transitions that can be chosen nondeterministically, then an implementation may have just one of the two transitions.

[1] From a historical perspective, it seems worth to mention that the term "Interactive Markov Chains" has first been coined by Conlisk to baptise a Markov chain variant that is at most mildly related to the model considered here [51].

Scheduling freedom: This is the classical use of nondeterminism in an interleaving semantics. Several processes run in parallel and there is a freedom in the choice of which process performs the next transition.

External environment: External actions represent interaction possibilities with some external process, or more generally an external environment, by means of synchronisation. The interaction capabilities of this environment then influence how the choice is determined.

In contrast, a stochastic process is *not* exhibiting nondeterministic behaviour, since each possible behaviour at any point in time has a specific probability. Therefore, if we want to combine Markov chains with interactive processes, we have to address the question how to deal with nondeterminism in this setting.

The issue of this chapter is to integrate CTMC and interactive processes. Since integration of DTMC and interactive processes has a richer tradition, dating back to [183, 158] and others, it appears to be helpful to analyse the solutions that have appeared in this setting.

4.1.1 Discrete-Time Interactive Markov Chains

In the literature that combines DTMC and interactive processes we may find two fundamentally distinct treatments of nondeterminism, the ***fully probabilistic*** and the *alternating* model. Other models, like 'reactive' and 'stratified' models [76] or 'simple' and 'general' model [173, 171] are variants (or combinations) of the above two approaches.[2]

Fully probabilistic approaches *replace* nondeterminism by (discrete) probability distributions. Whenever there are multiple interactive transitions out of a single state each of them has a certain probability to occur.

The alternating model differs substantially from the fully probabilistic model in that there is a clear distinction between nondeterminism and probability. It allows for *both* nondeterministic as well as probabilistic decisions. There are two disjoint kinds of states in this model, ***probabilistic states***, whose transitions define a probability distribution among successor states, and *nondeterministic states*, whose outgoing transitions are labelled with actions.

Fully Probabilistic Model. This model is also called the *generative* model [76]. Whenever there is a decision between different interactive transitions, this decision is taken probabilistically. Nondeterminism is completely ruled out by means of a probability distribution that assigns a specific probability to each possible action. Technically this is achieved by melting interactive transitions $\xrightarrow{\texttt{a}}$ and probabilistic transitions $\xrightarrow{p}$ into a single transition relation $\xrightarrow{p\ \ \texttt{a}}$ where p is the probability of interacting on action a.

[2] Although the simple model of Segala can be encoded into the alternating model (and hence we do not treat it separately here), it is a very elegant model and has appealing properties, see e.g. [172, 21].

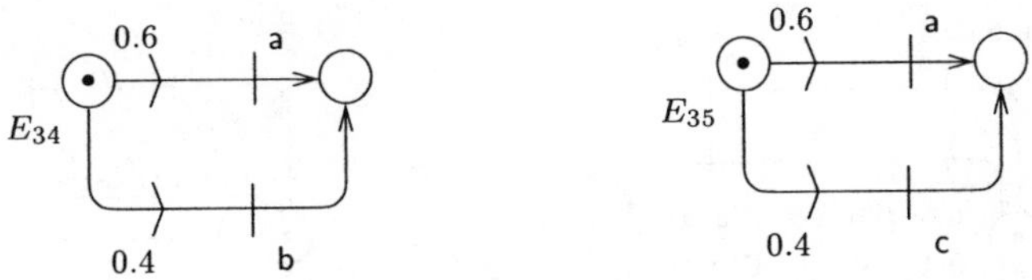

Figure 4.1. Two fully probabilistic transition systems

Example 4.1.1. Figure 4.1 shows two examples of the fully probabilistic model. The left one, E_{34}, chooses to perform a *with a probability of* 0.6 *and* b *with* 0.4 *probability.* E_{35} *is similar, but has the choice between* a *and* c.

This model has been extensively studied in the literature [69, 76, 133, 47, 50, 188, 88, 16]. In particular, since nondeterminism is absent, it is easily transferred into a DTMC by ignoring action labels of transitions. We shall not discuss the details of this model. We rather discuss an important implication that arises from this solution to avoid nondeterminism:

The exclusion of nondeterminism is not compatible with the concept of parallel composition we have introduced in Chapter 2. The reason is *scheduling freedom*: Parallel composition is realized by *interleaving* the different possibilities of each component. Nondeterminism is used to describe the freedom in the choice which component performs the next transition. Of course, the fully probabilistic model is not closed under such a kind of parallel composition since this would reintroduce nondeterminism. The solution that has been proposed by [69, 76] is introducing *synchrony.* This refers to a drastically different view on the evolution of processes. While we have assumed that actions that do not have to synchronise may occur independently and concurrently, synchrony means that all components have to interact (or explicitly signal that they intend to idle) each time, they proceed in *lockstep* [141]. In this way one avoids to make a scheduling decision since all components must perform a step.

Example 4.1.2. Synchronous parallel composition (denoted $\times$*) of* E_{34} *and* E_{35} *is depicted on the left in Figure 4.2. There are four different possibilities to proceed synchronously, and the respective probabilities are obtained by multiplying the individual probabilities. The action labels of the resulting transitions are also collected from the individual labels. A commutative function* $\aleph : Act \times Act \mapsto Act$ *has to be provided for this purpose. Remark that it is not possible to express that* E_{34} *and* E_{35} *only synchronise on a few specific actions, such as just action* a.

Some attempts have been made to avoid the synchrony assumption and to find an intuitive way of interleaving transitions by means of an *asynchronous* parallel composition operator [12, 152, 178, 84]. These operators are parametrised with probabilities that are used to determine a probability for each possible transition.

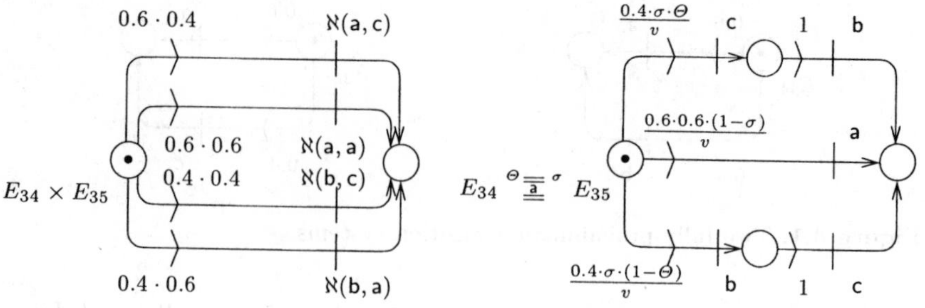

Figure 4.2. Different parallel compositions of fully probabilistic transition systems

Example 4.1.3. The approach of Baeten et al., *for instance, decorates parallel composition with two parameters σ and Θ [12]. Roughly speaking, $1-\sigma$ is the probability that a synchronisation takes place. The other parameter Θ is, again very roughly, the probability that the left component will perform a non-synchronising transition under the condition that no synchronisation takes place. It is a general form of parallel composition, where indeed synchronisation is assumed to be always possible, but it can be instantiated to the synchronisation style introduced in Chapter 2. This, however requires some normalisation, see [12] for the details. We may denote this operator as ${}^{\Theta}\!\equiv^{\sigma}$. Without going into the details, we depict in Figure 4.2 the fully probabilistic model obtained for $E_{34}\ {}^{\Theta}\!\underset{a}{\equiv}^{\sigma}\ E_{35}$ where $v = 0.4\cdot\sigma\cdot\Theta + 0.6\cdot 0.6\cdot(1-\sigma) + 0.4\cdot\sigma\cdot(1-\Theta)$ is a normalising denominator.*

This example allows one to catch a glimpse of the complications faced when aiming to define asynchronous parallel composition inside the fully probabilistic model. We consider all the existing approaches as unsatisfactory since neither of them possesses an intuitive interpretation. A detailed discussion of these operators and their problems appears in [55].

Another way to overcome the problems of parallel composition in this setting is to drop the assumption that concurrent interactive transitions may interleave. Resorting to, for instance, a *causality based semantics*, such as the one of Katoen *et al.* avoids the interpretation of parallelism by means of nondeterminism [127, 126]. However, such a semantics is far from resembling a DTMC. In order to derive a DTMC like model it is still necessary to decide whether to interleave the behaviour of components or to let them proceed synchronously.

Alternating Model. The alternating model has a distinct view on nondeterminism and probability, represented by two separate sets of states (nondeterministic and probabilistic states) and two separate transition relations, interactive and probabilistic transitions [183, 158, 87, 124]. Interactive transitions are only possible from nondeterministic states while probabilistic ones only

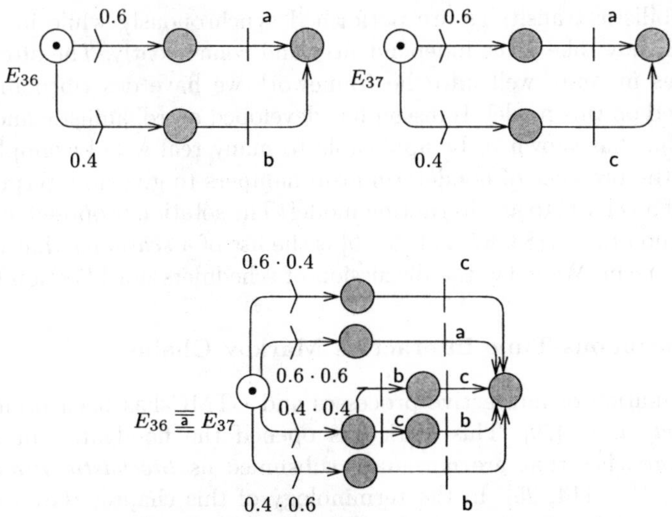

Figure 4.3. Two alternating transition systems and their parallel composition

emanate from probabilistic states. Probabilistic transitions define a probability distribution on successor states that cannot be influenced by the environment.

Example 4.1.4. Figure 4.3 shows two examples of the alternating model, similar to the fully probabilistic ones from above. The left one, E_{36}, has a (probabilistic) choice between two nondeterministic states. One of them performs a, *the other one performs* b. *E_{37} has a similar choice between* a *and* c. *We use* ● (○) *to indicate nondeterministic (probabilistic) states.*

This separation makes it possible to define *asynchronous* parallel composition because nondeterminism can be used to express scheduling freedom by interleaving of interactive transitions. Probabilistic transitions of different components are not interleaved. On a component level they describe a probability distribution on successor states. Since probabilistic decisions are assumed to be stochastically independent from each other (they are local to the component) the result of probabilistic decisions of two parallel components can be obtained by multiplying the respective individual probabilities. This can be represented by a synchronous probabilistic move in the parallel composition subsuming the probabilistic decisions of each component.

Example 4.1.5. On the right of Figure 4.3 we have depicted parallel composition of E_{36} and E_{37} when synchronising on a. *First both components may decide probabilistically where to move. In particular, with a probability of* $0.4 \cdot 0.4$ *both E_{36} and E_{37} decide to follow their respective lower branch. In this case, a state is reached where two actions,* b *and* c, *may occur independently and concurrently. They are interleaved nondeterministically.*

So, probabilistic transitions are performed synchronously while interactive transitions may take place independently and concurrently. The alternating model does fit quite well into the framework we have described in Chapter 2. Based on this model, Hansson has developed a probabilistic and timed calculus that has shown to be applicable to many real world examples [87]. However, the presence of nondeterminism hampers to give an interpretation in terms of a DTMC to an alternating model. The solution proposed by Vardi, Hansson and others [183, 87, 171, 5, 19] is the use of a *scheduler* that resolves nondeterminism. We delay the discussion of schedulers until Section 6.3.

4.1.2 Continuous-Time Interactive Markov Chains

The combination of interactive processes and CTMC has been brought up by Götz *et al.* in [79]. This work has opened the floodgates for a variety of approaches that are nowadays subsumed as *Stochastic Process Algebras* [42, 107, 114, 25]. In the terminology of this chapter they could be better characterised as *fully Markovian* approaches to the amalgamation of CTMC and interactive processes. Indeed, their common feature is that nondeterminism is *replaced* by probability distributions.

In this setting, probability distributions are continuous rather than discrete and they are drawn from the class of exponential distributions. This directly bridges to CTMC where Markovian transitions $\overset{\lambda}{-\!\!\square\!\!\rightarrow}$ describe exponentially distributed delays before moving to successor states. In the fully Markovian model, interactive transitions and Markovian transitions are melted in a single transition relation $\overset{\lambda}{-\!\!\square}\!\!\overset{\mathtt{a}}{\longrightarrow}$. The parameter λ represents an exponential distribution that describes the time needed *before* being able to move to the successor state by possible interaction on $\mathtt{a}$. There is an obvious correspondence to the *fully probabilistic* model of the discrete-time case: Nondeterminism is absent in this model because, implicitly, each transition possesses a probability to be chosen. To be specific, if several transitions emanate a single state P, say $P \overset{\lambda_i}{-\!\!\square}\!\!\overset{\mathtt{a}_i}{\longrightarrow} P_i$, then the probability to choose interaction on $\mathtt{a}_j$ is given by the fraction $\lambda_j / \sum \lambda_i$ (cf. property (D) on page 42).

Parallel Composition with Interleaving. Due to the analogy with the fully probabilistic model the probing question is whether the definition of parallel composition in the fully Markovian model suffers from similar problems. Indeed, severe problems arise, but not because of the issue of interleaving and nondeterminism. On the contrary, interleaving fits very well into the setting of continuous-time Markov chains.

Example 4.1.6. Assume that we are aiming to compose the two fully Markovian chains E_{40} and E_{41} of Figure 4.4 without any synchronisation. Using the general form of interleaving, the fully Markovian model depicted below would occur, where the rates $\phi_i(\lambda, \mu)$ should, if at all possible, *describe exponentially distributed probability distributions.*

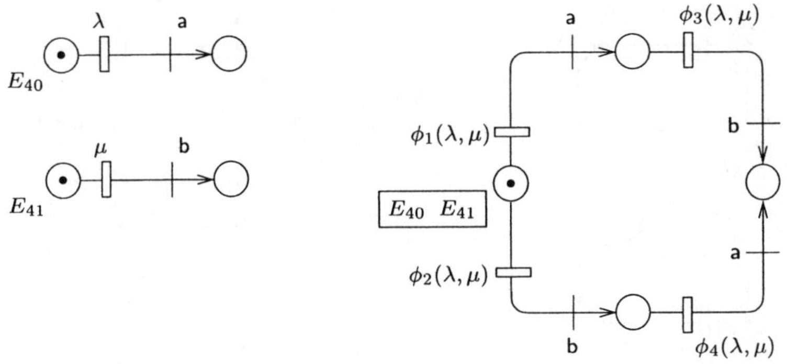

Figure 4.4. Asynchronous composition in the fully Markovian approach

In this example we have incorporated the hypothesis that it is indeed possible to represent non-synchronicity by interleaving. To verify this, we discuss constraints on the functions ϕ_i that are imposed due to a stochastic interpretation. First, the time until either of E_{40} or E_{41} decides to interact is given by the minimum of both distributions that amounts to a rate of $\lambda + \mu$ (cf. property (C) on page 41). From the interleaved chain $\boxed{E_{40}\ \ E_{41}}$ we derive $\phi_1(\lambda, \mu) + \phi_2(\lambda, \mu)$ as the rate until the first interaction takes place. As a consequence, the first constraint on ϕ_1 and ϕ_2 is

$$\phi_1(\lambda, \mu) + \phi_2(\lambda, \mu) = \lambda + \mu. \tag{4.1}$$

In addition, the probability that E_{40} decides earlier than E_{41} is given by $\lambda/(\lambda + \mu)$ since the distributions are stochastically independent from each other (property (D) on page 42). From the interleaved chain we infer

$$\frac{\phi_1(\lambda, \mu)}{\phi_1(\lambda, \mu) + \phi_2(\lambda, \mu)} = \frac{\lambda}{\lambda + \mu}. \tag{4.2}$$

It is now straightforward to deduce that $\phi_1(\lambda, \mu) = \lambda$, and symmetrically $\phi_2(\lambda, \mu) = \mu$.

Furthermore, the function ϕ_3 describes an exponentially distributed delay that should represent the distribution of the remaining delay of E_{41} after the move of E_{40}. As a consequence of the memoryless property (cf. property (B) on page 41) this is indeed again an exponential distribution, and its rate is $\phi_3(\lambda, \mu) = \mu$. Symmetrically, the only reasonable choice for ϕ_4 is $\phi_4(\lambda, \mu) = \lambda$.

To put it in a nutshell, non-synchronising transitions can be simply interleaved without adjusting any rate. Nondeterminism does not arise, since each transition has a probability to be chosen, and because the probability that both transitions are chosen at the same time is zero, due to the continuous nature of exponential distributions (property (E) on page 42).

Synchronisation of Distributions. Thus interleaving appears to be appropriate for non-synchronising transitions. Problems, however, arise in the context of synchronisation.

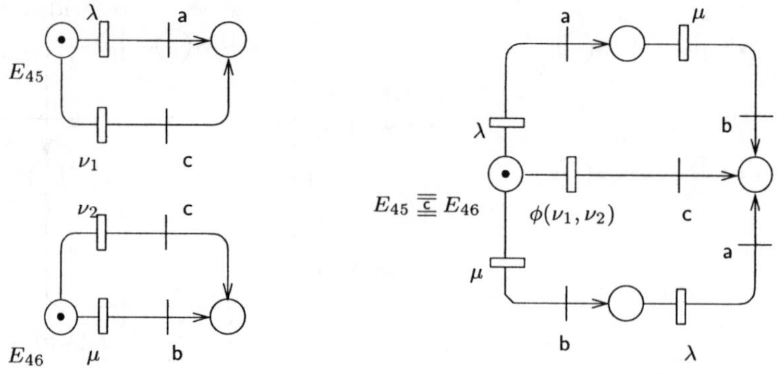

Figure 4.5. Composition with synchronisation in the fully Markovian approach

Example 4.1.7. Consider the composition of the two fully Markovian chains E_{45} and E_{46} by synchronising on c *depicted in Figure 4.5. Synchronisation is represented by a rate ϕ that should* somehow *depend on the rates ν_1 and ν_2. The other transitions are interleaved without synchronisation, using the rates determined before.*

Again, we have to address the question whether there is any reasonable choice for function ϕ. As before we can calculate that the synchronised chain $E_{45} \,\|\mathsf{c}\|\, E_{46}$ performs transition $\xrightarrow{\phi(\nu_1,\nu_2)\ \ \mathsf{c}}$ with probability $\phi(\nu_1, \nu_2)/(\lambda + \phi(\nu_1, \nu_2) + \mu)$. This is the probability that a synchronisation on action c occurs. From the individual components on the left we can calculate a constraint on this probability as follows. In order to be able to change state by synchronised interaction on c, both E_{45} and E_{46} have to decide in favour of their respective c-transition. In isolation, the probability to do so is $\nu_1/(\lambda + \nu_1)$ for E_{45} and $\nu_2/(\mu + \nu_2)$ for E_{46}. However, due to the continuous nature of the involved distributions, these decisions will not be taken at the same time (property (E) on page 42).

The probability that E_{45} decides between a and b earlier than E_{46} decides (between a and c) is given by $(\lambda + \nu_1)/(\lambda + \nu_1 + \mu + \nu_2)$, since both are concurrently heading for a decision (property (D) on page 42). Thus, the probability that an interaction on c occurs because first E_{45} and then E_{46} has decided accordingly amounts to

$$\frac{\nu_1}{\lambda + \nu_1} \; \frac{\lambda + \nu_1}{\lambda + \nu_1 + \mu + \nu_2} \; \frac{\nu_2}{\mu + \nu_2}. \tag{4.3}$$

By symmetry we derive

$$\frac{\nu_2}{\mu + \nu_2} \; \frac{\mu + \nu_2}{\lambda + \nu_1 + \mu + \nu_2} \; \frac{\nu_1}{\lambda + \nu_1} \tag{4.4}$$

as the probability that E_{46} is followed by E_{45}. So, the sum of (4.3) and (4.4) yields the probability that a synchronisation on action c occurs. As a result, we get the following constraint on function ϕ:

$$\frac{\nu_1}{\lambda+\nu_1+\mu+\nu_2}\,\frac{\nu_2}{\mu+\nu_2}+\frac{\nu_2}{\lambda+\nu_1+\mu+\nu_2}\,\frac{\nu_1}{\lambda+\nu_1} = \frac{\phi(\nu_1,\nu_2)}{\lambda+\phi(\nu_1,\nu_2)+\mu}. \tag{4.5}$$

After straightforward transformations we obtain that ϕ has to satisfy

$$\phi(\nu_1,\nu_2)=\frac{(\lambda+\nu_1)(\mu+\nu_2)(\lambda+\mu)}{(\lambda+\nu_1)(\mu+\nu_2)-\nu_1\nu_2}. \tag{4.6}$$

Thus, by means of this constraint, function ϕ is completely determined. Anyhow, an unfortunate observation is that the right hand side of this expression is *not* a function in the two parameters ν_1 and ν_2 of ϕ. It substantially depends on the values of λ and μ. A closer look into the train of thoughts that led to this dependency reveals the reason.

As in Section 3.3 we may associate with each transition a clock with an exponentially distributed expiration time. Then, the left hand side of equation (4.5) is essentially the probability that neither of the clocks associated with E_{45} $\xrightarrow{\lambda \quad \mathsf{a}}$ and E_{46} $\xrightarrow{\mu \quad \mathsf{b}}$ expires before the clocks of both E_{45} $\xrightarrow{\nu_1 \quad \mathsf{c}}$ and E_{46} $\xrightarrow{\nu_2 \quad \mathsf{c}}$ have expired. In other words, it is the probability that the *minimum* of two distributions (of λ and μ) is larger than the *maximum* of two other distributions (of ν_1 and ν_2). Under the assumption that $\phi(\nu_1,\nu_2)$ represents the maximum of the two exponential distributions, the right hand side of equation (4.5) *should indeed* describe exactly the required probability. But, there is a striking reason why problems occur anyhow: *The class of exponential distributions is not closed under maximum*, i.e., it is impossible to represent the maximum (as opposed to the minimum) of exponential distributions by an exponential distribution. So, the above assumption is invalidated because there is no function $\phi(\nu_1,\nu_2)$ that reflects the maximum of the distributions given by ν_1 and ν_2.

One may argue that using the maximum of distributions in the synchronisation case appears somewhat arbitrary. It is based on the idea that a synchronised state change is only possible if both associated distributions have expired. We insist on this interpretation like many others do [166, 10, 89, 38, 126, 160, 56] as a natural assumption. It is the only plausible choice if we accept that $\xrightarrow{\lambda \quad \mathsf{a}}$ has the meaning of describing that some time (given by λ) expires *before* interaction on a is possible. However, since the class of exponential distributions is not closed under maximum, none of the fully Markovian approaches employs maximum in case of synchronisation [79, 114, 107, 42, 25, 186]. So, how do these approaches solve the problem of synchronisation?

Passive Transitions. A solution common to all fully Markovian approaches is a second type of transitions where the rate is left unspecified. We use

$\stackrel{?}{\dashv}\!\xrightarrow{a}$ to represent such a transition and call it *passive* as opposed to *active* transitions. Passive transitions are assumed to simply wait for an active interaction partner. Synchronisation between both types of transitions is then easily defined by adopting the convention that the active partner determines the rate of the synchronised transition.

Bernardo *et al.* [25, 26, 33], as well as Wu *et al.* [186] require that *each synchronisation* involves at most one active transitions. While the former does not provide a recipe to achieve this, the latter approach assures this constraint by construction. Based on the model of Lynch and Tuttle [136] it distinguishes between input and output transitions and does not allow synchronisation among output transitions. Output transitions are active while input transitions are passive.[3]

For other fully Markovian approaches different extensions to the basic mechanism of active/passive synchronisation have been proposed. In Hillston's approach [114], also adopted by Priami [159], the unspecified rate '?' can be regarded as an arbitrarily large rate. Synchronisation is essentially realized (if abstracting from technical overhead) by a function $\phi(\nu_1, \nu_2) = \min(\nu_1, \nu_2)$. Note that this function does not calculate the minimum of *distributions* (which would give rise to $\nu_1 + \nu_2$) but the minimum of *rates*. If active and passive transitions are synchronised, function ϕ works as expected, since '?' is assumed to be arbitrarily large. Synchronisation of *active* transitions is also allowed. In this case the slower partner completely determines the rate of the synchronised transition. The term 'slower' refers to the fact that if $\nu_1 > \nu_2$, then, *in the average*, the clock of ν_1 expires before the clock of ν_2. However, there is always a nonzero probability that the other clock expires first, and this probability is not compensated in the approach. Hillston's justification to choose $\min(\nu_1, \nu_2)$ is an argument on the *mean durations*, given by $1/\nu_1$ and $1/\nu_2$. By selecting the smaller rate, the maximum of mean durations is selected. The resulting distribution can in general be much different from the maximum of the distributions.

The MTIPP approach [107, 78] is also based on the notion of active and passive transitions. Synchronisation is realized by multiplication: $\phi(\nu_1, \nu_2) = \nu_1\nu_2$, and passivity is expressed using rate 1, the neutral element of multiplication, as the unspecified rate '?'. Götz has discussed the benefits of this particular choice of ϕ in detail [77]. Apart from technical simplicity, it enables to scale rates (and implicitly mean durations) by weighting them with a scalar. Depending on the context, a scalar can be a particular probability or a factor indicating the relative 'speed' of this component, see also [100]. In this way, the concept of passivity is generalised in order to enhance modelling convenience. However, the distinction between rate and scalar is not made

[3] To be precise, the model of Wu *et al.* is not based on a transition system semantics. It uses so-called probabilistic behaviour maps as fully abstract models with respect to a notion of probabilistic testing. Negligently simplified, these maps are obtained from an embedded discrete-time Markov chain.

explicit. So, it is difficult to give an interpretation to specifications where, for instance, the parameter 1 might denote passivity, or denote a specific exponential distributed delay (with a mean duration of 1 time unit). And again, the product of rates can in general be much different from the maximum of the distributions.

Buchholz has taken the idea of scaling to the limit [42]. Both parameters occuring in a synchronisation play the roles of scalars. The actual rate is obtained by multiplication of both scalars with a third value that depends on the action label of synchronisation. This detour avoids different interpretations of parameters occuring in one specification.

Nondeterminism among Passive Transitions. From a stochastic point of view, neither of the existing approaches is completely satisfactory. In addition, the inclusion of passive transitions can easily lead to chains where *multiple* passive transitions emanate a single state. Since the probability of passive transitions is not known, it is not clear which passive transition will happen. So, *nondeterminism* appears to reenter through the back door. In order to avoid this, all existing approaches take a similar solution, though technically realized in different ways. Nondeterminism is treated as ***equi-probability.*** Whenever multiple passive transitions may independently synchronise with an active transition, all synchronisations are possible, and the resulting rates are identical.

From a conceptual point of view, this solution appears questionable. In particular, it excludes to use multiple passive transitions to represent ***implementation freedom*** in a specification. In our context, implementation freedom specifically implies that the actual passive transitions chosen by an implementation should not at all be based on an equi-probable choice.

Resumee. This discussion provides us with a good insight about the complications that occur when combining interactive processes and Markovian chains by means of a single transition relation $-\!\!\square\!\!\longrightarrow\!\!\!\!\rightarrow$. To sum up, fully Markovian process algebras suffer from the following problems:

- Synchronisation (of active transitions) is not adequately representable in the approach, because exponential distributions are not closed under maximum.
- Nondeterminism (among passive transitions) is treated as equi-probability.

Similar to the discrete-time case, a separation of concerns with respect to timing and interaction seems preferable and is good practice in many real-timed process algebras, for instance [146, 151]. The fact that a separation is favourable has already been highlighted by Hoare:

> The actual occurrence of each event in the life of an object should be regarded as an instantaneous or an atomic action without duration. [116, p.24]

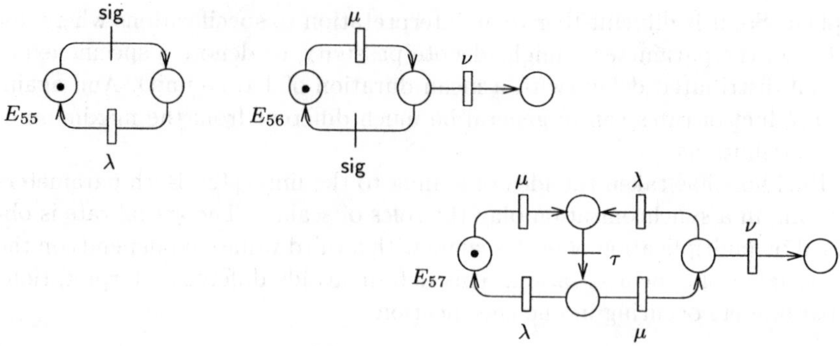

Figure 4.6. Three Interactive Markov Chains

Since in fully Markovian process algebras, all actions are inseparably linked to time consumption, it is impossible to follow this advice. The central formalism of this book will therefore be based on a different view. Similar to the alternating model in the discrete-time case we will use two *separate* relations, $\dashrightarrow$ representing the impact of stochastic time and $\twoheadrightarrow$ to represent the interaction potential of a chain. It will become clear that this solution

- avoids to reason about explicit synchronisation of distributions, because only interactive transitions may synchronise,
- implicitly realizes synchronisation as the maximum of distributions, because a synchronisation may only take place if all preceding delays have expired, and
- has a clear distinction between nondeterminism and probability.

4.2 Interactive Markov Chains

Interactive Markov Chains combine interactive processes and Markovian chains. On balance the result of our investigations in Section 4.1 has been that a separation of concerns is favourable in order to achieve a concise framework. So, IMC are defined by means of a twofold transition relation $\twoheadrightarrow$ and $\dashrightarrow$.

Definition 4.2.1. *An IMC transition system (IMCTS) is a quadruple* $(S, Act, \twoheadrightarrow, \dashrightarrow)$*, where*

- *S is a nonempty set of states,*
- *Act is a set of actions,*
- *$\twoheadrightarrow \subset S \times Act \times S$ is a set of interactive transitions, and*
- *$\dashrightarrow \subset S \times \mathbb{R}^+ \times S$ is a set of Markovian transitions.*

An Interactive Markov Chain is a quintuple $(S, Act, \twoheadrightarrow, \dashrightarrow, P)$*, where* $(S, Act, \twoheadrightarrow, \dashrightarrow)$ *is an IMC transition system and* $P \in S$ *is the initial state.*

Table 4.1. Structural operational rules for Markovian transitions

$$\frac{P \overset{\lambda}{\dashrightarrow} P'}{P \,\|_{a_1\ldots a_n}\, Q \overset{\lambda}{\dashrightarrow} P' \,\|_{a_1\ldots a_n}\, Q} \qquad \frac{Q \overset{\lambda}{\dashrightarrow} Q'}{P \,\|_{a_1\ldots a_n}\, Q \overset{\lambda}{\dashrightarrow} P \,\|_{a_1\ldots a_n}\, Q'}$$

$$\frac{P \overset{\lambda}{\dashrightarrow} P'}{\boxed{P\ \blacksquare_{a_1\ldots a_n}} \overset{\lambda}{\dashrightarrow} \boxed{P'\ \blacksquare_{a_1\ldots a_n}}}$$

Example 4.2.1. Figure 4.6 shows some examples of IMC. The first, E_{55}, repeatedly interacts on a signal sig *and takes some exponentially distributed rest after each interaction. E_{56} exhibits a similar behaviour. It first delays for an exponentially distributed time (given by rate μ). Afterwards it is able to interact on signal* sig *and return to its initial state. But if this signal does not occur within a certain time interval, E_{56} decides to terminate. That time interval is again exponentially distributed with rate ν. The third example, E_{57}, contains an internal transition together with some exponentially distributed delays.*

Theorem 4.2.1. *Each interactive process is (isomorphic to) an Interactive Markov Chain. Each Markovian chain is (isomorphic to) an Interactive Markov Chain.*

Proof. Immediate from the definitions (Definition 2.1.2 and 3.3.1). Note that transition relations are allowed to be empty.

In Chapter 2 we have seen that composition operators such as parallel composition and abstraction are crucial to build complex models in a stepwise, hierarchical manner. Since our ultimate goal is to use IMC as part of a compositional methodology, we have to discuss the way in which IMC can be composed. We therefore need to define parallel composition and abstraction of IMC. Surely, the result of composition should again lead to an IMC.

Definition 4.2.2. *Let P and Q be two IMC with state spaces S_P and S_Q. Parallel composition of P and Q on actions $a_1 \ldots a_n$ is an IMC $(S, Act, \longrightarrow, \dashrightarrow, P \,\|_{a_1\ldots a_n}\, Q)$, where*

- *$S := \{P' \,\|_{a_1\ldots a_n}\, Q' \mid P' \in S_P \wedge Q' \in S_Q\}$,*
- *Act is the union of all actions appearing in P or Q,*
- *$\longrightarrow$ is the least relation satisfying the first three rules in Table 2.1, and*
- *$\dashrightarrow$ is the least relation satisfying the first two rules in Table 4.1.*

According to this definition, Markovian transitions $\dashrightarrow$ are interleaved without adjusting rates. From the discussion in Section 4.1 it should be clear that, due to the memoryless property, recalculation of rates is not required. Interactive transitions are treated exactly as in Definition 2.1.3. They are interleaved

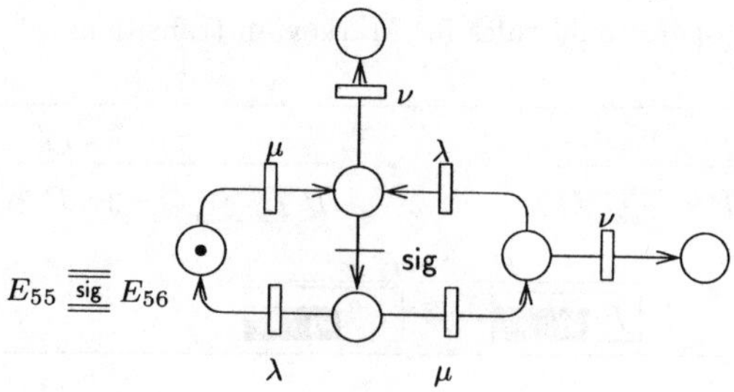

Figure 4.7. Parallel composition of two IMC

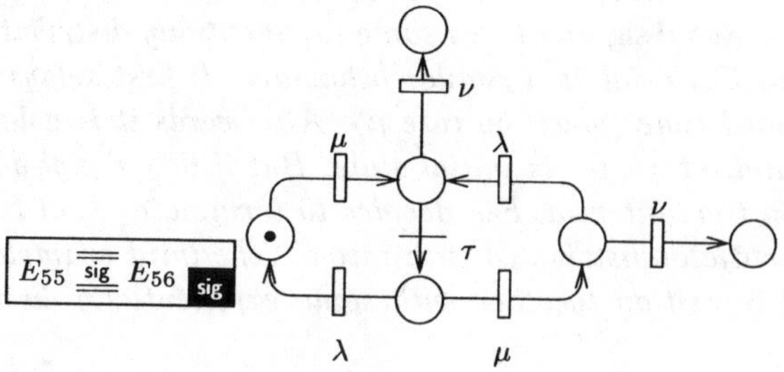

Figure 4.8. Abstraction applied to composed IMC

nondeterministically, if no interaction is forced, otherwise a synchronous state change occurs.

Example 4.2.2. In Figure 4.7 the result of composition of E_{55} *and* E_{56} *by synchronising on* sig *is depicted.*

Abstraction obviously has no impact on Markovian transitions. It is thus easily adapted from Definition 2.1.4.

Definition 4.2.3. *Let* P *be an IMC with state space* S_P*. Abstraction of actions* $\mathsf{a}_1 \dots \mathsf{a}_n$ *in* P *is an IMC* $\left(S, Act, \twoheadrightarrow, \dashrightarrow, \boxed{P \;\blacksquare_{\mathsf{a}_1 \dots \mathsf{a}_n}}\right)$*, where*

– $S := \left\{ \boxed{P' \;\blacksquare_{\mathsf{a}_1 \dots \mathsf{a}_n}} s \;\middle|\; P' \in S_P \right\}$,

– $\twoheadrightarrow$ *is the least relation satisfying the last two rules in Table 2.1, and*

– $\dashrightarrow$ *is the least relation satisfying the last rule in Table 4.1.*

Example 4.2.3. Figure 4.8 shows the result of internalising the interactive transition $\xrightarrow{\mathsf{sig}}$ *in* $E_{55} \overset{\mathsf{sig}}{=\!=} E_{56}$ *by means of abstraction.*

The following theorem is an immediate consequence of the above definitions.

Theorem 4.2.2. *IMC are closed under parallel composition and abstraction.*

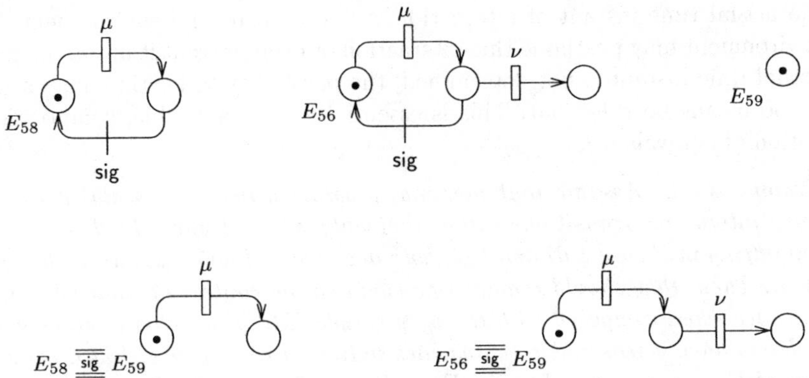

Figure 4.9. Some not equivalent IMC

4.3 Strong Bisimilarity

Interactive transitions and Markovian transitions coexist in Interactive Markov Chains. Meaningful equivalences for IMC should reflect their coexistence. Strong and weak bisimilarities will therefore be based on the respective notions for interactive processes and Markovian chains. Additionally, the interrelation of interactive and Markovian transitions has to be captured as well. This interrelation is not evident from the operational rules as introduce above.

Example 4.3.1. Consider, for instance, the internal transition in $\boxed{E_{55} \,\underline{\overline{\overline{\text{sig}}}}\, E_{56}\ \text{sig}}$ *(Figure 4.8). This transition is in conflict with an exponentially distributed delay, given by rate* ν. *From a stochastic perspective, the probability that this delay finishes* instantaneously *is zero, since (cf. property (A) on page 41)* $1 - e^{-\nu 0} = 0$. *On the other hand, an internal transition may happen instantaneously because nothing may prevent or delay it. This observation justifies to assume that an IMC that may perform an internal action is* not allowed *to let time pass.*

This assumption is usually called the *maximal progress assumption* and it is widely used in real time process algebra [151, 187, 93, 49].[4] As a consequence of maximal progress, our equivalence will equate $\boxed{E_{55} \,\underline{\overline{\overline{\text{sig}}}}\, E_{56}\ \text{sig}}$ to E_{57} depicted in Figure 4.6.

It is worth to emphasise that maximal progress is linked to *internal* transitions, but not to *external* transitions. An IMC that may interact with the environment does not prevent the passage of time. This is due to the fact that

[4] In the context of GSPN [4, 3] a similar assumption is present: immediate transition are assumed to have a higher priority level than (Markovian) timed transitions.

the actual time instant of interaction is governed by the environment. The environment may postpone this interaction or even prevent it at all. Since the actual time instant is not determined, the possibility to let time pass should by no means be ruled out. This is essential for a reasonable, compositional notion of equivalence.

Example 4.3.2. Assume that maximal progress would be associated to arbitrary interactive transitions rather that only internal ones. In this case E_{56} *(appearing in Figure 4.6) and* E_{58}*, both depicted in Figure 4.9, would be equivalent. Then, they should remain equivalent in the context of composition, for instance when composed with the very simple IMC* E_{59}*. This is necessary in order to have a reasonable equivalence notion, a congruence. But a comparison of* $E_{56} \underline{\overline{\text{sig}}} E_{59}$ *and* $E_{58} \underline{\overline{\text{sig}}} E_{59}$*, (Figure 4.9) reveals that this is not the case, since both give rise to different Markov chains.*

In order to formalise maximal progress, we distinguish IMC according to their ability to initially perform an internal action. We use $P \stackrel{\tau}{\not\longrightarrow}$ to denote the absence (and $P \stackrel{\tau}{\longrightarrow}$ to denote the presence) of such internal transitions and call them *stable* (respectively *unstable*) IMC. With this abbreviation we introduce strong bisimilarity based on Definition 2.2.2 and 3.5.1.

Definition 4.3.1. *An equivalence relation* $\mathcal{E}$ *on* $\mathcal{S}^{\text{all}}$ *is a strong bisimulation iff* $P \,\mathcal{E}\, Q$ *implies for all* $\mathsf{a} \in Act$ *and all equivalence classes* C *of* $\mathcal{E}$

1. $P \stackrel{\mathsf{a}}{\longrightarrow} P'$ *implies* $Q \stackrel{\mathsf{a}}{\longrightarrow} Q'$ *for some* Q' *with* $P' \,\mathcal{E}\, Q'$,
2. $P \stackrel{\tau}{\not\longrightarrow}$ *implies* $\gamma_{\text{M}}(P, C) = \gamma_{\text{M}}(Q, C)$.

Two processes P *and* Q *are strongly bisimilar, written* $P \sim Q$*, if they are contained in some strong bisimulation.*

This definition amalgamates strong bisimilarity for interactive processes and for Markovian chains. In order to compare the stochastic timing behaviour, the cumulative rate function γ_{M} is used, as motivated in Section 3.5. In addition, maximal progress is realized because the stochastic timing behaviour (evaluated by means of γ_{M}) is irrelevant for unstable expressions. Note that the first clause of the definition implies $Q \stackrel{\tau}{\not\longrightarrow}$ if $P \stackrel{\tau}{\not\longrightarrow}$, and vice versa.

Example 4.3.3. Figure 4.10 depicts a few examples of strongly bisimilar IMC. The first two, E_{60} *and* E_{61} *are bisimilar because the rate of reaching class cumulates to* 2ν *in either case.* E_{62} *falls into the same equivalence class, because of maximal progress: The Markovian transition labelled with rate* λ *is irrelevant, because the initial state is unstable. The same reasoning is applicable to* E_{63} *concerning the Markovian transition* $\stackrel{\mu}{\longrightarrow}$ *. So, all the four IMC are strongly bisimilar.*

Strong bisimilarity is well defined, as expressed in the following lemma. In addition, it is a substitutive relation.

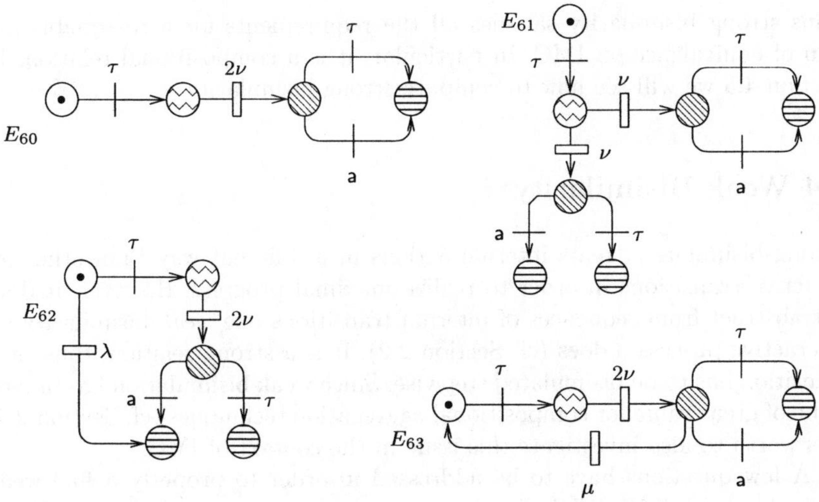

Figure 4.10. Some strongly bisimilar IMC

Lemma 4.3.1. *Strong bisimilarity*

- *is an equivalence relation on* $\mathcal{S}^{\text{all}}$,
- *is a strong bisimulation on* $\mathcal{S}^{\text{all}}$, *and*
- *is the largest strong bisimulation on* $\mathcal{S}^{\text{all}}$.

Theorem 4.3.1. *Strong bisimilarity is substitutive with respect to parallel composition and abstraction, i.e.,*

$$P_1 \sim P_2 \quad \textit{implies} \quad P_1 \;\overline{\overline{\underline{\underline{a_1 \ldots a_n}}}}\; P_3 \sim P_2 \;\overline{\overline{\underline{\underline{a_1 \ldots a_n}}}}\; P_3,$$

$$P_1 \sim P_2 \quad \textit{implies} \quad P_3 \;\overline{\overline{\underline{\underline{a_1 \ldots a_n}}}}\; P_1 \sim P_3 \;\overline{\overline{\underline{\underline{a_1 \ldots a_n}}}}\; P_2,$$

$$P_1 \sim P_2 \quad \textit{implies} \quad \boxed{P_1 \; a_1 \ldots a_n} \sim \boxed{P_2 \; a_1 \ldots a_n}\,.$$

Proof. See Appendix A.2.

Furthermore, strong bisimilarity turns out to conservatively extend the respective notions on interactive processes and Markovian chains. This answers indeed why we have 'recycled' the symbol $\sim$, that has been used in Section 2.2 already to denote strong bisimilarity on interactive processes.

Theorem 4.3.2. *Two interactive processes are strongly bisimilar according to Definition 4.3.1 if and only if they are strongly bisimilar according to Definition 2.2.2.*

Two Markovian chains are strongly bisimilar according to Definition 4.3.1 if and only if they are Markovian bisimilar according to Definition 3.5.1.

Proof. See Appendix A.3.

Thus strong bisimilarity satisfies all the requirements for a reasonable notion of equivalence on IMC. In particular, it is a compositional relation. In Section 4.5 we will see how to compute strong bisimilarity.

4.4 Weak Bisimilarity

Strong bisimilarity treats internal actions in a different way than other interactive transitions in order to realize maximal progress. However, it does not abstract from sequences of internal transitions like *weak* bisimilarity on interactive processes does (cf. Section 2.2). It is a strong relation, thus each transition has to be bisimulated stepwise. Since weak bisimulation has proven to be of great value for compositional aggregation techniques (cf. Section 2.4) it is worth to also investigate this issue in the context of IMC.

A few questions have to be addressed in order to properly define weak bisimulation on IMC. While the treatment of interactive transitions can follow the lines of Section 2.2, the treatment of Markovian transitions in a weak bisimulation has to be clarified. As discussed in Section 3.5, it is impossible to replace a sequence of Markovian transitions by a single Markovian transition without affecting the probability distribution of the total delay. So, we are forced to demand that Markovian transitions have to be bisimulated in the *strong* sense, using function γ_{M}, even for a weak bisimulation. However, we allow them to be preceded and followed by arbitrary sequences of internal interactive transitions.

These sequences are, according to Definition 2.2.4 given by $\stackrel{\tau}{\Longrightarrow}$, the reflexive and transitive closure of $\stackrel{\tau}{\longrightarrow}$. To incorporate these sequences into the definition of weak bisimulation is technically a bit involved, and there are various different strategies how to proceed. The variants we explored all led to the same notion of weak bisimilarity, and hence we focus on the strategy that appears most elegant.

For strong bisimilarity, γ_{M} has been used to cumulate rates of Markovian transitions that directly lead from a state P into a specific equivalence class C. We broaden this treatment in order to keep track of the impact of internal transitions that *follow* a Markovian transition: We cumulate all rates of Markovian transitions leading to states that can internally evolve into an element of class C. For this purpose, we define the *internal backward closure* C^τ as the set of processes that may internally evolve into an element of a set C, i.e., $C^\tau = \{P' \mid \exists P \in C : P' \stackrel{\tau}{\Longrightarrow} P\}$.

Example 4.4.1. Concerning the IMC E_{67} in Figure 4.11, the internal backward closure $^\tau$ *of the set*  *is the union of the sets*  *and* . $^\tau$ *is*  *and* $^\tau$ *is* .

The treatment of internal sequences *preceding* a Markovian transitions can follow the usual style (Definition 2.2.5), but with a slight exception. Since we

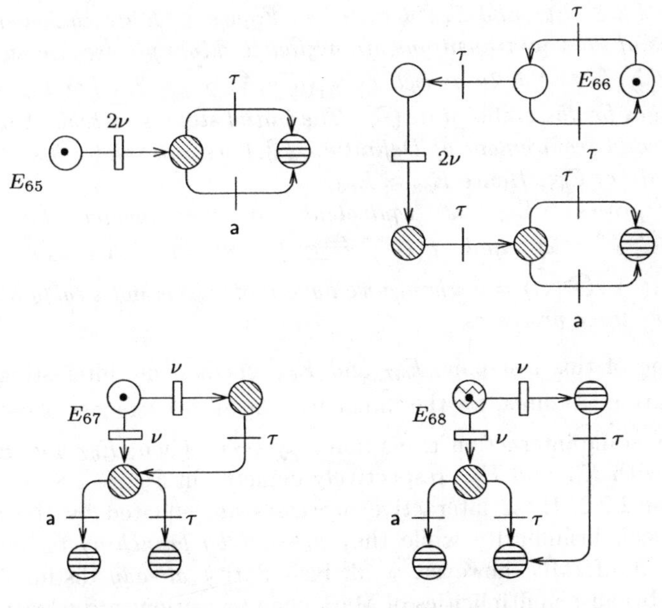

Figure 4.11. Some characteristic examples for weak bisimilarity

are aiming to incorporate maximal progress into our notion of weak bisimilarity, Markovian transitions *preceded* by a sequence of internal transitions are only relevant if they emanate a stable state P. Then the cumulative rate $\gamma(P, C^{\tau})$ should be taken into account. This requirement will be made more precise in the following definition.

Definition 4.4.1. *An equivalence relation $\mathcal{E}$ on $\mathcal{S}^{\mathrm{all}}$ is a weak bisimulation iff $P \; \mathcal{E} \; Q$ implies for all* $\mathsf{a} \in$ *Act*

1. $P \stackrel{\mathsf{a}}{\Longrightarrow} P'$ *implies* $Q \stackrel{\mathsf{a}}{\Longrightarrow} Q'$ *for some* Q' *with* $P' \; \mathcal{E} \; Q'$,
2. $P \stackrel{\tau}{\Longrightarrow} P'$ *and* $P' \stackrel{\tau}{\not\rightarrowtail}$ *imply* $Q \stackrel{\tau}{\Longrightarrow} Q'$ *for some* $Q' \stackrel{\tau}{\not\rightarrowtail}$ *with* $\gamma_{\mathrm{M}}(P', C^{\tau}) = \gamma_{\mathrm{M}}(Q', C^{\tau})$ *for all equivalence classes* C *of* $\mathcal{E}$.

Two processes P and Q are weakly bisimilar, written $P \approx Q$, if they are contained in some weak bisimulation.

Lemma 4.4.1. *Weak bisimilarity*

– *is an equivalence relation on* $\mathcal{S}^{\mathrm{all}}$,
– *is a weak bisimulation on* $\mathcal{S}^{\mathrm{all}}$, *and*
– *is the largest weak bisimulation on* $\mathcal{S}^{\mathrm{all}}$.

So, weak bisimilarity is a well defined equivalence. We illustrate the distinguishing power of $\approx$ by means of some examples.

Example 4.4.2. *E_{65} and E_{66} depicted in Figure 4.11 are equivalent because sequences of internal transitions are negligible. More precise, we have for E_{65} that γ_{M}(◯ , ▧$^{\tau}$) $= 2\nu$ as well as γ_{M}(◯ , ▤$^{\tau}$) $= 2\nu$. For E_{66} we get the same values for the* stable *state* ◯ *. The initial state is instable, but by means of the second requirement of Definition 4.4.1 we can relate this state to the initial state of E_{65}. Hence $E_{65} \approx E_{66}$.*

The process E_{67} is equivalent to the former two, because γ_{M}(◯ , ▧$^{\tau}$) $= 2\nu$ and γ_{M}(◯ , ▤$^{\tau}$) $= 2\nu$ in either case. In contrast, γ_{M}(⊜ , ▧$^{\tau}$) $= \nu$ whence we have that E_{68} is not weakly bisimilar to the former three processes.

The shape of this last pair, E_{67} and E_{68}, sheds some interesting light on our definition. Assume, for the moment, that all occurrences of $\overset{\nu}{\dashrightarrow}$ are replaced by some interaction transition, say $\xrightarrow{\mathsf{b}}$. Then, E_{67} and E_{68} would coincide with E_{16} and E_{19}, respectively depicted in Figure 2.8. As discussed in Section 2.2.2, these interactive processes are equated by the usual notion of weak bisimilarity while they are *not by branching* bisimilarity. In the context of IMC, however, weak bisimilarity *already* distinguishes E_{67} and E_{68}, because multiplicities of Markovian transitions are relevant. Specifically this fact is the reason why γ_{M}(◯ , ▤$^{\tau}$) $= 2\nu = \gamma_{M}$(⊜ , ▤$^{\tau}$) but γ_{M}(◯ , ▧$^{\tau}$) $\neq \gamma_{M}$(⊜ , ▧$^{\tau}$).

This observation suggests to investigate $\approx$ in terms of branching bisimulation. Indeed, it is in general possible to reformulate weak bisimulation such that Markovian transitions are treated in the same way as external transitions in branching bisimulation. This is particularly expressed by the following lemma, where the equivalence class C has replaced C^{τ}. The reader is invited to compare the second clause of this lemma to Lemma 2.2.4.

Lemma 4.4.2. *An equivalence relation $\mathcal{E}$ on $\mathcal{S}^{\text{all}}$ is a weak bisimulation iff $P\ \mathcal{E}\ Q$ implies for all $\mathsf{a} \in Act$ and for all equivalence classes C of $\mathcal{E}$,*

1. *$P \xrightarrow{\mathsf{a}} P'$ implies $Q \overset{\mathsf{a}}{\Longrightarrow} Q'$ for some Q' with $P'\ \mathcal{E}\ Q'$,*
2. *$P \not\xrightarrow{\tau}$ implies $\gamma_{M}(P, C) = \gamma_{M}(Q'', C)$ for some $Q'' \not\xrightarrow{\tau}$ such that $Q \overset{\tau}{\Longrightarrow} Q''$ and $P\mathcal{E}Q''$.*

Proof. See Appendix A.4.

We shall frequently use this reformulation in the sequel. In particular this lemma eases to show that weak bisimilarity is a compositional relation with respect to parallel composition and hiding.

Theorem 4.4.1. *Weak bisimilarity is substitutive with respect to parallel composition and abstraction, i.e.,*

$$
\begin{array}{lll}
P_1 \approx P_2 & \textit{implies} & P_1 \overset{a_1\ldots a_n}{\equiv} P_3 \approx P_2 \overset{a_1\ldots a_n}{\equiv} P_3, \\
P_1 \approx P_2 & \textit{implies} & P_3 \overset{a_1\ldots a_n}{\equiv} P_1 \approx P_3 \overset{a_1\ldots a_n}{\equiv} P_2, \\
P_1 \approx P_2 & \textit{implies} & \boxed{P_1 \; \blacksquare_{a_1\ldots a_n}} \approx \boxed{P_2 \; \blacksquare_{a_1\ldots a_n}}.
\end{array}
$$

Proof. See Appendix A.5.

As with strong bisimilarity, weak bisimilarity conservatively extends the notion of weak bisimilarity on interactive processes as well as the notion of Markovian bisimilarity on Markovian chains.

Theorem 4.4.2. *Two interactive processes are weakly bisimilar according to Definition 4.4.1 if and only if they are weakly bisimilar according to Definition 2.2.5.*

Two Markovian chains are weakly bisimilar according to Definition 4.4.1 if and only if they are Markovian bisimilar according to Definition 3.5.1.

Proof. See Appendix A.6.

As a whole, we have defined strong and weak bisimilarity by extending the respective definitions of Chapter 2 but incorporating the notion of maximal progress. Both relations are substitutive relations. We will apply these relations, weak bisimilarity in particular, to compositionally construct and aggregate Interactive Markov Chains. Practical applicability of this aggregation relies, as in previous chapters, on efficient algorithms to compute the respective relations.

4.5 Algorithmic Computation

In order to develop algorithms to compute strong and weak bisimilarity on IMC we build upon partition refinement techniques introduced in Section 2.3 and Section 3.6.

4.5.1 Strong Bisimilarity

To begin with, we show that strong bisimilarity can be characterised as a fixed-point of successively finer relations. This characterisation will then be turned into an algorithm.

Theorem 4.5.1. *Let P be an IMC with finite (reachable) state space S. Strong bisimilarity on S is the unique fixed-point of*

- $\smile_0 = S \times S.$
- $\smile_{k+1} = \smile_k \cap E_k$
  where $(P', Q') \in E_k$ *iff* $(\forall a \in Act)\ (\forall C \in S/_{\smile_k})$

Table 4.2. Algorithm for computing strong bisimilarity classes

Input: IMC transition system $(S, Act, \dashrightarrow, \rightarrowtail)$
Output: $S/\sim$
Method: $S\_Part := \Big\{\{P' \in S \mid P' \not\xrightarrow{\tau}\}\Big\} - \{\emptyset\};$
$U\_Part := \Big\{\{P' \in S \mid P' \xrightarrow{\tau}\}\Big\} - \{\emptyset\};$
$Spl := Act \times (S\_Part \cup U\_Part);$
While Spl not empty **do**
  Choose (a, C) **in** Spl;
  $Old := S\_Part \cup U\_Part;$
  $S\_Part := Refine(S\_Part, a, C);$
  $U\_Part := Refine(U\_Part, a, C);$
  $S\_Part := M\_Refine(S\_Part, C);$
  $New := (S\_Part \cup U\_Part) - Old;$
  $Spl := (Spl - \{(a, C)\}) \cup (Act \times New);$
od
Return $S\_Part \cup U\_Part$.

(*i*) $\gamma_O(P', a, C) = \gamma_O(Q', a, C)$,
(*ii*) *if* $P' \not\xrightarrow{\tau}$ *and* $Q' \not\xrightarrow{\tau}$, *then* $\gamma_M(P', C) = \gamma_M(Q', C)$.

Proof. See Appendix A.7.

Our abstract algorithm is based on this fixed-point characterisation; it is depicted in Table 4.2. The algorithm differs from the ones we have developed before. Initially, two partitions *S_Part* and *U_Part* are separated, containing stable, respectively unstable states. Only stable states are refined with respect to their cumulative rates, using *M_Refine* (cf. Section 3.6). In this way condition (*ii*) of Theorem 4.5.1 is assured. Refinement with respect to condition (*i*) is performed before by function *Refine*, as in Table 2.3. Just because the state space is split into two separate partitionings initially, function *Refine* is invoked twice.

Example 4.5.1. We illustrate the algorithm in Table 4.2 by means of an example. We aim to verify the equivalence classes of E_{61} *depicted in Figure 4.10. We initialise the algorithm by collecting stable states in S_Part* := {}, *and unstable states in U_Part* = {}. *This situation is represented in the first column of Figure 4.12. The set Spl of splitters contains four elements,* (a, ) (a, 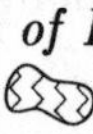) (τ, ) (τ, 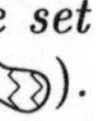).

We start partition refinement by choosing the splitter (a, ) *and refining the partition of stable states. We compute* $\gamma_O(\_, a,$ $)$ = `false` *for all states in* . *Thus*  *requires no splitting, function Refine*(*S_Part*, a, ) *will return S_Part unchanged. We then proceed computing* $\gamma_O(\_, a,$ $)$ *for each of the states in* . *One of the states returns* `false` *while the others return*

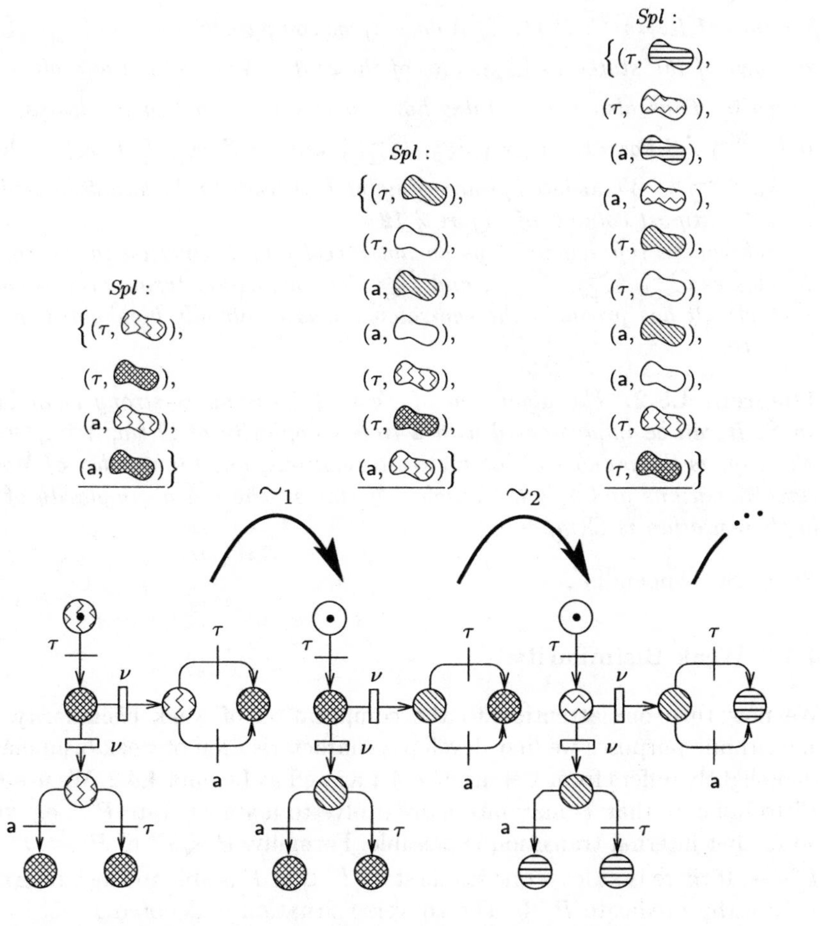

Figure 4.12. Partition refinement for strong bisimilarity

true. *As a result, we have to refine partition into two partitions, thus $U\_Part := \{$, $\}$.*

Afterwards, we consider Markovian transitions by means of function $M\_Refine(S\_Part,$ $)$. We compute the values of $\gamma_M(\_,$ $)$ for each of the states in , the only partition in S_Part. All states return 0, thus the set of stable states still requires no splitting.

We complete the first iteration of refinement by adding new splitters to Spl (and removing splitter (a,)). This leads to the situation depicted in the second column of this Figure.

We may then choose a splitter (a,). Refinement with respect to interactive transitions (i.e., $Refine(S\_Part, a,$ $)$ as well as

Refine(U_Part, a,) *does not lead to any further splitting. But then, function M_Refine(S_Part,*) *does. If we compute the values of* $\gamma_M(\_, \ )$ *for each of the states in* , *one of the states returns* 2ν *while all others return* 0. *Thus also stable states have to be split into two partitions,* *and* . *S_Part becomes* { , } *and* (*S_Part* ∪ *U_Part*) − *Old is* { , }. *We update Spl accordingly. This leads to the situation depicted in the rightmost column of Figure 4.12.*

Subsequent refinement steps do not reveal any distinction inside each of the classes , , , *and* . *The algorithm terminates when Spl is empty. It has produced the equivalence classes already highlighted in Figure 4.10.*

Theorem 4.5.2. *The algorithm of Table 4.2 computes strong bisimilarity on S. It can be implemented with a time complexity of* $\mathcal{O}((m_I + m_M)\log n)$ *where* m_I *is the number of interactive transitions,* m_M *the number of Markovian transitions and n is the number of states. The space complexity of this implementation is* $\mathcal{O}(m_I + m_M)$.

Proof. See Appendix A.8.

4.5.2 Weak Bisimilarity

We now turn our attention to the computation of weak bisimilarity. For algorithmic purposes we first develop a characterisation of weak bisimulation that slightly differs from Definition 4.4.1 as well as Lemma 4.4.2. We use $P \searrow_{\tau} P'$ to indicate that P may internally evolve to a stable state P', i.e., where no further internal transition is possible. Formally, $P \searrow_{\tau} P'$ iff $P \overset{\tau}{\Longrightarrow} P'$ and $P' \not\xrightarrow{\tau}$. If there is at least one stable state P' that P is able to reach internally we use the predicate $P \searrow_{\tau}$. The converse situation is denoted $P \not\searrow_{\tau}$. In this case P has no possibility to internally evolve into a stable state. We shall call such a chain *time-divergent* because of the notion of maximal progress. From a timing perspective, a time-divergent chain is forced to follow the maximal progress assumption without being able to reach a state where time may pass again. So, a time divergent chain prevents the passage of time. If $P \searrow_{\tau}$ holds, on the contrary, P will be said to be *time-convergent.*

Lemma 4.5.1. *An equivalence relation* $\mathcal{E}$ *on* $\mathcal{S}^{\text{all}}$ *is a weak bisimulation iff* $P \ \mathcal{E}\ Q$ *implies for all* $\mathsf{a} \in Act$ *and all equivalence classes* C *of* $\mathcal{E}$

1. $\gamma_O(P, \mathsf{a}, C) = \gamma_O(Q, \mathsf{a}, C)$,
2. $P \searrow_{\tau} P'$ *and* $P\mathcal{E}P'$ *imply* $Q \searrow_{\tau} Q'$ *for some* Q' *such that* $Q\mathcal{E}Q'$ *and* $\gamma_M(P', C) = \gamma_M(Q', C)$, *and*
3. $P \searrow_{\tau}$ *iff* $Q \searrow_{\tau}$.

Proof. See Appendix A.9.

In previous chapters we have derived an algorithm to compute the respective bisimulation from a characterisation as a fixed-point of successively finer relations. This is also possible here, but it is not necessary and technically fairly involved. We therefore prefer to directly present the algorithm to compute weak bisimilarity, and to provide a rigorous proof that this algorithm computes weak bisimilarity indeed.

The basic idea is as follows: The third clause of Lemma 4.5.1 is incorporated into the initial partitioning where we will separate time-convergent from time-divergent states. How to refine with respect to the first clause is known from Section 2.3 already. To refine with respect to the second clause of this lemma is the crucial difficulty of our algorithm. It is important to observe that according to Lemma 4.5.1 time-divergent states do not have to be refined according to this clause. Since we will separate time-convergent from time-divergent states initially, we can concentrate on an efficient way to ensure the second clause of Lemma 4.5.1 only for *time-convergent* states.

The global structure of the algorithm is represented in Table 4.3. In the initialisation phase, we first compute the weak transition relation $\Longrightarrow$. We then partition the whole state space into time-convergent and time-divergent states contained in *TC_Part*, respectively *TD_Part*. The set of splitters, *Spl* is initialised accordingly. In the body of the algorithm, *TC_Part* and *TD_Part* are refined with respect to splitters (a, C), in order to ensure the first clause of Lemma 4.5.1. The function $\mathit{M\_Refine}(\mathit{TC\_Part}, C)$ afterwards refines the current partitioning of *time-convergent* states with respect to C. The rest of the algorithm is as before. So, only function $\mathit{M\_Refine}(\mathit{TC\_Part}, C)$ remains to be explained. It is defined as

$$\mathit{M\_Refine}(\mathit{Part}, C) :=$$
$$\left(\bigcup_{X \in \mathit{Part}} \left(\mathit{M\_Spread}(X, C) \cup \{\mathit{Rest}(X, C)\} \right) \right) - \{\emptyset\}.$$

This function, the crucial part of the algorithm, refines each class X into multiple classes, by means of a function *M_Spread*. For some reasons, explained later on, the result of *M_Spread* does not necessarily contain all the states of X. The missing states are collected in $\mathit{Rest}(X, C)$, defined below.

For a class X, function $\mathit{M\_Spread}(X, C)$ splits the *stable* states inside this class, according to their rates of moving into class C. Additionally, each resulting new class is possibly enriched with some unstable states of X by means of a function *Enrich*,

$$\mathit{M\_Spread}(X, C) :=$$
$$\bigcup_{v \in \mathbb{R}^+} \left\{ \mathit{Enrich}\Big(X, \{P \in X \mid P \not\xrightarrow{\tau} \text{ and } \gamma_{\mathrm{M}}(P, C) = v\}\Big) \right\}.$$

For class X and a new (sub-)class, say Y, function $\mathit{Enrich}(X, Y)$ inserts exactly those unstable states of X to this new class Y that can internally evolve *only* to stable states contained in class Y, i.e.,

Table 4.3. Algorithm for computing weak bisimilarity classes

Input: IMC transition system $(S, Act, \longrightarrow\!\!\!\!\!\rightarrow, \longrightarrow\!\!\!\!\!\Box\!\!\!\!\rightarrow)$
Output: $S/\approx$
Method: Compute $\Longrightarrow$ from $\longrightarrow\!\!\!\!\!\rightarrow$;
$TC\_Part := \{\{P' \in S \mid P' \searrow^{\tau}\}\} - \{\emptyset\}$;
$TD\_Part := \{\{P' \in S \mid P' \not\searrow^{\tau}\}\} - \{\emptyset\}$;
$Spl := Act \times (TC\_Part \cup TD\_Part)$;
While Spl not empty **do**
  Choose (a, C) **in** Spl;
  $Old := TC\_Part \cup TD\_Part$;
  $TC\_Part := \mathit{IRefine}(TC\_Part, \mathsf{a}, C)$;
  $TD\_Part := \mathit{IRefine}(TD\_Part, \mathsf{a}, C)$;
  $TC\_Part := \mathit{IM\_Refine}(TC\_Part, C)$;
  $New := (TC\_Part \cup TD\_Part) - Old$;
  $Spl := (Spl - \{(\mathsf{a}, C)\}) \cup (Act \times New)$;
od
Return $TC\_Part \cup TD\_Part$.

$$Enrich(X, Y) := \{P \in X \mid P \searrow^{\tau} P' \text{ implies } P' \in Y\}.$$

Unstable states that may internally and nondeterministically evolve to stable states of different new classes, are not added to any of the new classes. So, $M\_Spread(X, C)$ may in general contain less states than contained in X. All these missing states are collected in a new class $Rest(X, C)$, defined as

$$Rest(X, C) := X - \{P \in Y \mid Y \in M\_Spread(X, C)\}.$$

This explains how a single class X is split into multiple classes by comparing rates of stable states into class C. Unstable states are either inserted into one of the new classes or collected in $Rest(X, C)$. All classes X of time-convergent states are treated in this way, by means of function $\mathit{IM\_Refine}$.

Example 4.5.2. We aim to explain the algorithm in Table 4.3 by means of an example, taken from Figure 4.11. We focus on E_{66}. *During initialisation, we compute* $\Longrightarrow$, *TC_Part and TD_Part. The latter is empty while the former contains all the states,* *TC_Part* = {}. *Spl is initialised with two splitters,* (a, ) *and* (τ, ). *This situation is depicted in Figure 4.13. Loops of weak internal transitions are omitted.*

Choosing splitter (τ, 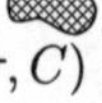) *we enter the body of the algorithm. Refinement with* $\mathit{IRefine}(TC\_Part, \tau, C)$ *provides no finer partitioning of TC_Part. Since TD_Part is empty, we proceed with* $\mathit{IM\_Refine}$(*TC_Part*, 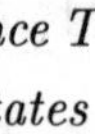). *Since TC_Part contains just* , *we compute M_Spread*(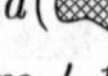, ). *Only two states in*  *are stable, marked* E_{80} *and* E_{81} *in Figure 4.13. We compute* $\gamma_{\mathrm{M}}(E_{80}$, ) = 2ν *and* $\gamma_{\mathrm{M}}(E_{81}$, ) = 0. *Thus,*

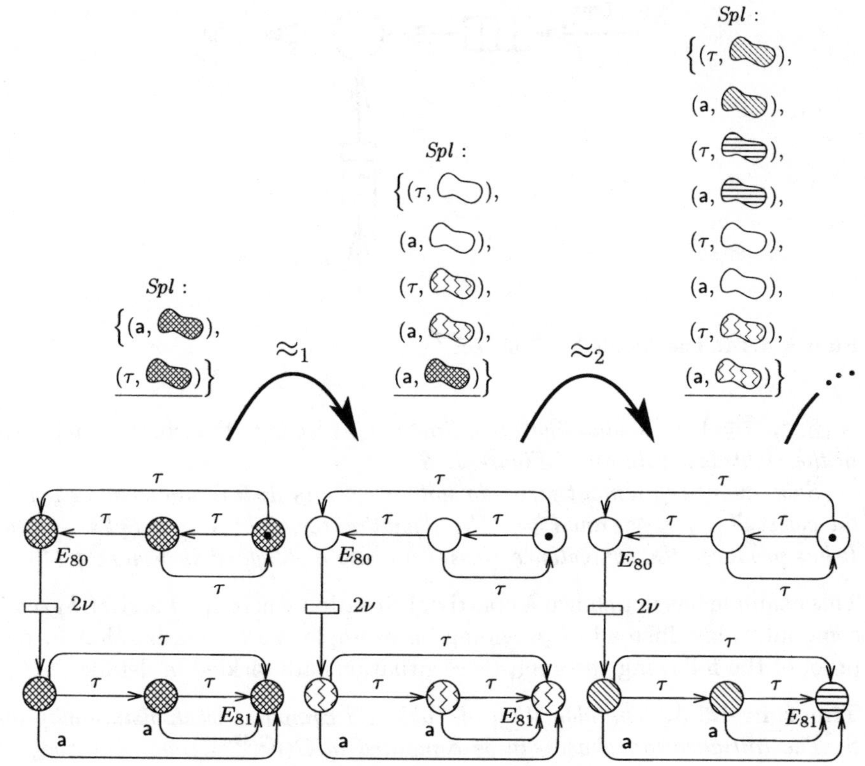

Figure 4.13. Partition refinement for weak bisimilarity

$$M\_Spread(\;, \;) = \{Enrich(\;, \{E_{80}\}), Enrich(\;, \{E_{81}\})\}.$$

Since each unstable state is inserted into either $\{E_{80}\}$ *or* $\{E_{81}\}$, $M\_Spread($, $)$ *returns two partitions, say* and , *covering the whole state space. Thus* $Rest($, $) = \emptyset$. *We complete the first iteration of refinement by adding four new splitters to Spl (and removing splitter* $(\tau,$ $))$. *This leads to the situation depicted in the second column of this Figure.*

We may then choose the next splitter, say (a,). *Refinement with respect to interactive transitions (i.e.,* $I\!Refine(TC\_Part,$ a, $))$ *splits the class* into two classes indicated by and in the third column of Figure 4.13. Then,* $I\!M\_Refine(TC\_Part,$ $)$ *provides no refinement. Remark that* $M\_Spread($, $) = \{\emptyset\}$. *Thus* $Rest($, $) =$. *As a whole,* $TC\_Part = \{$, , $\}$ *at the end of this iteration. Since New*

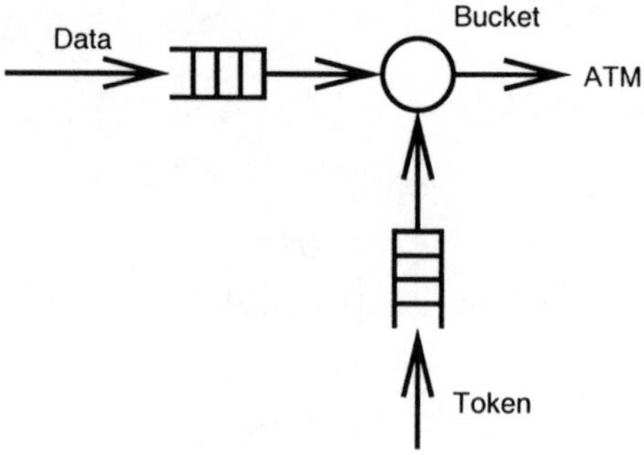

Figure 4.14. The simple leaky bucket

is {, }, *we update Spl accordingly. This leads to the situation depicted in the rightmost column of Figure 4.13.*

Subsequent refinement steps do not reveal any distinction inside each of the classes 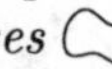, , *and* 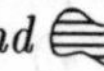. *The algorithm terminates when Spl is empty. It has produced the equivalence classes already highlighted in Figure 4.11.*

This example has not shown a nontrivial situation where *Rest* and *M_Spread* come into play. Instead of providing an example, we emphasise that in the proof of the following theorem, these situations are tackled in detail.

Theorem 4.5.3. *The algorithm of Table 4.3 computes weak bisimilarity on* S. *The initialisation phase can be computed in* $\mathcal{O}(n^{2.376})$ *time. The body of the algorithm can be implemented such that it requires* $\mathcal{O}(n\ (m_I'' + m_M))$ *time where* n *is the number of states,* m_M *is the number of Markovian transitions and* m_I'' *is the number of interactive transitions after transitive closure of internal transitions. The space complexity of this implementation is* $\mathcal{O}(m_I'' + m_M)$.

Proof. See Appendix A.10.

4.6 Application Example: Leaky Bucket

In this section we describe a first, simple application example of IMC. The *leaky bucket* is an access control mechanism used in ATM networks [8]. The basic idea behind the leaky bucket principle is *traffic shaping.* Before establishing a connection, sender and receiver negotiate with the net provider in order to properly describe the characteristics of a connection. In particular, they may negotiate a constant bucket rate. As data cells arrive at the ATM switch they flow into a 'bucket' which drains at bucket rate. If the cells are arriving faster than the bucket is draining eventually the bucket will overflow.

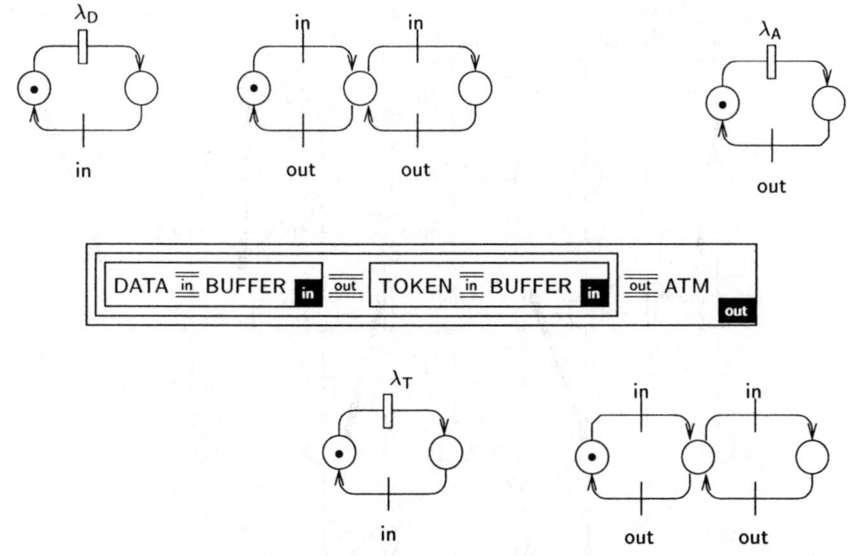

Figure 4.15. Hierarchical structure of the simple leaky bucket specification

Several variants of this mechanism have appeared in the literature, and we refer to [138] for an introduction into the topic. We consider a slight variant of the basic principle, taken from [130]. As sketched in Figure 4.14 we view the leaky bucket as consisting of two queues. One of them buffers data cells that are waiting for transmission over the net. In abstract terms, they simply need some service and they have to wait until service is available. But, in order to be allowed for service, a data cell requires a *token*. If no token is available the cell has to wait. Tokens are buffered in a second queue. We assume that tokens and cells possess exponentially distributed inter-arrival times, given by, say λ_D and λ_T. The service time is given by rate λ_A.

If we assume that each of the queues consist of two places, we can describe the whole system as the following composition of five Interactive Markov Chains, depicted in Figure 4.15,

DATA ‖in‖ BUFFER in ‖out‖ TOKEN ‖in‖ BUFFER in ‖out‖ ATM out .

The structure of all components of this chain has appeared before. The two chains named BUFFER are indeed replica of the interactive process E_6 that has appeared in Figure 2.4. The other three chains are variants of E_{58} (Figure 4.9). As informally discussed above, data cells are iteratively produced by the chain DATA and inserted (by action in) into the private buffer if space is

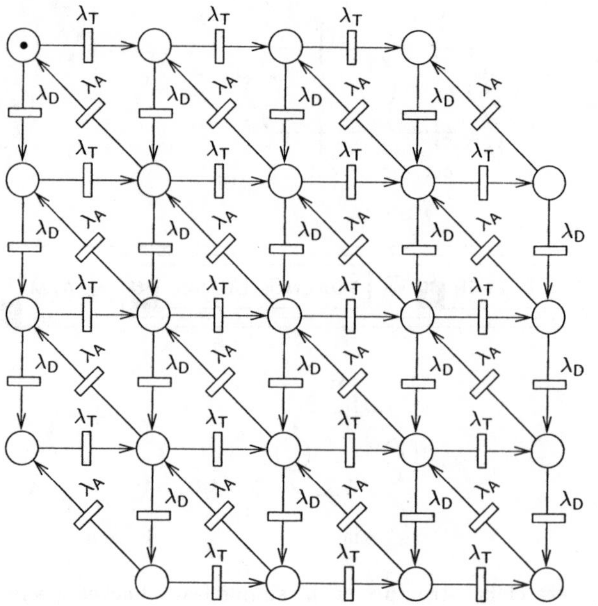

Figure 4.16. Lumped Markov chain of the simple leaky bucket

available.[5] Similarly, tokens are produced by TOKEN and handed over to the private buffer. A cell will be transmitted whenever three components are able to synchronise on action out: the chain ATM and both buffers. This realizes the constraint that data cells can only be served if a token is available.
In order to explore the behaviour of this specification, we iteratively apply Definition 2.1.3 and Definition 2.1.4. This leads to an IMC with 72 states and 244 transitions. By applying the algorithms sketched in Section 4.5 this IMC can be aggregated to 51 states, in the case of strong bisimilarity, and 23 states in the case of weak bisimilarity. Quite important, the result of factoring with respect to weak bisimilarity is a (lumped) Markovian chain, it does not contain any interactive transitions. It is depicted in Figure 4.16. Performance analysis of the simple leaky bucket system can therefore proceed as outlined in Section 3.4.

Since we have consequently used abstraction we may, as in Section 2.4, apply *compositional aggregation* to generate the state space and hence the Markovian chain. Compositional aggregation in the setting of IMC relies on Theorem 4.4.1, the congruence result. To be specific, we can aggregate the state space of $\boxed{\text{DATA} \;\|_{\text{in}}\; \text{BUFFER} \;\blacksquare_{\text{in}}}$ which has 6 states into an equivalent IMC, say D, with 4 states. The same is true for $\boxed{\text{TOKEN} \;\|_{\text{in}}\; \text{BUFFER} \;\blacksquare_{\text{in}}}$, we

[5] Note that DATA (and TOKEN) may hold an additional data cell (respectively a token) when the respective buffer is full.

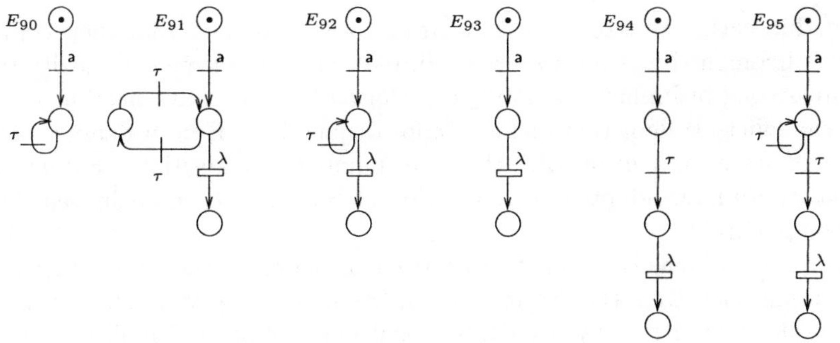

Figure 4.17. Distinguishing examples for weak bisimilarity

denote the aggregated IMC T. Due to the congruence result we have

Analysing the right hand side leads to an IMC with 32 states. This IMC, in turn, can be aggregated to the Markovian chain depicted in Figure 4.16. So, to put it in a nutshell, we can perform the same kind of compositional aggregation that we have outlined in Section 2.4 also in the context of Interactive Markov Chains.

The benefits of compositional aggregation become more impressing if we increase the buffer sizes to 100 places each. Then, the original model has a size of 81608 states while the one obtained by means of compositional aggregation possesses 20808 states. Because of the congruence property, aggregation of either of the two IMC leads to the same result, a Markovian chain consisting of 10607 states.

4.7 Discussion

In this chapter we have introduced IMC, the central formalism of this book. It arises as an integration of interactive processes and continuous-time Markov chains. Our decision to maintain two separate transition relations, $\longrightarrow$ and $-\!\!\square\!\!\rightarrow$, has been justified by a discussion of the shortcomings of other approaches where a single relation $-\!\!\square\!\!\longrightarrow$ is used. After introducing parallel composition and abstraction on IMC we have defined strong and weak bisimilarity by extending the respective definitions of Chapter 2 but incorporating the notion of *maximal progress*. In order to efficiently compute these relations we have developed algorithms that are more involved than the ones presented in earlier chapters. Anyhow, it is worth to remark that their computational complexity is *not increased*. In particular, the time complexity to compute strong bisimilarity is (still) $\mathcal{O}(m \log n)$ since we have been able to adopt the efficient algorithm of Paige and Tarjan [154]. Here n denotes the number of states and m denotes the total number of transitions. For weak bisimilarity we have adopted the algorithm of Bouali [30], and achieved the same complexity.

The notion of weak bisimilarity is central in order to abstract from internal computation such that the system can be regarded as a Markovian chain. We have illustrated this by a compositional analysis of our small example specifying the leaky bucket principle. More impressing examples will follow in Chapter 6.

Our notion of weak bisimilarity is closely related to the one used in the dissertation of Michael Rettelbach [163]. This is not surprising, since both are based on previous joint work [108]. Indeed, it can be shown that the relation in [108] coincides with our notion. The difference to Rettelbach's definition essentially lies in the treatment of loops of internal transitions. As highlighted in Section 4.5, such a loop may cause *time-divergence*, if it is not possible to escape this loop by an internal transition. The reason is our *maximal progress* assumption. An IMC with such a loop is forced to follow the maximal progress assumption without actually progressing (thus essentially preventing the passage of time). The same is true in Rettelbach's work, *except* for loops of length one.

Example 4.7.1. Consider the IMC E_{92} depicted in Figure 4.17. Rettelbach does equate this IMC with E_{93} but not with E_{91} or E_{90}. The latter two are considered to be time-divergent, while E_{92} is not. Our notion of bisimilarity, on the contrary, distinguishes E_{92} from E_{93} but equates E_{92} with E_{90} and E_{91}. Both relations equate E_{93}, E_{94} and E_{95}.

The issue of loops of internal transitions will be a crucial aspect in the next chapter, where we are developing an equational theory of IMC under weak bisimilarity.

5. Algebra of Interactive Markov Chains

In this chapter we will develop an algebra of Interactive Markov Chains. What turns Interactive Markov Chains into an algebra? An algebra usually consists of a set of operations on a given carrier set together with equational laws that characterise these operations. A well-know algebra is the algebra of natural numbers where addition and multiplication satisfy associativity and commutativity laws.

To begin with, we show how to specify IMC in a purely syntactic way by means of a *language*, and this language shall be the carrier set of our *algebra of Interactive Markov Chains*. We will investigate strong and weak bisimilarity in this context, introducing the notion of *weak congruence*, a slight refinement of weak bisimilarity. Then, we tackle the issue of a *sound and complete equational theory* for strong bisimilarity and weak congruence. Nontrivial problems will have to be solved in order to establish an equational treatment of time-divergence and maximal progress. Indeed, our solution solves an open problem for timed process calculi in general, and we highlight that it can be adapted to solve a similar open problem for process calculi with priority.

In addition, we introduce further operators that exemplify two different directions of extensions to Interactive Markov Chains. One operator is introduced to enhance specification convenience. A second operator is introduced for more pragmatic reasons, namely to diminish the infamous state space explosion problem for specifications that exhibit symmetries.

5.1 Basic Language

The carrier set of an *algebra of Interactive Markov Chains* will be a set of expressions that are basically syntactic descriptions of IMC. For this purpose we shall introduce a language and show that this language is expressive enough to generate all IMC. The language is given by a grammar consisting of terminal symbols and non-terminal symbols. Some of these terminal symbols play the role of composition operators, additional to parallel composition and abstraction. So, we shall define a formal semantics for this language by giving a meaning to each of these additional operators, again in a structural operational style.

We assume a countable set of *variables* $\mathcal{V}$ that will be used to express repetitive behaviour. As in earlier chapters, we suppose a set of actions *Act* containing a distinguished internal action τ and let $\mathbb{R}^+$ denote the set of nonnegative reals. We use λ, μ, ... to range over $\mathbb{R}^+$ and a, b, ... for elements of *Act*.

Definition 5.1.1. *Let $\lambda \in \mathbb{R}^+$, $\mathsf{a} \in Act$ and $X \in \mathcal{V}$. We define the language* IML *as the set of expressions given by the following grammar.*

$$\mathcal{E} ::= 0 \quad | \quad \mathsf{a}.\mathcal{E} \quad | \quad (\lambda).\mathcal{E} \quad | \quad \mathcal{E} + \mathcal{E} \quad | \quad X \quad | \quad \underline{_{X:=}\,\mathcal{E}} \quad | \quad \perp$$

We use E, E_1, E_2, F, ... to range over arbitrary expressions of IML. The intuitive meaning of the language constructs is as follows.

- The terminal symbol 0 describes a *terminated* behaviour that cannot engage in any interaction and is also unable to perform internal actions.
- The expression $\mathsf{a}.E$ may interact on action a and afterwards behave as expression E. We say that E is *action prefixed* by a.
- The expression $(\lambda).E$, a *delay prefixed* expression, describes a behaviour that will behave as expression E after a delay that is governed by an exponentially distribution with a mean duration of $1/\lambda$ time units.
- The expression $E + F$ describes two alternatives. It may either exhibit the behaviour of expression E or the behaviour of expression F. The terminal symbol $+$ is called the *choice operator*. Dependent on the possibilities of E and F, the choice between E and F is taken either nondeterministically, or it may be taken probabilistically, in particular if both E and F consist of delay prefixes.
- The expression $\underline{_{X:=}\,E}$ describes a recursively defined behaviour. Assuming that the variable X appears somewhere inside expression E, the meaning is as follows. Whenever the variable X is encountered during the evolution of the expression, the expression will reinitialise its behaviour to $\underline{_{X:=}\,E}$. (If variable X does not occur inside E, the expression $\underline{_{X:=}\,E}$ simply behaves as E.)
- The symbol $\perp$ is intended to represent an *ill-defined* behaviour. This notion will be explained later on.

We will formalise this intuitive interpretation by means of structural operational rules, as in Section 2.1. We define a semantics for each expression of IML, by mapping the complete language IML onto a transition system. The state space of this transition system is implicitly defined. It is given by the set of expressions according to Definition 5.1.1. Since each expression $E \in$ IML appears somewhere in this transition system its semantics is determined by the state space reachable from this expression. Not surprising, we define two transition relations, one for actions and one to represent the impact of time. We will provide rules for both action prefix ($\mathsf{a}.$) and delay prefix ($(\lambda).$), as well as choice ($+$) and recursion ($\underline{_{X:=}}$). Termination (0) and ill-definedness ($\perp$) do not require specific rules.

Table 5.1. Operational semantic rules for IML

$\dfrac{}{\mathsf{a}.E \xrightarrow{\mathsf{a}} E}$	$\dfrac{}{(\lambda).E \overset{\lambda}{\dashrightarrow} E}$
$\dfrac{E \xrightarrow{\mathsf{a}} E'}{E + F \xrightarrow{\mathsf{a}} E'}$	$\dfrac{E \overset{\lambda}{\dashrightarrow} E'}{E + F \overset{\lambda}{\dashrightarrow} E'}$
$\dfrac{F \xrightarrow{\mathsf{a}} F'}{E + F \xrightarrow{\mathsf{a}} F'}$	$\dfrac{F \overset{\lambda}{\dashrightarrow} F'}{E + F \overset{\lambda}{\dashrightarrow} F'}$
$\dfrac{E\{\underline{X{:=}\,E}\,/X\} \xrightarrow{\mathsf{a}} E'}{\underline{X{:=}\,E} \xrightarrow{\mathsf{a}} E'}$	$\dfrac{E\{\underline{X{:=}\,E}\,/X\} \overset{\lambda}{\dashrightarrow} E'}{\underline{X{:=}\,E} \overset{\lambda}{\dashrightarrow} E'}$

Definition 5.1.2. *The* action transition *relation* $\longrightarrow \subset \mathsf{IML} \times Act \times \mathsf{IML}$ *is the least relation and the* Markovian transition *relation* $\dashrightarrow \subset \mathsf{IML} \times \mathbb{R}^+ \times \mathsf{IML}$ *is the least* multi-*relation given by the rules in Table 5.1, where* $E\{F/X\}$ *denotes simultaneous substitution of each occurrence of variable* X *inside expression* E *by expression* F.

Note that the rules for Markovian transitions define a *multi*-relation instead of an ordinary relation. This ensures that multiple transitions are generated for expressions like $(\lambda).0 + (\lambda).0$. Whenever there are k different proof trees for a transition $\overset{\lambda}{\dashrightarrow}$ the multi-relation will contain k such transitions. The need to represent multiplicities stems from our interpretation of the choice operator '+' in the presence of delays. The delay until the choice is actually decided is governed by an exponential distribution given by the sum of rates (cf. property (C) on page 41). As a consequence, the behaviour described by $(\lambda).0 + (\lambda).0$ should be the same as that of $(2\lambda).0$. For the former expression we represent this delay by means of two separate transitions labelled λ which is technically simple using a multi-relation [77, 114].

Definition 5.1.3. *A variable* X *occurs* free *in an expression* E *if it occurs inside* E *outside the scope of any binding* $\underline{X{:=}\ldots}$. *We let* $\mathcal{V}(E)$ *denote the set of variables occuring free in expression* E. *If* $\mathcal{V}(E) = \emptyset$ *we say that* E *is a closed expression, otherwise we call it an open expression. The set of closed expressions will be denoted* IMC, *ranged over by* P, P_1, Q, *and so on.*

Example 5.1.1. The expressions 0, $\bot{+}\mathsf{a}.0$, $\underline{Y{:=}\,\mathsf{a}.Y} + 0$, *and* $\underline{X{:=}\,X + \underline{Y{:=}\,\mathsf{a}.Y}}$ *are closed expressions. On the contrary, the expressions* X, $\underline{Z{:=}\,Z + (\lambda).Y}$, *and* $\underline{Z{:=}\,Y + \mathsf{a}.(\lambda).X}$ *are open, since each of them contains at least one free variable.*

The behaviour of open expressions is only partially specified. They become closed expressions, by adding further recursion operators, such as $\underline{X{:=}\,E}$ if

$\mathcal{V}(E) = \{X\}$. Closed expressions are, as suggested by the name IMC, closely related to IMC. Indeed, we shall prove in Section 5.4 that any closed and *well-defined* expression corresponds to an IMC, and vice versa. The notion of well-definedness is formally introduced in the following definition. It is of crucial importance for the whole chapter.

Definition 5.1.4. *The set of* well-defined *expressions* $\mathsf{IML}_{\downarrow}$ *is the smallest subset of* IML *such that*

- $\mathcal{V} \subseteq \mathsf{IML}_{\downarrow}$ *and* $0 \in \mathsf{IML}_{\downarrow}$,
- *if* $E \in \mathsf{IML}$, *then* $\mathsf{a}.E \in \mathsf{IML}_{\downarrow}$ *and* $(\lambda).E \in \mathsf{IML}_{\downarrow}$,
- *if* $E \in \mathsf{IML}_{\downarrow}$ *and* $F \in \mathsf{IML}_{\downarrow}$, *then* $E + F \in \mathsf{IML}_{\downarrow}$,
- *if* $E\{\underline{{\scriptstyle X:=}\,E}\,/X\} \in \mathsf{IML}_{\downarrow}$, *then* $\underline{{\scriptstyle X:=}\,E} \in \mathsf{IML}_{\downarrow}$.

The complementary set containing all ill-defined *expressions will be denoted* $\mathsf{IML}_{\uparrow}$. *We write* $E{\downarrow}$ $(E{\uparrow})$ *if* $E \in \mathsf{IML}_{\downarrow}$ $(E \in \mathsf{IML}_{\uparrow})$.

Intuitively, an expression is well-defined if its *initial* behaviour is determined (or if it is just a variable). In particular, $\perp$ and expressions like $\underline{{\scriptstyle X:=}\,X}$ are ill-defined.

Example 5.1.2. The expressions 0, $\underline{{\scriptstyle Y:=}\,\mathsf{a}.Y} + 0$, X, *and* $\underline{{\scriptstyle Z:=}\,Y + \mathsf{a}.(\lambda).X}$ *are well-defined. Note that the former two expressions are closed, while the latter two are not closed, since they contain free variables.*

The expressions $\perp + \mathsf{a}.0$, $\underline{{\scriptstyle Z:=}\,Z + (\lambda).Y}$, *and* $\underline{{\scriptstyle X:=}\,X + \underline{{\scriptstyle Y:=}\,\mathsf{a}.Y}}$ *are ill-defined. The former two expressions are closed, while the latter one is an open expression. If* E *is ill-defined, then* $E + F$ *and* $F + E$ *are always ill-defined, for arbitrary* F. *Note that well-definedness only refers to the initial possibilities of an expression. In particular,* $\tau.(\perp + \mathsf{a}.0)$, $\mathsf{b}.\underline{{\scriptstyle Z:=}\,Z + (\lambda).Y}$, *and* $(\lambda).\underline{{\scriptstyle X:=}\,X + \underline{{\scriptstyle Y:=}\,\mathsf{a}.Y}}$ *are all well-defined.*

5.2 Strong Bisimilarity and Weak Congruence

The fact that the language IML contains new composition operators urges us to re-investigate the notions of strong and weak bisimilarity, in order to ensure that we are still dealing with a proper substitutive notion of equivalence. Indeed, the fact that IML contains open and ill-defined expressions implies that IML is a bit more expressive than IMC. We will hence redefine strong and weak bisimilarity for this language. For expressions E that are stable ($E \overset{\tau}{\not\twoheadrightarrow}$) as well as well-defined ($E{\downarrow}$) we use the shorthand notation $E_{\surd}{\downarrow}$.

Definition 5.2.1. *An equivalence relation* $\mathcal{E}$ *on* IMC *is a strong bisimulation iff* $P\ \mathcal{E}\ Q$ *implies for all* $\mathsf{a} \in Act$,

1. $P \overset{\mathsf{a}}{\longrightarrow\!\!\!\!\!\!+} P'$ *implies* $Q \overset{\mathsf{a}}{\longrightarrow\!\!\!\!\!\!+} Q'$ *for some* Q' *with* $P'\ \mathcal{E}\ Q'$,
2. $P_{\surd}{\downarrow}$ *implies that* $Q_{\surd}{\downarrow}$ *and that* $\gamma_{\mathrm{M}}(P, C) = \gamma_{\mathrm{M}}(Q, C)$ *for all* $C \in \mathsf{IMC}/\mathcal{E}$.

Two closed expressions P and Q are strongly bisimilar, written $P \sim Q$, if they are contained in some strong bisimulation.

This definition is a slight but conservative extension of Definition 4.3.1. In the second clause, $\surd\!\downarrow$ has replaced $\overset{\tau}{\not\rightarrow}$, and we require that stable and well-defined expressions are only related to stable and well-defined expressions. Furthermore, we do not compare the stochastic behaviour of ill-defined expressions. A reason for this reformulation is best explained by means of an example, other reasons will appear later in this chapter.

Example 5.2.1. The ill-defined expression $\underline{{}_{X:=}X + (\lambda).0}$ has an infinite number of $\overset{\lambda}{\dashrightarrow}$ transitions, since we can generate arbitrarily many terms $(\lambda).0$ as summands, and hence infinitely different proof trees of the following form:

$$\cfrac{\cfrac{\cfrac{\cfrac{(\lambda).0 \overset{\lambda}{\dashrightarrow} 0}{\vdots}}{\left(\underline{{}_{X:=}X + (\lambda).0} + (\lambda).0\right) + (\lambda).0 \overset{\lambda}{\dashrightarrow} 0}}{\underline{{}_{X:=}X + (\lambda).0} + (\lambda).0 \overset{\lambda}{\dashrightarrow} 0}}{\underline{{}_{X:=}X + (\lambda).0} \overset{\lambda}{\dashrightarrow} 0}$$

Our restriction to well-defined expressions avoids the need to calculate (and compare) infinite sums of rates. We redefine weak bisimilarity in the same way, based on Definition 4.4.1. Recall that for a set of states C, C^τ denotes the internal backward closure $\{P' \mid \exists P \in C : P' \overset{\tau}{\Longrightarrow} P\}$.

Definition 5.2.2. *An equivalence relation $\mathcal{E}$ on* IMC *is a weak bisimulation iff $P\ \mathcal{E}\ Q$ implies for all* a $\in$ *Act,*

1. *$P \overset{a}{\longrightarrow} P'$ implies $Q \overset{a}{\Longrightarrow} Q'$ for some Q' with $P'\ \mathcal{E}\ Q'$,*
2. *$P \overset{\tau}{\Longrightarrow} P'$ and $P'\surd\!\downarrow$ imply $Q \overset{\tau}{\Longrightarrow} Q'$ for some $Q'\surd\!\downarrow$ such that $\gamma_M(P', C^\tau) = \gamma_M(Q', C^\tau)$ for all $C \in \mathsf{IMC}/\mathcal{E}$.*

Two closed expressions P and Q are weakly bisimilar, written $P \approx Q$, if they are contained in some weak bisimulation.

Similar to strong bisimilarity, the only change concerns the second clause of this definition. According to this improved clause (where $\surd\!\downarrow$ has replaced $\overset{\tau}{\not\rightarrow}$), the stochastic behaviour is only relevant if we compare two expressions which both may internally evolve to stable and well-defined expressions.

It can be shown that $\approx$ ($\sim$, respectively) itself satisfies the definition of weak (strong) bisimulation. Either of them is therefore the largest such relation. It is worth to remark that both relations have been defined on closed expressions, a subset of IML. Anyway we may 'lift' these relations to also cover open expressions. Two open expressions are considered equivalent if and only if they behave equivalent whenever arbitrary closed expressions are substituted for their free variables. This is expressed in the next definition.

Definition 5.2.3. *Let $\mathcal{R}$ be a relation on* IMC × IMC *and let E and F be two expressions of* IML *with a vector of free variables $\boldsymbol{X} = (X_1, \ldots, X_n)$. We define $E \; \mathcal{R} \; F$ iff $E\{(\boldsymbol{P}/\boldsymbol{X})\} \; \mathcal{R} \; F\{(\boldsymbol{P}/\boldsymbol{X})\}$ holds for arbitrary vectors $\boldsymbol{P} = (P_1, \ldots, P_n)$ of closed expressions of* IMC, *where $\{(\boldsymbol{P}/\boldsymbol{X})\}$ denotes the simultaneous substitution of each X_i by P_i.*

Now that we have defined strong and weak bisimilarity for the language IML, we turn our attention towards their compositional properties with respect to the operators of the language.

Theorem 5.2.1. *Strong bisimilarity is a congruence with respect to all operators of* IML. *In other words, if E_1, E_2 and E_3 are expressions of* IML *and* $\mathsf{a} \in Act$, $\lambda \in \mathbb{R}^+$, *and* $X \in \mathcal{V}$, *then*

- $E_1 \sim E_2$ *implies* $\mathsf{a}.E_1 \sim \mathsf{a}.E_2$,
- $E_1 \sim E_2$ *implies* $(\lambda).E_1 \sim (\lambda).E_2$,
- $E_1 \sim E_2$ *implies* $E_1 + E_3 \sim E_2 + E_3$ *and* $E_3 + E_1 \sim E_3 + E_2$,
- $E_1 \sim E_2$ *implies* $\underline{X{:=}\,E_1} \sim \underline{X{:=}\,E_2}$.

Weak bisimilarity is a congruence with respect to all operators of IML except the choice operator '+'.

Proof. The proof proceeds along the lines of the proof of Theorem 5.2.2, the sketch of which is included as Appendix B.1.

The fact that weak bisimilarity is not substitutive with respect to choice is a well-known deficiency inherited from non-stochastic weak bisimilarity, see [145] for a counterexample. In order to rectify this situation we crystallise a specific congruence contained in $\approx$.

Definition 5.2.4. *P and Q are weakly congruent, written $P \simeq Q$, iff for all* $\mathsf{a} \in Act$ *and all* $C \in$ IMC/$\approx$

1. $P \overset{\mathsf{a}}{\longrightarrow} P'$ *implies* $Q \overset{\tau}{\longrightarrow}{}^{*} \overset{\mathsf{a}}{\longrightarrow} \overset{\tau}{\longrightarrow}{}^{*} Q'$ *for some Q' with $P' \approx Q'$,*
2. $Q \overset{\mathsf{a}}{\longrightarrow} Q'$ *implies* $P \overset{\tau}{\longrightarrow}{}^{*} \overset{\mathsf{a}}{\longrightarrow} \overset{\tau}{\longrightarrow}{}^{*} P'$ *for some P' with $P' \approx Q'$,*
3. $P{\surd}{\downarrow}$ *(or $Q{\surd}{\downarrow}$) implies* $\gamma_{\mathrm{M}}(P, C) = \gamma_{\mathrm{M}}(Q, C)$,
4. $P{\surd}{\downarrow}$ *iff* $Q{\surd}{\downarrow}$.

Weak congruence and weak bisimilarity only differ in the treatment of initial internal steps of P and Q. Weak bisimilarity requires that an internal transition $\overset{\tau}{\longrightarrow}$ is simulated by a weak transition $\overset{\tau}{\Longrightarrow}$, which includes the possibility that no internal transition has to be carried out (cf. Definition 2.2.4). For initial behaviours, weak congruence strengthens this requirement. It requires that an internal transition $\overset{\tau}{\longrightarrow}$ has to be matched by $\overset{\tau}{\longrightarrow}{}^{*} \overset{\tau}{\longrightarrow} \overset{\tau}{\longrightarrow}{}^{*}$, i.e., by at least on internal transition $\overset{\tau}{\longrightarrow}$. This small change simplifies the conditions imposed on (initial) Markovian transitions, and it is enough to fix the congruence problem with respect to choice: in contrast to weak bisimilarity, weak congruence is a proper substitutive relation with respect to all language operators (using the lifting according to Definition 5.2.3).

Theorem 5.2.2. *Weak congruence is a congruence with respect to all operators of* IML*: If E_1, E_2 and E_3 are expressions of* IML *and* $\mathsf{a} \in Act$, $\lambda \in \mathbb{R}^+$, $X \in \mathcal{V}$, *then*

- $E_1 \simeq E_2$ *implies* $\mathsf{a}.E_1 \simeq \mathsf{a}.E_2$,
- $E_1 \simeq E_2$ *implies* $(\lambda).E_1 \simeq (\lambda).E_2$,
- $E_1 \simeq E_2$ *implies* $E_1 + E_3 \simeq E_2 + E_3$ *and* $E_3 + E_1 \simeq E_3 + E_2$,
- $E_1 \simeq E_2$ *implies* $\underline{X := E_1} \simeq \underline{X := E_2}$.

Proof. See Appendix B.1.

Indeed, weak congruence is unique in the sense that it turns out to be the *coarsest* congruence contained in weak bisimilarity, as a consequence of the following lemma.

Lemma 5.2.1. $E_1 \simeq E_2$ *iff, for each* $E_3 \in$ IML, $E_1 + E_3 \approx E_2 + E_3$ *and* $E_3 + E_1 \approx E_3 + E_2$.

As a result, we have obtained two substitutive equivalence notions on IML: strong bisimilarity and weak congruence, a distinguished subset of weak bisimilarity. In the next section we will develop an equational theory for both congruences. The interrelation between these relations is expressed in the following lemma.

Lemma 5.2.2. $\sim \subset \simeq \subset \approx$.

5.3 Equational Theory of Strong Bisimilarity and Weak Congruence

The issue of this section is to develop an equational theory of strong bisimilarity, respectively weak congruence, for IML. An equational theory characterises the impact of the language operators by means of a set $\mathcal{A}$ of *equational laws.*

Example 5.3.1. An example of an equational law is the commutativity law $E + F = F + E$. The intuitive meaning of this law is as follows: Whenever a pattern of the form $E + F$ can be found in an expression, it can be replaced by $F + E$. E and F play the roles of meta variables *and can be instantiated by arbitrary expressions of* IML*. In this way we may transform* $(\lambda).(\mathsf{a}.0+\perp)$ *into* $(\lambda).(\perp+\mathsf{a}.0)$*: We instantiate* $E \equiv \mathsf{a}.0$ *and* $F \equiv \perp$ *and afterwards substitute* $F + E$ *for* $E + F$.

Formally, an equational law is a pair of expressions of IML connected with the symbol '=', where either of the expressions may contain some meta variables such as E, F, and so on. In technical terms, a law (or a set of laws) induces an equivalence on IML, or more precisely, a *congruence*, since we are allowed to replace sub-expressions inside larger expressions, as in the above example.

The question arises, in what sense such an induced congruence is related to the congruences we have defined on the semantics of IML, i.e., strong bisimilarity and weak congruence. We are aiming to provide laws that are *sound* with respect to, say, strong bisimilarity. A law is sound with respect to an equivalence if any application of the law does not alter the equivalence class of the expression. The converse direction is called *completeness*. A set of laws is complete with respect to an equivalence, if two expressions can be transformed into each other by (iterative) application of laws, whenever they are equivalent.

Example 5.3.2. The law $0 = 0$ *is sound for any equivalence relation on* IML. *However, this law is, as it stands alone, far from providing a complete set of laws for any nontrivial equivalence. On the other hand, a law* $E = F$ *is complete for any equivalence relation on* IML, *but it is sound only for the trivial relation* IML $\times$ IML *that equates all expressions.*

So, our ultimate goal is to provide sets of laws that are *sound as well as complete* with respect to strong bisimilarity, respectively weak congruence. We shall say that such a set *axiomatises* the respective congruence.

In order to reach this goal, we can rely on experience gained in the context of interactive processes. Milner has pioneered this topic by developing axiomatisations for regular CCS [142, 144]. Regular CCS arises from IML by disallowing $\perp$ and delay prefixing, $(\lambda).E$. Before we introduce Milner's axiomatisation, we need to introduce different notions of *guardedness*.

Definition 5.3.1. *We define the following notions of guardedness:*

- *A variable* X *is* weakly guarded *in an expression* E, *if every free occurrence of* X *in* E *is contained in a prefixed sub-expression, i.e., an expression of the form* $\mathsf{a}.F$ *or* $(\lambda).F$. *In this case,* a, *respectively* (λ) *is called a* guard *of* X.
- *A variable* X *is* strongly guarded *in expression* E *if each free occurrence of* X *is weakly guarded by a guard different from* τ.
- *A variable* X *is said to be* fully unguarded *in expression* E *if it is not weakly guarded. It is called* partially guarded *if it is not strongly but weakly guarded.*
- *An* expression E *is said to be* strongly (weakly) guarded *if, for every sub-expression of the form* $\underline{X{:=}\,E'}$, *the variable* X *is strongly (weakly) guarded in* E'.

Example 5.3.3. X *is weakly guarded in* $(\lambda).X$, $Y+\perp$, $\underline{X{:=}\,X}$, *and* $\tau.(X+0)$, *but neither strongly guarded in the latter, nor in* $\tau.(\underline{X{:=}\,X}+X)$. *Furthermore,*

Table 5.2. The set $\mathcal{A}_{\sim}^{\mathrm{CCS}}$ axiomatises strong bisimilarity on CCS

(C)	$E+F$	$=$	$F+E$	
(A)	$(E+F)+G$	$=$	$E+(F+G)$	
(I)	$E+E$	$=$	E	
(N)	$E+0$	$=$	E	
$(_{:=}1)$	$\underline{_{X:=}E}$	$=$	$\underline{_{Y:=}E\{Y/X\}}$	provided Y is not free in $\underline{_{X:=}E}$.
$(_{:=}2)$	$\underline{_{X:=}E}$	$=$	$E\left\{\underline{_{X:=}E}/X\right\}$	
$(_{:=}3)$	$\underline{_{X:=}E}$	$=$	$\underline{_{X:=}X+E}$	
$(_{:=}4)$	$F=E\{F/X\}$ implies $F=\underline{_{X:=}E}$			provided X is weakly guarded in E.

X is fully unguarded in $X+\mathsf{a}.X$, $\underline{_{Y:=}Y+X}$, and $\underline{_{Z:=}(\lambda).\underline{_{X:=}\underline{_{Y:=}X}}+X}$, but not in $\underline{_{Z:=}(\lambda).\underline{_{X:=}\underline{_{Y:=}X}}}$.

The expressions 0, $\mathsf{a}.\underline{_{X:=}\tau.X}$ and $\underline{_{Y:=}\tau.\underline{_{X:=}Y+(\lambda).X}}$ are weakly guarded, and only the first of them is also strongly guarded. The other two are hence partially guarded. Examples for fully unguarded expressions are $\underline{_{Z:=}(\lambda).\underline{_{X:=}\underline{_{Y:=}X}}+X}$ and $\underline{_{Z:=}(\lambda).\underline{_{X:=}\underline{_{Y:=}X}}}$.

5.3.1 Strong Bisimilarity

We are now ready to introduce an equational theory for strong bisimilarity on IML. Table 5.2 lists equational laws axiomatising strong bisimilarity on regular CCS. The first four laws state that the choice operator is commutative (C), associative (A), idempotent (I), and that 0 is the neutral element of choice (N). The other four laws handle recursion, and we refer to [142] for a detailed explanation. In short, law $(_{:=}1)$ states that bound variables can be renamed if no additional bindings are introduced. Law $(_{:=}2)$ is immediate from the structure of the operational rules for recursion, and law $(_{:=}3)$ is a means to remove ill-definedness. The effect of law $(_{:=}4)$ will be extensively discussed below.

Our axiomatisation is based on this system, but with a few modifications. The first question we pose concerns soundness of these laws. Indeed, two of the laws, (I) and $(_{:=}3)$, turn out to be *not* sound with respect to strong bisimilarity on IML. We will show later (Theorem 5.3.1) that all other laws are sound however.

Example 5.3.4. Law (I) allows one to equate $(\lambda).0+(\lambda).0$ and $(\lambda).0$, but $(\lambda).0+(\lambda).0 \sim (\lambda).0$ does not hold according to Definition 5.2.1. Similarly $\underline{_{X:=}X+0}$ can be equated to $\underline{_{X:=}0}$ by means of law $(_{:=}3)$, but these two expressions are not strongly bisimilar, since the former is ill-defined, while the latter is well-defined (and both are stable).

Law ($_{:=}$3) is therefore not sound since it makes it possible to equate well-defined and ill-defined expressions. The purpose of this law in [142] is indeed to remove ill-definedness. But we have decided to treat ill-definedness differently in order to avoid the calculation of infinite sums. Another striking reason why we deviate from Milner's treatment will become evident later. Anyhow, we have to avoid a law like ($_{:=}$3). We replace it by a law that explicitly indicates ill-definedness, using the symbol $\bot$:

$$(_{:=}5) \quad \underline{{}_{X:=}\bot + E} \quad = \quad \underline{{}_{X:=}X + E}$$

We also require to avoid the general idempotency of choice. This is needed because of the presence of stochastic time. Delay rate quantities have to be cumulated according to property (C) of exponential distributions (cf. page 41) in order to represent the stochastic timing behaviour of expressions like $(\lambda).E + (\lambda).E$. As a consequence, idempotency does not hold for delay prefixed expressions, while it does hold for action prefixed ones. We thus have to split law (I) into several laws, ($I1$) – ($I3$). Together with other laws, they form the set $\mathcal{A}_\sim$ and are depicted in Table 5.3.

As we will see (Corollary 5.3.1), this set $\mathcal{A}_\sim$ is sound and complete for IML modulo strong bisimilarity. Law ($I1$) axiomatises property (C) (and is the reason why (I) is invalidated in general). The laws ($I2$) and ($I3$) state that idempotency is valid for action prefixed expressions as well as in the presence of ill-definedness. Note that, according to the second clause of Definition 5.2.1, ill-definedness or instability implies that delay rate quantities are irrelevant. This fact is indeed also the justification for the law ($\bot$1), as well as law ($\bot$2). The latter two are crucial for an equational treatment of maximal progress. The characteristic law for maximal progress, $(\lambda).E + \tau.F = \tau.F$ [151, 187, 49], is derivable from these two laws. We shall write $\mathcal{A} \vdash E = F$ whenever $E = F$ may be proved from the set of laws $\mathcal{A}$.

Lemma 5.3.1. $\mathcal{A}_\sim \vdash (\lambda).E + \tau.F = \tau.F.$

Proof. Use ($\bot$1) and ($\bot$2) (twice) to equate

$$\mathcal{A}_\sim \vdash (\lambda).E + \tau.F = (\lambda).E + \bot + \tau.F = \bot + \tau.F = \tau.F.$$

In this proof, as well as in the sequel, commutativity (C) and associativity (A) are used without explicit reference. After these remarks we are now tackling the issues of soundness and completeness of $\mathcal{A}_\sim$. First, we state that each of the laws of $\mathcal{A}_\sim$ is sound with respect to strong bisimilarity.

Theorem 5.3.1. *For arbitrary expressions E and F it holds that*

$$\mathcal{A}_\sim \vdash E = F \qquad \textit{implies} \qquad E \sim F.$$

The proof requires a tedious case analysis and is omitted.

We are now addressing the question whether this set is also complete, i.e., enough to allow the deduction of all semantic equalities. We closely follow

Table 5.3. The set $\mathcal{A}_\sim$ axiomatises strong bisimilarity on IML

(C)	$E+F$	$=$	$F+E$	
(A)	$(E+F)+G$	$=$	$E+(F+G)$	
$(I1)$	$(\lambda).E+(\mu).E$	$=$	$(\lambda+\mu).E$	
$(I2)$	$\mathsf{a}.E+\mathsf{a}.E$	$=$	$\mathsf{a}.E$	
$(I3)$	$E+E{+}\bot{+}\bot$	$=$	$E{+}\bot$	
(N)	$E+0$	$=$	E	
$(\bot 1)$	$(\lambda).E{+}\bot$	$=$	$\bot$	
$(\bot 2)$	$\bot{+}\tau.E$	$=$	$\tau.E$	
$({:=}1)$	$\underline{X{:=}\,E}$	$=$	$\underline{Y{:=}\,E\{Y/X\}}$	provided Y is not free in $\underline{X{:=}\,E}$.
$({:=}2)$	$\underline{X{:=}\,E}$	$=$	$E\left\{\underline{X{:=}\,E}/X\right\}$	
$({:=}4)$	$F = E\{F/X\}$ implies		$F = \underline{X{:=}\,E}$	provided X is weakly guarded in E.
$({:=}5)$	$\underline{X{:=}\bot{+}E}$	$=$	$\underline{X{:=}\,X+E}$	

the lines of Milner and use sets of mutually recursive defining equations to capture the impact of recursion. The main use of the recursion operator $\underline{X{:=}\,\ldots}$ is to provide a means to specify multiply recursive dependencies in just a single expression of the language, as in $\underline{X{:=}\,\mathsf{in}.\underline{Y{:=}\,\mathsf{in.out}.Y+\mathsf{out}.X}}$. By means of a set of mutually recursive defining equations the same situation can be encoded as

$$\begin{aligned} X &:= \mathsf{in}.Y \\ Y &:= \mathsf{in.out}.Y+\mathsf{out}.X. \end{aligned}$$

Such sets of defining equations will be employed as a means to prove completeness. In addition, these sets are also much more convenient to work with when specifying complex dependencies. We will therefore use sets of defining equations extensively in later chapters, when we discuss applications of Interactive Markov Chains.

Definition 5.3.2.

1. *An* equation set *(ES)* $\{\boldsymbol{X} := \boldsymbol{F}\}$ *is a finite non empty sequence (a vector) of declarations* $(X_1 := F_1, \ldots, X_n := F_n)$*, where the* X_i*s are pairwise distinct variables and the* F_i *are expressions of* IML.
2. *An equation set* $S = \{\boldsymbol{X} := \boldsymbol{F}\}$ *is weakly guarded if each* X_j *is weakly guarded in each* F_i*. We shall call* $\boldsymbol{X}$ *the* formal *variables of* S*. All other variables occuring in any of the expressions* F_i *will be called free variables of* S*.*
3. *An equation set* $\{\boldsymbol{X} := \boldsymbol{F}\}$ *is a standard equation set (SES) iff each* F_i *is of the form:*

$$F_i \equiv \sum_{j=1}^{r(i)} \mathsf{a}_{i,j}.X_{f(i,j)} + \sum_{k=1}^{s(i)} (\lambda_{i,k}).X_{g(i,k)} + \sum_{l=1}^{t(i)} W_{h(i,l)} + \sum_{m=0}^{u(i)} \perp .$$

where the variables W_k are free, i.e., $\boldsymbol{W}$ is disjoint from $\boldsymbol{X}$. An empty sum denotes 0.

4. *For a given set of laws $\mathcal{A}$, a vector $\boldsymbol{E}$ $\mathcal{A}$-provably satisfies an ES $\{\boldsymbol{X} := \boldsymbol{F}\}$ iff for each E_i it holds that $\mathcal{A} \vdash E_i = F_i\{\boldsymbol{E}/\boldsymbol{X}\}$.*
5. *An expression E $\mathcal{A}$-provably satisfies $\{\boldsymbol{X} := \boldsymbol{F}\}$ iff there exists a vector $\boldsymbol{E}$ which $\mathcal{A}$-provably satisfies $\{\boldsymbol{X} := \boldsymbol{F}\}$ and $\mathcal{A} \vdash E = F_1$,*

For convenience, we introduce some further abbreviations, that are needed later on, especially in the proofs in Appendix B. They reflect the syntactic shape of an equation set in a transition oriented notation.

Definition 5.3.3. *Let $S = \{\boldsymbol{X} := \boldsymbol{F}\}$ be some ES with free variables $\boldsymbol{W}$. We define*

1. $X_i \xrightarrow{\mathsf{a}}\!\!\!\!\rightarrow_S X_j$ *iff* $\mathsf{a}.X_j$ *appears as a summand in* F_i*,*
2. $X_i \rhd_S W_k$ *iff* W_k *appears as a summand in* F_i*,*

Our proof of completeness consists of three main steps. First, we show that each expression provably satisfies some weakly guarded SES. Then, we verify that two separate weakly guarded SES, each provably satisfied by some expression, can be merged into a single weakly guarded SES, provably satisfied by both expressions, if both expressions are strongly bisimilar. This will be the crucial part of the proof. In the last step we show that two expressions can be equated by means of $\mathcal{A}_\sim$ if both provably satisfy the same weakly guarded SES.

This strategy turns out to work for weakly guarded expressions, but not for fully unguarded ones. The latter however can be transformed into the former, as expressed by the following lemma.

Lemma 5.3.2. *For each expression E, there exists a weakly guarded expression F such that $\mathcal{A}_\sim \vdash E = F$.*

Proof. By induction on the structure of E. The only interesting case is recursion where ($_{:=}$5) is applied to remove fully unguardedness (by transformation into $\perp$).

Theorem 5.3.2. *For each weakly guarded expression E, there is some weakly guarded SES in the free variables of E that E $\mathcal{A}_\sim$-provably satisfies.*

Proof. See Appendix B.2.

We proceed by verifying that, whenever two weakly guarded expression are strongly bisimilar, and each of them provably satisfies a weakly guarded SES, then these SES can be merged into a single weakly guarded SES. This is expressed by Theorem 5.3.3.

Theorem 5.3.3. *Let E and F be two strongly bisimilar expressions, i.e., $E \sim F$. Furthermore let E (F, respectively) $\mathcal{A}_\sim$-provably satisfy the weakly guarded SES S_1 (S_2). Then there is some weakly guarded SES S, that both E and F $\mathcal{A}_\sim$-provably satisfy.*

Proof. See Appendix B.3.

As a last last step we show that two expressions can be equated by means of $\mathcal{A}_\sim$ if both provably satisfy the same SES.

Theorem 5.3.4. *If two expressions E and F $\mathcal{A}_\sim$-provably satisfy a single weakly guarded SES in the free variables of E and F, then $\mathcal{A}_\sim \vdash E = F$.*

Proof. A direct adaption of the proof of [142, Theorem 5.7].

We now have the necessary means to prove completeness for arbitrary expressions.

Theorem 5.3.5. *For arbitrary expressions E and F it holds that*

$$E \sim F \qquad \textit{implies} \qquad \mathcal{A}_\sim \vdash E = F.$$

Proof. Lemma 5.3.2 implies the existence of weakly guarded expressions E' and F' with $\mathcal{A}_\sim \vdash E = E'$ and $\mathcal{A}_\sim \vdash F = F'$. Soundness of the laws gives $E' \sim F'$. Using Theorem 5.3.2, there exist two weakly guarded SES S_1 and S_2, such that E' $\mathcal{A}_\sim$-provably satisfies S_1 and F' $\mathcal{A}_\sim$-provably satisfies S_2. Theorem 5.3.3 implies the existence of some weakly guarded SES S that both E' and F' $\mathcal{A}_\sim$-provably satisfy. Now, Theorem 5.3.4 implies that S has a unique solution. Therefore, $\mathcal{A}_\sim \vdash E' = F'$ and hence $\mathcal{A}_\sim \vdash E = F$.

So, summarising Theorem 5.3.1 and Theorem 5.3.5, we have achieved that $\mathcal{A}_\sim$ axiomatises strong bisimilarity.

Corollary 5.3.1. *For arbitrary expressions E and F it holds that*

$$E \sim F \qquad \textit{if and only if} \qquad \mathcal{A}_\sim \vdash E = F.$$

5.3.2 Weak Congruence

In this section we develop a set of equational laws that is sound and complete with respect to weak congruence. As in the preceding section, we follow the lines of Milner [144]. Table 5.4 presents a set of laws, $\mathcal{A}_{\simeq}^{\mathrm{CCS}}$. This set is sound and complete for weak congruence on CCS. At first glance, the upper part of these laws seems to be literally copied from the set $\mathcal{A}_{\sim}^{\mathrm{CCS}}$ (Table 5.2). This fact should not be surprising, because strong bisimilarity is a subset of weak congruence (cf. Lemma 5.2.2) and therefore every pair that can be proven to be strongly bisimilar has to be weakly congruent as well. This is a striking reason why the axiomatisation of weak congruence is an extension of the axiomatisation of strong bisimilarity.

However, there is an irritating difference between Table 5.2 and the upper half of Table 5.4. Law ($:=$4) has been changed to law ($:=$4′), the alteration being just that the word 'weakly' has been replaced by 'strongly'. Indeed, ($:=$4) is more than just a law, it is the essence of the so-called *recursive specification principle* [14]. It states that a recursively defining equation, such as $X := E$, possesses a unique solution. 'To have a solution' refers to the fact that there is an expression F such that if X is replaced by F on both sides of the equation, the resulting expression, i.e., F and $E\{F/X\}$, are congruent. The solution is said to be unique, if all such F fall into the same congruence class.

Example 5.3.5. The recursively defining equation $X := X$ has (infinitely) many solutions, such as $\mathsf{a}.0$ *and* 0*, that are neither weakly congruent nor strongly bisimilar. It does not possess a unique solution at all.*

The equation $X := \mathsf{a}.X$ (with a *different from τ), on the other hand, has also infinitely many solutions, such as* $\underline{{}_{Y:=}\,\mathsf{a}.\mathsf{a}.Y}$ *and* $\underline{{}_{Z:=}\,\mathsf{a}.Z+0}$*, but they are all strongly bisimilar to each other, as well as weakly congruent. So, this equation possesses a unique solution with respect to either congruence.*

Now, the equation $X := \tau.X$ has a unique solution if we are interested in strong *bisimilarity, where τ is treated in the same way as* a*. But with respect to* weak *congruence, there are expressions, such as* $\tau.\mathsf{a}.0$ *and* $\tau.0$ *that are not weakly congruent, but either of them is a solution of the above equation. So, this equation does not possess a unique solution with respect to weak congruence, in contrast to strong congruence.*

From this example it should be clear that ($:=$4) is not valid for expressions where X appears fully unguarded in E, such as $\underline{{}_{X:=}\,X}$. It is sound for strong bisimilarity if X is (at least) weakly guarded in E (cf. Theorem 5.3.1). Weak congruence additionally requires to restrict this law to strongly guarded expressions in order to exclude partially guarded expressions like $\underline{{}_{X:=}\,\tau.X}$.

In this setting, such partially guarded expressions require a treatment that differs from the strong case. For this reason, Milner introduces the laws ($:=$6) and ($:=$7), depicted in the lower part of $\mathcal{A}^{\mathsf{CCS}}_{\simeq}$. These laws make it possible to transform any partially guarded expression into a strongly guarded one, that is, of course, weakly congruent to the original expression.

The laws (τ1)-(τ3) describe different impacts of internal actions on expressions. It is worth to emphasise that we obtain an axiomatisation of branching bisimulation by removing law (τ3) from $\mathcal{A}^{\mathsf{CCS}}_{\simeq}$.[1]

After this discussion of Milner's axiomatisation of weak congruence for CCS, let us now focus on the goal of this section, an axiomatisation of weak congruence for IML. Unsurprisingly, we will base this axiomatisation on the lower part of $\mathcal{A}^{\mathsf{CCS}}_{\simeq}$, together with the set $\mathcal{A}_{\sim}$ that axiomatises strong bisimilarity on IML.

[1] To be precise, the axiomatisation of branching bisimulation congruence replaces the laws (τ1) – (τ3) by the single law $\mathsf{a}.(E+F) = \mathsf{a}.(\tau.(E+F)+F)$ from which (τ1) and a guarded version of (τ2) can be derived.

Table 5.4. The set $\mathcal{A}_{\simeq}^{\mathrm{CCS}}$ axiomatises weak congruence on CCS

(C)	$E+F$	$=$	$F+E$	
(A)	$(E+F)+G$	$=$	$E+(F+G)$	
(I)	$E+E$	$=$	E	
(N)	$E+0$	$=$	E	
$({:=}1)$	$\underline{X{:=}\,E}$	$=$	$\underline{Y{:=}\,E\{Y/X\}}$	provided Y is not free in $\underline{X{:=}\,E}$.
$({:=}2)$	$\underline{X{:=}\,E}$	$=$	$E\left\{\underline{X{:=}\,E}\,/X\right\}$	
$({:=}3)$	$\underline{X{:=}\,E}$	$=$	$\underline{X{:=}\,X+E}$	
$({:=}4')$	$F=E\{F/X\}$ implies F	$=$	$\underline{X{:=}\,E}$	provided X is *strongly* guarded in E.
$({:=}6)$	$\underline{X{:=}\,\tau.(X+E)+F}$	$=$	$\underline{X{:=}\,\tau.X+E+F}$	
$({:=}7)$	$\underline{X{:=}\,\tau.X+E}$	$=$	$\underline{X{:=}\,\tau.E}$	
$(\tau 1)$	$\mathsf{a}.\tau.E$	$=$	$\mathsf{a}.E$	
$(\tau 2)$	$E+\tau.E$	$=$	$\tau.E$	
$(\tau 3)$	$\mathsf{a}.(E+\tau.F)+\mathsf{a}.F$	$=$	$\mathsf{a}.(E+\tau.F)$	

Again, the first question that arises concerns soundness of the CCS laws for our variant of weak congruence. Indeed, the law $({:=}7)$ turns out to be problematic. This has to do with the notion of *time-divergence* already discussed in Section 4.5. Recall that an expression is time-divergent if it cannot evolve to a stable expression by means of internal steps only. Weak congruence inherits from weak bisimilarity that time-divergent expressions are separated from time-convergent ones. We have argued that this is a highly desirable feature, since time-divergent expressions are forced to perform an infinite number of internal steps, due to maximal progress, without actually letting time progress. But law $({:=}7)$ would allow us to blur this distinction between time-convergence and time-divergence.

Example 5.3.6. Consider the time-divergent expression $\underline{X{:=}\,\tau.X+(\lambda).0}$. *An application of law* $({:=}7)$ *turns this expression into the time-convergent expression* $\underline{X{:=}\,\tau.(\lambda).0}$. *These two expressions are not weakly congruent.*

As a consequence, law $({:=}7)$ is not sound with respect to $\simeq$ on IML. In order to motivate an alternative law, we point out *the* crucial particularity of our definition of weak bisimilarity, and thus of weak congruence: Definition 5.2.2 treats time-divergence and ill-definedness in exactly the same way. In order to equate two closed (or open, applying Definition 5.2.3) expressions, the second clause of this definition allows two possibilities. Either both expressions are time-divergent *or* ill-defined, or both expressions are time-convergent *and* well-defined. It is therefore possible, and also quite intuitive, to regard time-divergence as a specific case of ill-definedness, and to use the symbol $\perp$ to indicate it.

Table 5.5. The set $\mathcal{A}_{\simeq}$ axiomatises weak congruence on IML

(C)	$E+F$	$=$	$F+E$	
(A)	$(E+F)+G$	$=$	$E+(F+G)$	
$(I1)$	$(\lambda).E+(\mu).E$	$=$	$(\lambda+\mu).E$	
$(I2)$	$\mathsf{a}.E+\mathsf{a}.E$	$=$	$\mathsf{a}.E$	
$(I3)$	$E+E+\perp+\perp$	$=$	$E+\perp$	
(N)	$E+0$	$=$	E	
$(\perp 1)$	$(\lambda).E+\perp$	$=$	$\perp$	
$(\perp 2)$	$\perp+\tau.E$	$=$	$\tau.E$	
$({:=}1)$	$\underline{X{:=}\,E}$	$=$	$\underline{Y{:=}\,E\{Y/X\}}$	provided Y is not free in $\underline{X{:=}\,E}$.
$({:=}2)$	$\underline{X{:=}\,E}$	$=$	$E\left\{\underline{X{:=}\,E}/X\right\}$	
$({:=}4')$	$F=E\{F/X\}$ implies F	$=$	$\underline{X{:=}\,E}$	provided X is *strongly* guarded in E.
$({:=}5)$	$\underline{X{:=}\perp+E}$	$=$	$\underline{X{:=}\,X+E}$	
$({:=}6)$	$\underline{X{:=}\,\tau.(X+E)+F}$	$=$	$\underline{X{:=}\,\tau.X+E+F}$	
$({:=}8)$	$\underline{X{:=}\,\tau.X+E}$	$=$	$\underline{X{:=}\,\tau.(\perp+E)}$	
$(\tau 1)$	$\mathsf{a}.\tau.E$	$=$	$\mathsf{a}.E$	
$(\tau 2)$	$E+\tau.E$	$=$	$\tau.E$	
$(\tau 3)$	$\mathsf{a}.(E+\tau.F)+\mathsf{a}.F$	$=$	$\mathsf{a}.(E+\tau.F)$	
$(\tau 4)$	$(\lambda).\tau.E$	$=$	$(\lambda).E$	

We will replace $({:=}7)$ by a law $({:=}8)$ that equates $\underline{X{:=}\,\tau.X+E}$ with $\underline{X{:=}\,\tau.(\perp+E)}$ instead of $\underline{X{:=}\,\tau.E}$. This is the crucial aspect of our set of equations $\mathcal{A}_{\simeq}$, depicted in Table 5.5. The upper part of these laws is essentially $\mathcal{A}_{\sim}$, but with $({:=}4)$ changed to $({:=}4')$, as in the case of $\mathcal{A}_{\simeq}^{\text{CCS}}$. In the lower part of $\mathcal{A}_{\simeq}$ we find law $({:=}6)$ and law $({:=}8)$.

The remaining laws are Milner's laws $(\tau 1)$-$(\tau 3)$, together with a law $(\tau 4)$ which is an obvious adaption of $(\tau 1)$. It is an important observation that no adaption of $(\tau 3)$ is included in $\mathcal{A}_{\simeq}$, such as

$$(\lambda).(E+\tau.F)+(\lambda).F=(\lambda).(E+\tau.F).$$

Indeed, this law is not required and, additionally, not even sound for $\simeq$ on IML. This is basically a consequence of Lemma 4.4.2 saying that $\approx$ treats Markovian transitions in the same way as external transitions in branching bisimilarity. Be reminded that law $(\tau 3)$ is the distinguishing law between weak and branching bisimilarity.

After this explanatory introduction we are now tackling the issues of soundness and completeness of $\mathcal{A}_{\simeq}$.

Theorem 5.3.6. *For arbitrary expressions E and F it holds that*

$$\mathcal{A}_{\simeq} \vdash E = F \qquad \textit{implies} \qquad E \simeq F.$$

Proof. See Appendix B.4.

So, $\mathcal{A}_{\simeq}$ is sound with respect to weak congruence. In order to verify completeness, we will, once again, follow closely the lines of Milner [144]. The proof is a bit more involved than the one for strong bisimilarity. In a preprocessing step, Lemma 5.3.3, we first show that we can transform an arbitrary expression into a *strongly* guarded one.

Lemma 5.3.3. *For each expression E, there exists a strongly guarded expression F such that* $\mathcal{A}_{\simeq} \vdash E = F$.

Proof. See Appendix B.5.

The fact that we transform each expression into a *strongly* guarded one differs from the treatment in Section 5.3.1, where only *weak* guardedness was achieved by Lemma 5.3.2 in a preprocessing step.

The reason to require *weak* guardedness has been the last step of the completeness proof, Theorem 5.3.4. This theorem essentially extends ($_{:=}4$) from a single recursively defining equation to sets of equations. These sets have to be *weakly* guarded (Definition 5.3.2). Since weak congruence forces us to use law ($_{:=}4'$) instead of law ($_{:=}4$), we need *strong* guardedness to ensure that multiple recursively defining equations have a unique solution. This fact will be shown in Theorem 5.3.9. The notion of strong guardedness of equation sets is based on Definition 5.3.3.

Definition 5.3.4. *Let* $S = \{\boldsymbol{X} := \boldsymbol{F}\}$ *be some ES with free variables* $\boldsymbol{W}$. *S is strongly guarded, if it contains no cycle* $X_i \xrightarrow{\tau}{}^{*}_{S} X_i$.

In order to show completeness of $\mathcal{A}_{\simeq}$, we need some intermediate transformations, similar to those required in the strong case.

Theorem 5.3.7. *For each strongly guarded expression E there is some strongly guarded SES in the free variables of E that E* $\mathcal{A}_{\simeq}$*-provably satisfies.*

Proof. The construction in Appendix B.2 provides a strongly guarded SES *S* if expression *E* is strongly guarded.

So we can associate a strongly guarded SES with each strongly guarded expression. We now show how each strongly guarded SES can be transformed into a strongly guarded and *saturated* SES. The (syntactic) notion of saturation corresponds to the (semantic) notion of weak transitions. It is based on Definition 5.3.3.

Definition 5.3.5. *An SES* $S = \{\boldsymbol{X}_i := \boldsymbol{F}_i\}$ *with free variables* $\boldsymbol{W}$ *and formal variables* $\boldsymbol{X}$ *is saturated iff for all* X_i, X_j *and all* W_k *it holds that*

1. $X_i \overset{\tau}{\longrightarrow}\!\!\!\!\!\!\rightarrow^{*}_{S} \overset{a}{\longrightarrow}\!\!\!\!\!\!\rightarrow_{S} \overset{\tau}{\longrightarrow}\!\!\!\!\!\!\rightarrow^{*}_{S} X_j$ *implies* $X_i \overset{a}{\longrightarrow}\!\!\!\!\!\!\rightarrow_{S} X_j$, *and*
2. $X_i \overset{\tau}{\longrightarrow}\!\!\!\!\!\!\rightarrow^{*}_{S} \rhd_S W_k$ *implies* $X_i \rhd_S W_k$.

Lemma 5.3.4. *If a strongly guarded expression E $\mathcal{A}_{\simeq}$-provably satisfies some strongly guarded SES S, then there is some strongly guarded and saturated SES S' that E $\mathcal{A}_{\simeq}$-provably satisfies.*

Proof. See the proof of [144, Lemma 3.1], or [134] for a more detailed one.

We proceed by verifying that, whenever two strongly guarded expression are weakly congruent and each of them provably satisfies a strongly guarded and saturated SES, then these SES can be merged into a single strongly guarded SES. This is expressed by Theorem 5.3.8.

Theorem 5.3.8. *Let E and F be two weakly congruent expressions, i.e., $E \simeq F$. Furthermore let E $\mathcal{A}_{\simeq}$-provably satisfy the SES S_1 and F $\mathcal{A}_{\simeq}$-provably satisfy the SES S_2, where both S_1 and S_2 are strongly guarded and saturated. Then there is some guarded SES S, that both E and F $\mathcal{A}_{\simeq}$-provably satisfy.*

Proof. See Appendix B.6.

As a last step we show, as in the strong case (Theorem 5.3.4), that whenever two expressions provably satisfy the same SES, they can be equated by means of the laws.

Theorem 5.3.9. *If two expressions E and F $\mathcal{A}_{\simeq}$-provably satisfy a single strongly guarded SES in the free variables of E and F, then $\mathcal{A}_{\simeq} \vdash E = F$.*

Proof. Rework of the proof of [144, Theorem 4.2].

We are now in the position to put all these results together to prove completeness for arbitrary expressions.

Theorem 5.3.10. *For arbitrary expressions E and F it holds that*

$$E \simeq F \qquad \textit{implies} \qquad \mathcal{A}_{\simeq} \vdash E = F.$$

Proof. Assume $E \simeq F$. Lemma 5.3.3 implies the existence of strongly guarded expressions E' and F' with $\mathcal{A}_{\simeq} \vdash E = E'$ and $\mathcal{A}_{\simeq} \vdash F = F'$. Soundness of the laws gives $E' \simeq F'$. Using Theorem 5.3.7, there exist two strongly guarded SES S_1 and S_2, such that E' $\mathcal{A}_{\simeq}$-provably satisfies S_1 and F' $\mathcal{A}_{\simeq}$-provably satisfies S_2. Due to Lemma 5.3.4 we can assume that both SES are saturated. Then, Theorem 5.3.8 implies the existence of some strongly guarded SES S that both E' and F' $\mathcal{A}_{\simeq}$-provably satisfy. Now, Theorem 5.3.9 implies that S has a unique solution. Therefore, $\mathcal{A}_{\simeq} \vdash E' = F'$ and hence $\mathcal{A}_{\simeq} \vdash E = F$.

Corollary 5.3.2. *For arbitrary expressions E and F it holds that*

$$E \simeq F \qquad \textit{if and only if} \qquad \mathcal{A}_{\simeq} \vdash E = F.$$

So, we have, in summary, achieved that $\mathcal{A}_{\simeq}$ axiomatises weak congruence, in addition to the result of Corollary 5.3.1, saying that $\mathcal{A}_{\sim}$ axiomatises strong bisimilarity. We have thus tackled the problem of sound and complete equational theories of strong bisimilarity and weak congruence on IML successfully.

This result is made possible due to our deliberate treatment of time-divergence and ill-definedness. In fact, we are now in the position to completely understand why the separation into well-defined and ill-defined expressions is crucial. In Section 4.3 we have argued that the notion of *maximal progress* arises naturally whenever a decision between Markovian transitions and internal action transitions has to be taken. With the syntactic means of IML, this has been expressed in Lemma 5.3.1 as

$$(\lambda).E + \tau.F = \tau.F.$$

The notion of maximal progress is widely used in many timed process algebras, e.g. [151, 187, 93, 49]. However, complete equational theories for weak congruences for such algebras have been an open problem. The reason is that Milner's law ($_{:=}7$) is easily shown to contradict maximal progress.

Example 5.3.7. In our setting, a counterexample is $\underline{_{X:=}\, \tau.X + (\lambda).0}$. *This expression is time-divergent, it can perform an infinite number of steps without evolving to a stable state. Law* ($_{:=}7$) *equates this expression to* $\underline{_{X:=}\, \tau.(\lambda).0}$, *while maximal progress leads to* $\underline{_{X:=}\, \tau.X}$. *The former is time-convergent, while the latter is not. Obviously, they are not equivalent.*

So, the presence of maximal progress and time-divergence hampers the treatment of expressions that are usually covered by ($_{:=}7$). We have overcome this problem by means of three ingredients:

- We have incorporated the notion of well-definedness into the definition of (strong bisimilarity and) weak congruence.
- We have introduced a distinguished symbol $\perp$ that indicates ill-definedness.
- We have replaced law ($_{:=}7$) by law ($_{:=}8$), together with some other laws. The laws make use of the symbol $\perp$ to represent ill-definedness. Time divergence is treated as ill-definedness.

From a practical point of view, the symbol $\perp$ can be seen as an auxiliary construct needed during the process of equating two expressions. To make this view more precise, we introduce a restricted language, $\mathsf{IML}^{\not\perp}$ where the symbol $\perp$ is absent.

Definition 5.3.6. *Let* $\lambda \in \mathbb{R}^+$, $\mathsf{a} \in Act$ *and* $X \in \mathcal{V}$. *We define the language* $\mathsf{IML}^{\not\perp}$ *as the set subset of* IML *given by the following grammar.*

$$\mathcal{E} ::= \; 0 \;\mid\; \mathsf{a}.\mathcal{E} \;\mid\; (\lambda).\mathcal{E} \;\mid\; \mathcal{E}+\mathcal{E} \;\mid\; X \;\mid\; \underline{_{X:=}\,\mathcal{E}}$$

The subset of $\mathsf{IML}^{\not\perp}$ *of closed guarded expressions will be denoted* $\mathsf{IMC}^{\not\perp}$. *Furthermore, we use* $\mathsf{IMC_L}$ *to denote the subset of weakly guarded expressions of* $\mathsf{IMC}^{\not\perp}$.

Though this language does not contain the symbol $\perp$, its axiomatisation still relies on the use of this symbol.

Example 5.3.8. The expressions $\underline{X := \tau.X + \tau.(\lambda).0}$ *and* $\tau.(\lambda).0$ *are both contained in* $\mathsf{IML}^{\not\perp}$ *(even in* $\mathsf{IMC_L}$*), and they are weakly congruent. To establish this property makes the symbol* $\perp$ *appear and vanish again inside the derivation (that uses the laws (:=8), (:=2), (⊥2), and (τ2)):*

$$\mathcal{A}_{\simeq} \vdash \underline{X := \tau.X + \tau.(\lambda).0} \;=\; \underline{X := \tau.(\perp + \tau.(\lambda).0)} \;=\; \tau.(\perp + \tau.(\lambda).0) \;=\; \tau.\tau.(\lambda).0 = \tau.(\lambda).0.$$

So, in order to equate expressions from this restricted language, the laws involving the symbol $\perp$ are still necessary. Since the subset $\mathsf{IMC_L}$, the subset of closed and weakly guarded expressions of $\mathsf{IML}^{\not\perp}$, has a specific importance, we restate Corollary 5.3.1 and Corollary 5.3.2 for $\mathsf{IMC_L}$.

Corollary 5.3.3. *For arbitrary expressions E and F of* $\mathsf{IMC_L}$ *it holds that* [2]

1. $E \sim F$ *if and only if* $\mathcal{A}_{\sim} \vdash E = F$, *and*
2. $E \simeq F$ *if and only if* $\mathcal{A}_{\simeq} \vdash E = F$.

In fact, the set $\mathsf{IMC_L}$ has a direct correspondence to Interactive Markov Chains as they have been defined in Section 4.2. It plays the role of the set $\mathcal{S}^{\text{all}}$, already introduced in Section 2.1 as the state space of 'an immense transition system' containing all transition systems we have discussed. In other words, any interactive processes, and any Interactive Markov Chains can be generated by some expression taken from $\mathsf{IMC_L}$.

Formally, we say that an expression E *generates* some IMC P (modulo strong bisimilarity) if there is a strong bisimulation containing the pair (E, P) of initial states. The notion of strong bisimulation used may be instantiated to either Definition 5.2.1 or Definition 4.3.1. The reason is that both definitions agree on $\mathsf{IMC_L}$, because the elements of $\mathsf{IMC_L}$ must be well-defined by definition.[3]

Theorem 5.3.11. *Each expression from* $\mathsf{IMC_L}$ *generates some IMC with a finite state space. Furthermore, each IMC with a finite state space is generated by some expression from* $\mathsf{IMC_L}$.

So, IMC and $\mathsf{IMC_L}$ are equipotent,[4] and we may therefore call the expressions of $\mathsf{IMC_L}$ *chains* in the sequel. The language $\mathsf{IMC_L}$ is the core of a *specification*

[2] Since $\mathsf{IMC_L}$ does not contain fully unguarded expressions, law (:=5) is superfluous for a complete theory, while it is still needed to completely axiomatise $\mathsf{IML}^{\not\perp}$ and $\mathsf{IMC}^{\not\perp}$.

[3] We are a bit sloppy here, since strong bisimulation according to Definition 4.3.1 is defined on a transition relation, while the semantics of E is a multi-transition relation.

[4] Similar results are possible with respect to *weak congruence* and *weak bisimilarity.* IMC and IMC are equipotent modulo *weak congruence*, if the latter is restricted to weakly guarded expressions. Without any restriction, IMC and IMC are equipotent modulo *weak bisimilarity.*

language for IMC. In order to allow the use of defining equations for the specification of IMC, we say that a standard equation set $\{\boldsymbol{X} := \boldsymbol{F}\}$ generates an IMC, if there is a chain $P \in \mathsf{IMC_L}$ that generates this IMC and P $\mathcal{A}_{\sim}$-provably satisfies $\{\boldsymbol{X} := \boldsymbol{F}\}$. From the restrictions imposed on P it is clear that each F_i does neither contain $\perp$ nor free variables.

Example 5.3.9. The equation (set)

$$E_1 := \mathsf{in.out}.E_1$$

generates the IMC E_1 (Figure 2.1). E_{59} (Figure 4.9) is generated by

$$E_{59} := 0,$$

and E_{63} (Figure 4.10) by

$$E_{63} = \tau.(2\nu).((\mu).E_{63} + \tau.0 + \mathsf{a}.0).$$

Let us now draw the complete picture of results achieved so far. In summary, we have defined a specification language $\mathsf{IMC_L}$ to generate IMC in a compositional way, using the operators, and possibly using defining equations. We have lifted strong and weak bisimilarity to this language. Weak bisimilarity needed a slight refinement, resulting in weak congruence, in order to preserve substitutivity. We have obtained sound and complete equational theories for strong bisimilarity and weak congruence (Corollary 5.3.3). The equational theories use an additional symbol $\perp$ to properly handle maximal progress and time-divergence.

5.4 Parallel Composition and Abstraction

In the previous section, we have tackled the problem of sound and complete equational theories of strong bisimilarity and weak congruence on IML, and we have clarified the relation between IMC and IML. We shall now discuss the operators we have defined in Chapter 4, abstraction and parallel composition in the context of our specification language. It turns out that these operators are *not elementary*, because they can be encoded using the operators of IML. This may seem a bit surprising (it is a standard phenomenon in process algebra though), but it just reflects that IMC are closed under these two operators (Theorem 4.2.2) and the fact that each IMC is generated by some chain of $\mathsf{IMC_L}$ (Theorem 5.3.11).

To establish this property formally, we present a set of additional laws, that allows us one to rewrite parallel composition as well as abstraction into the basic operators of IML. Table 5.6 lists the necessary laws. Law (X) is the IMC version of the well-known *expansion law.* It states that non-synchronising actions of components can be simply interleaved. Either the

Table 5.6. The set $\mathcal{A}_X$ allows rewriting of parallel composition and abstraction

(X)	$P \;\|\!\|_{a_1\ldots a_n}\; Q$	$=$	$\sum(\lambda_i).(P_i \;\|\!\|_{a_1\ldots a_n}\; Q) + \sum_{a_j \notin \{a_1\ldots a_n\}} a_j.(P_j \;\|\!\|_{a_1\ldots a_n}\; Q) +$ $\sum(\mu_k).(P \;\|\!\|_{a_1\ldots a_n}\; Q_k) + \sum_{b_l \notin \{a_1\ldots a_n\}} b_l.(P \;\|\!\|_{a_1\ldots a_n}\; Q_l) +$ $\sum_{a_j = b_l \in \{a_1\ldots a_n\}} a_j.(P_j \;\|\!\|_{a_1\ldots a_n}\; Q_l)$	$\overbrace{\sum(\lambda_i).P_i + \sum a_j.P_j}^{P}$ $\underbrace{\sum(\mu_k).Q_k + \sum b_l.Q_l}_{Q}$
$(H1)$	$\boxed{0}_{a_1\ldots a_n}$	$=$	0	
$(H2)$	$\boxed{a.P}_{a_1\ldots a_n}$	$=$	$a.\boxed{P}_{a_1\ldots a_n}$	provided $a \notin \{a_1 \ldots a_n\}$
$(H3)$	$\boxed{a.P}_{a_1\ldots a_n}$	$=$	$\tau.\boxed{P}_{a_1\ldots a_n}$	provided $a \in \{a_1 \ldots a_n\}$
$(H4)$	$\boxed{(\lambda).P}_{a_1\ldots a_n}$	$=$	$(\lambda).\boxed{P}_{a_1\ldots a_n}$	
$(H5)$	$\boxed{P+Q}_{a_1\ldots a_n}$	$=$	$\boxed{P}_{a_1\ldots a_n} + \boxed{Q}_{a_1\ldots a_n}$	

left ($a_j \notin \{a_1 \ldots a_n\}$), or the right component ($b_l \notin \{a_1 \ldots a_n\}$) performs a non-synchronising action. In case of synchronisation ($a_j = b_l \in \{a_1 \ldots a_n\}$), both partners evolve further. As discussed extensively in Section 4.1, the memoryless property (cf. property (B) on page 41) of exponential distributions implies that delay rate quantities can be interleaved as well, without any adjustment of distributions.

The laws $(H1)$ – $(H5)$ are very simple. They say that abstraction distributes over termination, over choice, and over (action and delay) prefix, where, according to $(H3)$, action a is internalised if it appears in the set $\{a_1 \ldots a_n\}$ of actions. With these laws, parallel composition and abstraction can be shifted arbitrarily deep into a specification, until either 0 or some variable X is reached. In Theorem 5.4.1 we establish that this is enough to ensure completeness for a language $\mathsf{IMC_{XL}}$ that includes abstraction and parallel composition of chains.

Definition 5.4.1. *Let* $\mathsf{IMC_{XL}}$ *be the set of expressions given by the following grammar, where* $P \in \mathsf{IMC_L}$ *and* $a_1, \ldots, a_n \in Act \setminus \{\tau\}$.

$$\mathcal{E} \;::=\; P \;\mid\; \boxed{\mathcal{E}}_{a_1,\ldots,a_n} \;\mid\; \mathcal{E} \;\|\!\|_{a_1,\ldots,a_n}\; \mathcal{E}$$

The action transition *relation* $\longrightarrow \subset \mathsf{IMC_{XL}} \times Act \times \mathsf{IMC_{XL}}$ *is the least relation and the* Markovian transition *relation* $\dashrightarrow \subset \mathsf{IMC_{XL}} \times \mathbb{R}^+ \times \mathsf{IMC_{XL}}$ *is the least multi-relation given by the rules in Table 2.1, Table 4.1, and Table 5.1.*

Theorem 5.4.1. *For arbitrary expressions* $P \in \mathsf{IMC_{XL}}$ *and* $Q \in \mathsf{IMC_{XL}}$ *it holds that*

1. $P \sim Q$ *if and only if* $(\mathcal{A}_\sim \cup \mathcal{A}_X) \vdash P = Q$, *and*
2. $P \simeq Q$ *if and only if* $(\mathcal{A}_\simeq \cup \mathcal{A}_X) \vdash P = Q$.

The proof is omitted since it exhibits no differences to the usual proofs (see for instance [107]).

5.5 Time Constraints and Symmetric Composition

In this section, we discuss two additional operators that exemplify two different strands of extensions to Interactive Markov Chains. One operator which we call the *elapse* operator, is introduced to enhance specification convenience. With this operator, *time constraints* can be inserted into a specification in a compositional way. Roughly speaking, it allows a constraint-oriented specification style for IMC [184]. The shape of the second operator to be introduced is not primarily driven by the desire to enhance modelling convenience. Instead, the reason to introduce *symmetric composition* is a pragmatic one, namely to diminish the state space explosion problem for specifications that exhibit symmetries.

5.5.1 Time Constraints

We aim to introduce an operator that allows one to add time constraints between certain interactions of an Interactive Markov Chain. To fix terminology, a time constraint is a delay that necessarily has to elapse between two kinds of interactions, unless some interaction of a third kind occurs in the meanwhile.

Example 5.5.1. The leaky bucket principle, investigated in Section 4.6, can be seen as a rudimentary example, where time constraints are imposed on certain interactions. For instance, we have incorporated a time constraint between successive out *actions by a small chain* ATM := (λ_{A}).out.ATM. *By synchronising* ATM *and*

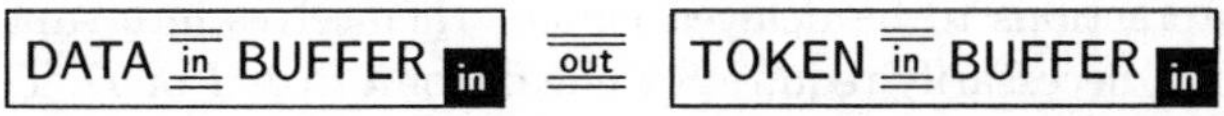

on action out*, successive* out *actions are separated by delays drawn from an exponential distribution with rate* λ_{A}*. In this simple example, interruption of the delay by some interaction of a different kind is not possible.*

In order to facilitate the definition of such (and much more involved) time constraints, we introduce a dedicated operator, the *elapse* operator. The operator has the following five parameters:

- an IMC P is the subject of some time constraint,
- an IMC Q determines the duration of this time constraint,
- a set of actions $D = \{\mathsf{d}_1, \ldots, \mathsf{d}_k\}$ determines which interactions of P should be delayed,

- a set of actions $S = \{s_1, \ldots, s_l\}$ determines when the delay (governed by Q) starts, and
- a set of actions $I = \{i_1, \ldots, i_m\}$ determines which interactions may interrupt the delay.

Though not formally required, it is only meaningful to employ an absorbing Markov chain as Q. The time until absorption will then govern the duration of the time constraint. Clearly, a Markov chain is any expression $Q \in \mathsf{IMC_L}$ that does not involve action prefix. If Q has the possibility of termination, it is an absorbing Markov chain, and the time until termination (i.e., until the state 0 is reached) describes some *phase-type* distribution. This is an important aspect here, since phase-type distributions can approximate arbitrary distributions arbitrarily closely [53, 150]. In other words, we can impose an arbitrarily distributed time constraint, by choosing the appropriate absorbing Markov chain for Q.

As a syntactical notation of the elapse operator, we have chosen the following notation, that appears in two variants,

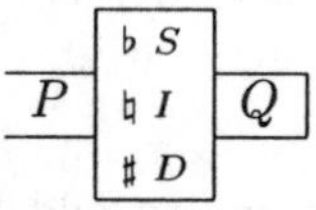

$$\boxed{P}\ \boxed{\begin{array}{l}\flat\ S \\ \natural\ I \\ \sharp\ D\end{array}}\ \boxed{Q} \qquad \text{and} \qquad \boxed{P}\ \boxed{\begin{array}{l}\flat\ (S,Q) \\ \natural\ I \\ \sharp\ D\end{array}}\ \boxed{Q'} \ .$$

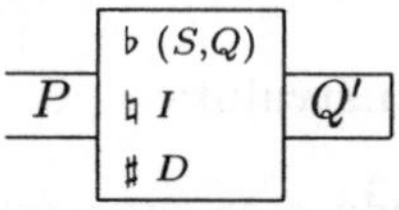

The left variant is used when the time constraint is *inactive*, i.e., when no interaction listed in D has to be delayed, because, for instance, no interaction listed in S has occured yet. If some interaction of S occurs, the elapse operator changes its appearance to the right variant. We say that the time constraint is *active* now. Q' indicates a running phase-type distribution of the absorbing Markov chain Q. Now interactions listed in D are prevented, until Q' has evolved to 0. Then interactions in D are enabled again. If one of them happens, or if some of the interrupting interactions from I occurs during the running time constraint, the elapse operator is de-activated again, by changing back to the left variant. For obvious reasons, we require that I and D are disjoint (interactions to be delayed cannot themselves interrupt the delay). But we do not necessarily require S to be disjoint from I or D. This opens up the possibility to re-initialise the phase-type distributions, if an interaction from $S \cap I$ occurs, or to immediately activate a new time constraint, if an interaction from $S \cap D$ occurs.

This informal explanation is formally realised by the operational rules given in Table 5.7 and Table 5.8; we explain the semantics rule by rule.

- The first rule in Table 5.7 says that an inactive time constraint does not affect, and is not affected by any interaction except those in S.
- The second rule handles the activation of a time constraint. In order to remember the distribution, Q is copied into (S, Q).
- The third rule says that an active time constraint does not affect, and is not affected by any interaction except those that may interrupt the constraint, and those that are to be delayed.

Table 5.7. Operational rules for action transitions of the elapse operator

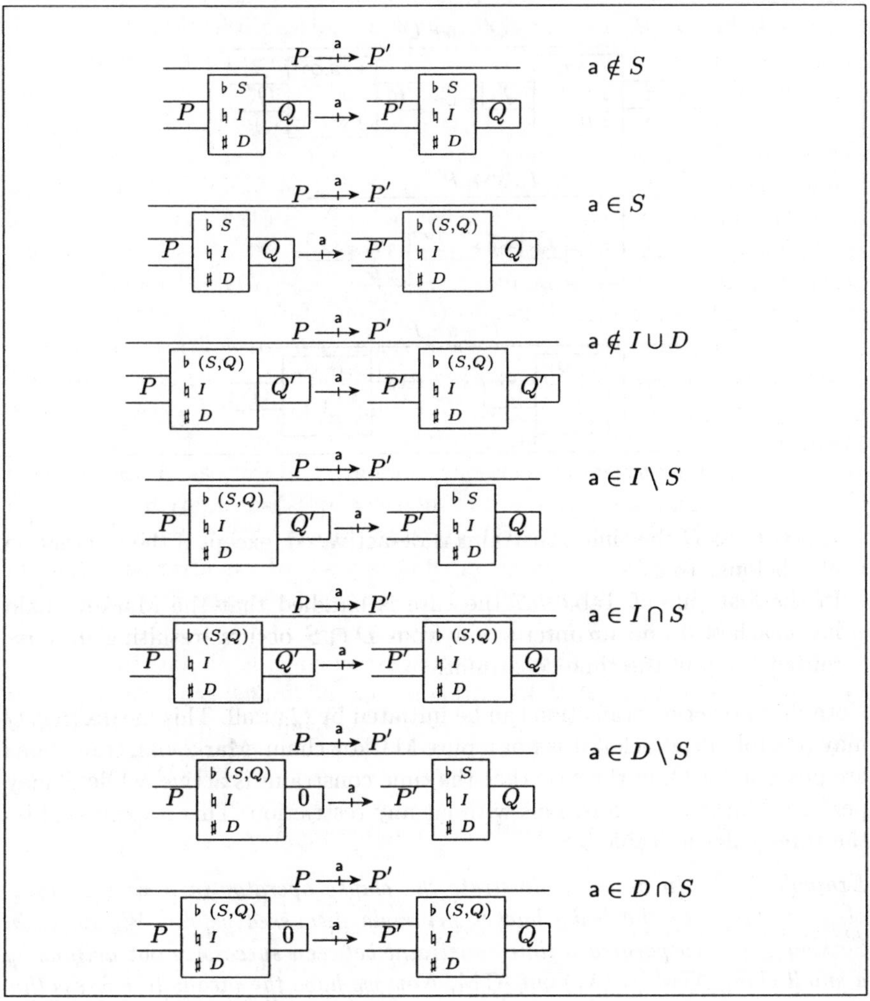

- The fourth rule realizes an interrupt of the time constraint. The time constraint is de-activated.
- In the fifth rule, re-initialisation of an active time constraint is realized. This requires an action that belongs to I as well as S.
- The sixth rule handles the case where the phase-type distribution is elapsed, the Markov chain has reached 0.[5] Upon the occurrence of an inter-

[5] Some reader may find it inelegant to rely on the fact that a terminated Markov chain has a specific syntactic form, namely 0. It seems to be more appropriate to employ an explicit termination signal there, such as δ as it is known from

Table 5.8. Operational rules for Markovian transitions of the elapse operator

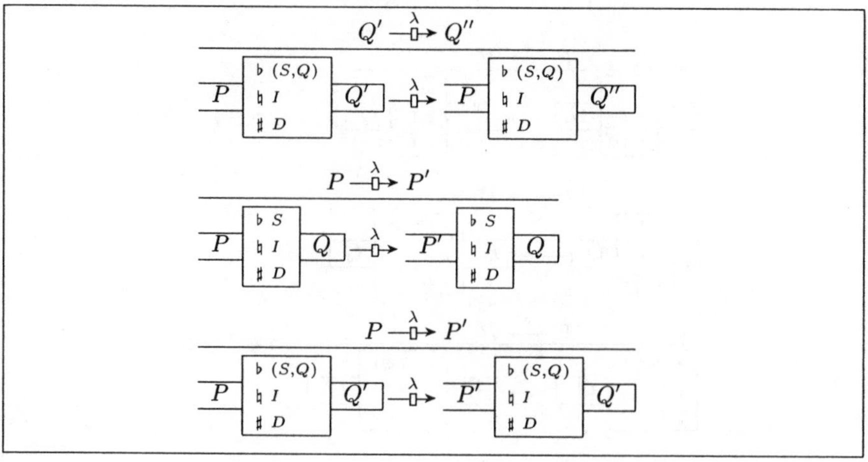

action from D the time constraint is de-activated, except if the interaction also belongs to S.

– In the last rule of Table 5.7 the case is handled that the Markov chain has reached 0 and an interaction from $D \cap S$ occurs, resulting in a re-initialisation of the time constraint.

Note that no action transition can be initiated by Q at all. This means that Q may possibly deadlock if it is not a pure Markov chain. Markovian transitions are possible for Q, in the case that the time constraint is active, while P may perform Markovian transitions without any restriction. This is expressed by the three rules in Table 5.8.

Example 5.5.2. In order to illustrate the elapse operator in a simple example, we return to the leaky bucket principle discussed before. We have, in Section 4.6, incorporated a time constraint between successive out *actions by a small chain* ATM := (λ_T).out.ATM*. Now we have the means to express this time constraint explicitly, using the elapse operator, as*

In a similar way, we could also replace DATA *and* TOKEN *by appropriate time constraints. Furthermore, we can employ different time constraints easily,*

LOTOS. We have chosen the syntactic solution here only for simplicity, since the whole operator is already somewhat complex. To include a signal δ does not pose substantial problems.

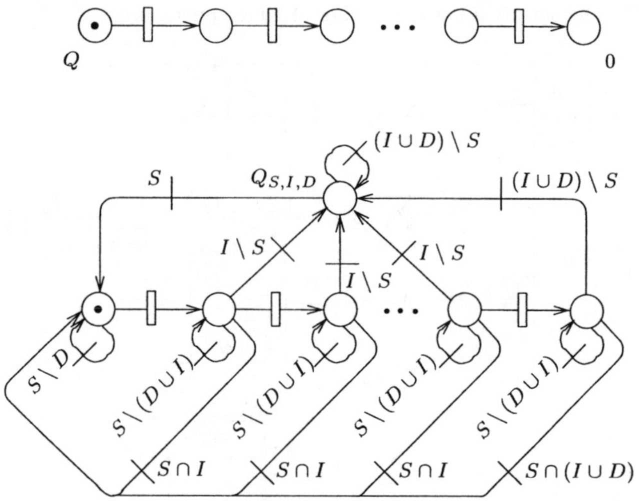

Figure 5.1. Time constraints can be added by means of parallel composition

such as an Erlang$_4$ distribution with mean duration $1/\lambda_A$ by changing the Markov chain that governs the time constraint between successive out *actions to*

From a slightly different point of view, this example indicates that time constraints governed by a certain MC Q can also be incorporated into a specification *without* using the dedicated elapse operator, namely by parallel composition with an IMC, say $Q_{S,I,D}$, of a certain shape. This is the reason why synchronisation with ATM on action out has been used in Section 4.6 for adding the time constraint to the leaky bucket system. For the general case, the shape of such an IMC $Q_{S,I,D}$ is sketched in Figure 5.1. In the essence, some loops and additional action transitions are attached to the MC Q (here an Erlang distribution). In this figure we have taken the liberty to abbreviate a set of parallel action transitions $\xrightarrow{a}$ by a single transition $\xrightarrow{A}$, where A subsumes the respective actions a.

If such an IMC $Q_{S,I,D}$ is synchronised with another IMC P on the union of S, I and D, the effect is exactly the same as specifying 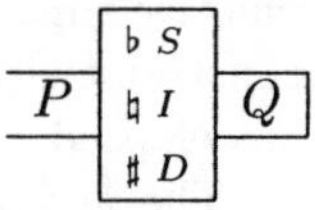

(formally, both are strongly bisimilar). In other words, the elapse operator

Table 5.9. The set $\mathcal{A}_\sharp$ allows rewriting of time constraints

$$(\natural)\quad P\,[\flat S,\ \natural I,\ \sharp D]\,Q = \sum(\lambda_i).P_i\,[\flat S,\ \natural I,\ \sharp D]\,Q\ +$$

$$\sum_{a_j\notin S} a_j.\overline{P_j}\,[\flat S,\ \natural I,\ \sharp D]\,Q\ +\ \sum_{a_j\in S} a_j.\overline{P_j}\,[\flat (S,Q),\ \natural I,\ \sharp D]\,Q$$

$$(\flat)\quad P\,[\flat (S,Q'),\ \natural I,\ \sharp D]\,Q = \sum(\lambda_i).\overline{P_i}\,[\flat (S,Q'),\ \natural I,\ \sharp D]\,Q\ +$$

$$\sum(\mu_k).\overline{P}\,[\flat (S,Q'),\ \natural I,\ \sharp D]\,Q_k\ +$$

$$\sum_{a_j\notin I\cup D} a_j.\overline{P_j}\,[\flat (S,Q'),\ \natural I,\ \sharp D]\,Q\ +$$

$$\sum_{a_j\in I\setminus S} a_j.\overline{P_j}\,[\flat S,\ \natural I,\ \sharp D]\,Q'\ +$$

$$\sum_{a_j\in I\cap S} a_j.\overline{P_j}\,[\flat (S,Q'),\ \natural I,\ \sharp D]\,Q' \qquad \text{provided } Q\not\equiv 0$$

$$(\sharp)\quad P\,[\flat (S,Q'),\ \natural I,\ \sharp D]\,0 = \sum(\lambda_i).\overline{P_i}\,[\flat (S,Q'),\ \natural I,\ \sharp D]\,0\ +$$

$$\sum_{a_j\notin I\cup D} a_j.\overline{P_j}\,[\flat (S,Q'),\ \natural I,\ \sharp D]\,0\ +$$

$$\sum_{a_j\in (I\cup D)\setminus S} a_j.\overline{P_j}\,[\flat S,\ \natural I,\ \sharp D]\,Q'\ +$$

$$\sum_{a_j\in (I\cup D)\cap S} a_j.\overline{P_j}\,[\flat (S,Q'),\ \natural I,\ \sharp D]\,Q'$$

$$P \overset{\textstyle\frown}{\underset{\textstyle\smile}{\overset{\displaystyle \Sigma(\lambda_i).P_i+\Sigma a_j.P_j}{}}}$$

$$\underbrace{\underbrace{\Sigma(\mu_k).Q_k+\Sigma b_l.Q_l}}_{Q}$$

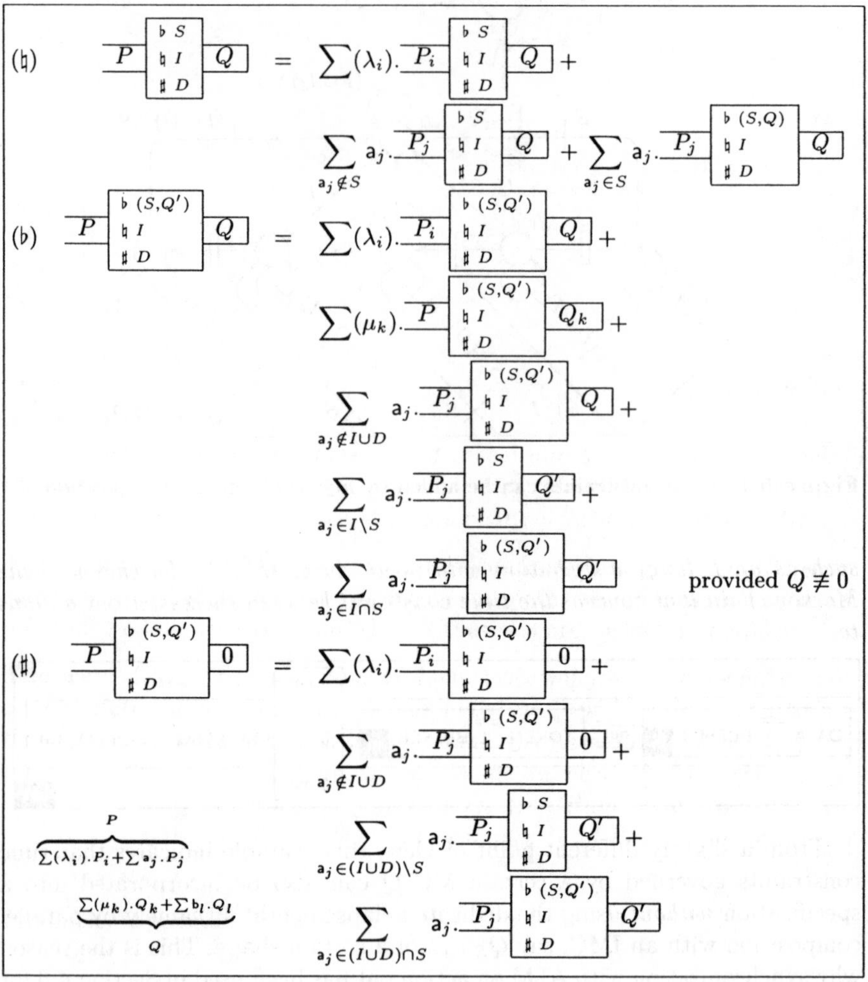

is not essential, it can be encoded by means of parallel composition. But, since parallel composition itself is not essential (Theorem 5.4.1), we may also give a direct encoding of the elapse operator in terms of the basic operators. This is done in Table 5.9, where a kind of expansion law for time constraints is given, consisting of three separate laws. Law ($\natural$) handles the case of an inactive time constraint, while law ($\flat$) describes what can happen if the time constraint is active. Law ($\sharp$) describes the possibilities if the time constraint has elapsed.

The algebraic properties of the elapse operator are as expected. Strong bisimilarity and weak congruence are also congruences with respect to this operator and a sound and complete axiomatisation can be achieved by adding $\mathcal{A}_\sharp$ to $\mathcal{A}_\sim$, respectively $\mathcal{A}_{\simeq}$, as formally established below (Theorem 5.5.4).

One may wonder whether the incorporation of a time constraint via the elapse operator influences the functional side of the behaviour of an interactive process P (which would be considered undesirable). To address this question, let us consider delays as internal moves, i.e., we consider a pragmatic abstraction that maps each (λ) prefix occuring in Q into τ, treating the passage of time as an internal system activity. Under this abstraction, the semantics of the elapse operator stays entirely in the ordinary setting of interactive processes, and we are able to compare the original process to the one resulting after adding the constraint. It is a simple exercise to verify that P and $P \boxed{\begin{smallmatrix}\flat S \\ \natural I \\ \sharp D\end{smallmatrix}} \boxed{Q}$ are equivalent according to ordinary weak (or branching) bisimulation (Definition 2.2.5), whatever the parameters of the constraint (Q, S, I and D) may be, under the above abstraction. As a consequence, we may say that the use of the elapse operator does *not* alter the functional behaviour modulo weak bisimulation.

5.5.2 Symmetric Composition

We now introduce another composition operator. The goal of this operator is different from an enhancement of modelling convenience which has been the motivation for the elapse operator. The main achievement of symmetric composition is to enhance the possibilities of analysing complex models. It avoids an *exponential blow-up* of the state space of a model, that occurs if many identical replica of some component appear in the context of parallel composition.

Example 5.5.3. A small example of a blow-up caused by parallel composition of identical replicas appeared in Figure 2.5. The chain $\boxed{E_1 \;\; E_1}$ *describes a two-place buffer as a composition of two one-place buffers. It requires four states, two of which can be identified by strong bisimilarity. As a consequence, only three states are necessary to represent the behaviour of a two place buffer, (cf.* E_6 *in Figure 2.5). If we increase the number of buffer places to, say, hundred places, the possible reduction is more dramatic. A parallel composition of hundred one-place buffers requires* 2^{100} *states, while a strongly bisimilar behaviour can be expressed by means of 101 states, as well.*

For this buffer example, it is well known how to achieve a compact state space directly (by enumerating the different filling degrees of the buffer). However, as soon as the component to be replicated is slightly more complex or involves synchronisation, it is a difficult and error-prone task to manually determine the reduced state space. It is possible to apply the bisimulation algorithms of Section 2.3, respectively Section 4.5, in order to aggregate the state space, but this requires (compositional) construction of the state space beforehand.

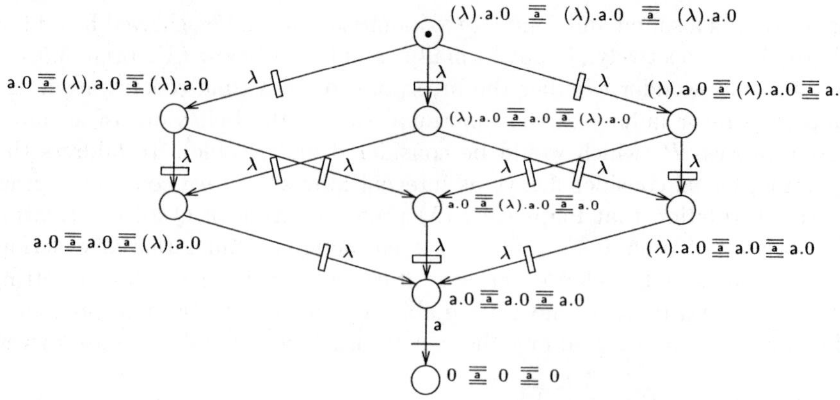

Figure 5.2. Parallel composition of three identical replicas of (λ).a.0

The symmetric composition operator shall avoid this exponential blow-up without any necessity for manual transformations or bisimulation algorithms. Symmetric composition of n identical replicas of P will be denoted by $\{n!P\}A$. These n identical replicas evolve independently by spending time or performing actions not belonging to the synchronisation set A. Actions contained in this set have to be performed synchronously by all replicas. We will define the semantics of symmetric composition in terms of structural operational rules. We then proceed by showing that this semantics is correct with respect to the usual semantics of parallel composition. Furthermore it turns out that, for a restricted class of processes, the semantics is minimal with respect to the number of states. Before introducing the semantics of symmetric composition formally, we discuss the effect of symmetric composition for a specific example.

Example 5.5.4. We use a composition of three replicas of the chain (λ).a.0, *all synchronised on action* a. *The state space obtained when relying on the operator* $\|_{\text{a}}\|$ *is depicted in Figure 5.2. In this transition system, the states in the second (as well as in the third) row are strongly bisimilar, since they are just permutations of each other.*

Figure 5.3 shows the transition system obtained for symmetric composition of three replicas of (λ).a.0. *The chain* {3!(λ).a.0}{a} *will first spend some time, since all three chains spend time. The time until it changes to another behaviour is given by the minimum of three exponential distributions, all with rate* λ. *Since the minimum of exponential distributions is given by the sum of the rates, the rate until a state change happens is* 3λ. *Then, one of the replicas will change its behaviour to* a.0. *The other two will stay in their current state. We denote this situation by the term* {2 ! (λ).a.0 , 1 ! a.0}{a}. *Note that this term contains a set of two different expressions with their respective multiplicities. What is the subsequent behaviour of this term? Performing action*

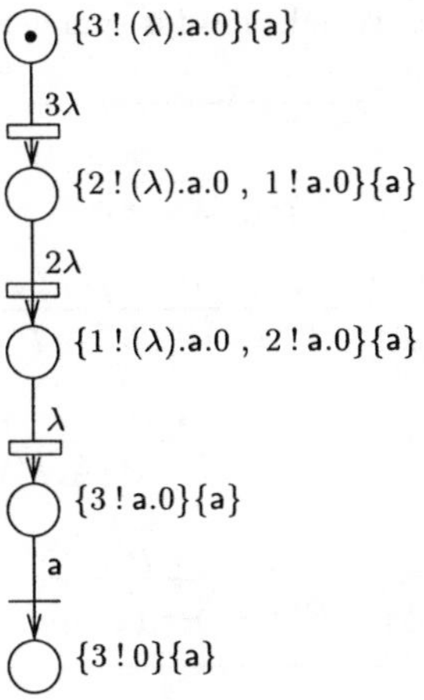

Figure 5.3. Symmetric composition of three identical replicas of (λ).a.0

a *is not possible, because only one out of three replicas is ready to perform it. Therefore, in complete analogy to the above, this process will evolve to* $\{1\,!\,(\lambda).\mathsf{a}.0\ ,\ 2\,!\,\mathsf{a}.0\}\{\mathsf{a}\}$ *with rate* 2λ*. This process is still not able to perform action* a*. It will hence evolve to* $\{3\,!\,\mathsf{a}.0\}\{\mathsf{a}\}$ *with rate* λ*. Now, all chains are ready to perform action* a*, leading to* $\{3\,!\,0\}\{\mathsf{a}\}$.

From this example it is not surprising that we are essentially dealing with multisets of chains. The bridge between the notation $\mathcal{M} = \{n_1\,!\,P_1, \ldots, n_m\,!\,P_m\}$ and a standard multiset over chains, i.e., a function (also denoted $\mathcal{M}$) from chains to $\mathbb{N}$ is the following definition:

$$\mathcal{M}(P) = k > 0 \quad \text{if and only if} \quad (k\,!\,P) \in \mathcal{M}.$$

With these notational preliminaries we are ready to define the semantics of symmetric composition. It is given by the set of rules given in Table 5.10. $\varnothing$ denotes an empty multiset (different from $\emptyset$, denoting an empty *set*). The rules make use of a standard multiset operation, namely insertion ($\oplus$) of elements to a multiset. It is defined as

$$\mathcal{M} \oplus P := \mathcal{M}' \text{ with } \mathcal{M}'(P') := \textbf{ if } (P' \equiv P) \textbf{ then } \mathcal{M}(P') + 1 \textbf{ else } \mathcal{M}(P').$$

The first two rules handle synchronisation, where all replicas have to change state on the occurrence of an action contained in A. The necessary preconditions that all elements in the multiset are able to perform an action are checked element-wisely, until the remaining multiset is empty. This effect is best explained by means of our example.

Example 5.5.5. The proof tree for the transition $\{3\,!\,\mathsf{a}.0\}\{\mathsf{a}\} \xrightarrow{\mathsf{a}} \{3\,!\,0\}\{\mathsf{a}\}$, *is constructed as follows:*

Table 5.10. Operational rules for symmetric composition

$$\frac{}{\varnothing A \xrightarrow{\texttt{a}} \varnothing A} \qquad \texttt{a} \in A$$

$$\frac{\mathcal{M}A \xrightarrow{\texttt{a}} \mathcal{M}'A \qquad P \xrightarrow{\texttt{a}} P'}{(\mathcal{M} \oplus P)A \xrightarrow{\texttt{a}} (\mathcal{M}' \oplus P')A} \qquad \texttt{a} \in A$$

$$\frac{P \xrightarrow{\texttt{a}} P'}{(\mathcal{M} \oplus P)A \xrightarrow{\texttt{a}} (\mathcal{M} \oplus P')A} \qquad \texttt{a} \notin A$$

$$\frac{P \xrightarrow{\lambda} P'}{(\mathcal{M} \oplus P)A \xrightarrow{n\lambda} (\mathcal{M} \oplus P')A} \qquad \mathcal{M}(P) = n-1$$

$$\cfrac{\cfrac{\cfrac{\cfrac{}{\varnothing\{\texttt{a}\} \xrightarrow{\texttt{a}} \varnothing\{\texttt{a}\}} \quad \cfrac{}{\texttt{a.0} \xrightarrow{\texttt{a}} \texttt{0}}}{\{1\,!\,\texttt{a.0}\}\{\texttt{a}\} \xrightarrow{\texttt{a}} \{1\,!\,\texttt{0}\}\{\texttt{a}\}} \quad \cfrac{}{\texttt{a.0} \xrightarrow{\texttt{a}} \texttt{0}}}{\{2\,!\,\texttt{a.0}\}\{\texttt{a}\} \xrightarrow{\texttt{a}} \{2\,!\,\texttt{0}\}\{\texttt{a}\}} \quad \cfrac{}{\texttt{a.0} \xrightarrow{\texttt{a}} \texttt{0}}}{\{3\,!\,\texttt{a.0}\}\{\texttt{a}\} \xrightarrow{\texttt{a}} \{3\,!\,\texttt{0}\}\{\texttt{a}\}}$$

Note that, according to the first rule, $\varnothing A$ is able to interact on any action listed in A, without changing state. This is required by the left precondition of the second rule in order to let a synchronisation happen. The latter two rules of symmetric composition handle asynchronous action transitions, respectively Markovian transitions, in a straightforward way. As we have seen in the example, the multiplicity of possible Markovian transitions has to be taken into account.

Example 5.5.6. For instance, $\{2\,!\,(\lambda).\texttt{a.0}\ ,\ 1\,!\,\texttt{a.0}\}\{\texttt{a}\} \xrightarrow{2\lambda} \{1\,!\,(\lambda).\texttt{a.0}\ ,\ 2\,!\,\texttt{a.0}\}\{\texttt{a}\}$ *can be derived as follows:*

$$\cfrac{\cfrac{}{(\lambda).\texttt{a.0} \xrightarrow{\lambda} \texttt{a.0}}}{\{2\,!\,(\lambda).\texttt{a.0}\ ,\ 1\,!\,\texttt{a.0}\}\{\texttt{a}\} \xrightarrow{2\lambda} \{1\,!\,(\lambda).\texttt{a.0}\ ,\ 2\,!\,\texttt{a.0}\}\{\texttt{a}\}}$$

Now, once we have discussed the effect of these laws, it is worth to observe in what sense symmetric composition is consistent with parallel composition. More precise, it is important to establish that the transition system obtained by applying the above rules is different, but equivalent (i.e., strongly bisimilar) to the transition system obtained by applying parallel composition. This central property of symmetric composition will be expressed in Theorem 5.5.1. Before we are able to establish this result, we have to be a bit more precise about the chains we allow to appear inside a multiset.

The complete language of IMC we are going to consider, $\mathsf{IMC_{XXL}}$, consists of IMC, together with the four operators introduced in the previous sections: symmetric and parallel composition, time constraints, and abstraction. It is defined on the basis of $\mathsf{IMC_L}$ and as a superset of $\mathsf{IMC_{XL}}$.

Definition 5.5.1. *Let* $\mathsf{IMC_{XXL}}$ *be the set of expressions given by the following grammar where* $P, P' \in \mathsf{IMC_L}$, *and* $S, I, D, A, \{\mathsf{a}_1, \ldots, \mathsf{a}_n\} \subseteq Act \setminus \{\tau\}$.

$$\mathcal{E} \ ::= \ P \mid \boxed{\mathcal{E}\ \mathsf{a_1\ldots a_n}} \mid \mathcal{E} \overline{\overline{\mathsf{a_1\ldots a_n}}} \mathcal{E} \mid$$

$$\mathcal{E}\ \boxed{\begin{array}{l}\flat\, S\\ \natural\, I\\ \sharp\, D\end{array}}\ \boxed{P} \mid \mathcal{E}\ \boxed{\begin{array}{l}\flat\,(S,P')\\ \natural\, I\\ \sharp\, D\end{array}}\ \boxed{P} \mid \mathcal{MA}$$

$$\mathcal{M} \ ::= \ \mathcal{M} \oplus \mathcal{E} \mid \varnothing$$

The language $\mathsf{IMC_{XXL}}$ comprises arbitrary multisets, where each of the elements of a multiset is built according to the first row of the grammar. Symmetric composition is a special case, where $\mathcal{M}$ is just $\{n\,!\,\mathcal{E}\}$ for $n \in \mathbb{N}$. This syntax is not explicitly included in the grammar of $\mathsf{IMC_{XXL}}$, we use it as a shorthand for a multiset obtained by adding the same expression $\mathcal{E}$ to $\varnothing$ exactly n times.[6]

Definition 5.5.2. *The* action transition *relation* $\dashrightarrow\!\!\!\!\!\rightarrow \subset \mathsf{IMC_{XXL}} \times Act \times \mathsf{IMC_{XXL}}$ *is the least relation and the* Markovian transition *relation* $\dashrightarrow\!\!\!\!\!\rightarrow \subset \mathsf{IMC_{XXL}} \times \mathbb{R}^+ \times \mathsf{IMC_{XXL}}$ *is the least multi-relation given by the rules in Table 2.1, Table 4.1, Table 5.1, Table 5.7, Table 5.8, and Table 5.10.*

Since symmetric composition is intended to be a specific reformulation of parallel composition, it is not surprising that we can find an expansion law to encode symmetric composition into the basic language. However, in order to make the connection to parallel composition explicit, Table 5.11 provides a direct encoding of symmetric composition in terms of parallel composition. Indirectly (using the expansion law (X)), this gives an encoding of symmetric composition into the basic language. The set $\mathcal{A}_!$ contains two laws having a clear correspondence to the operational rules of Table 5.10. Law $(\oplus)$ states that symmetric composition can be rewritten to parallel composition element-wisely. Law $(\varnothing)$ handles the empty multiset, as it is done by the first operational rule. In fact, the soundness of these two laws is the key to establish that symmetric composition is consistent with parallel composition.

---

[6] In technical terms, define $\{1\,!\,P\}$ as $(\varnothing \oplus P)$, and for $n \geq 0$ define $\{n+1\,!\,P\}$ as $(\{n\,!\,P\} \oplus P)$. One may demand that a user of the language is only allowed to use the notation $\{n\,!\,P\}$ instead of specifying arbitrary multisets. With this restriction, the implementation of symmetric parallel composition by means of a multiset is completely hidden from the user of the language. However, since this constraint is not required for any technical reason, we do not restrict the use of multisets.

Table 5.11. The set $\mathcal{A}_!$ allows rewriting of symmetric composition

$$(\oplus)\quad (\mathcal{M} \oplus P)A \;=\; \mathcal{M}A \overline{\overline{\;\mathsf{a}_1 \ldots \mathsf{a}_m\;}} P \qquad \text{provided } \{\mathsf{a}_1 \ldots \mathsf{a}_m\} = A$$

$$(\varnothing)\quad \varnothing A \;=\; \sum_{\mathsf{a} \in A} \mathsf{a}.(\varnothing A)$$

Theorem 5.5.1. *For $n > 0$ and $P \in \mathsf{IMC}_{\mathsf{XXL}}$ it holds that*

$$\underbrace{P \overline{\overline{\;\mathsf{a}_1 \ldots \mathsf{a}_m\;}} P \ldots \overline{\overline{\;\mathsf{a}_1 \ldots \mathsf{a}_m\;}} P}_{n \text{ times}} \quad \textit{is strongly bisimilar to } \{n \,!\, P\}\{\mathsf{a}_1 \ldots \mathsf{a}_m\}.$$

Proof. A side result of the proof of Theorem 5.5.4.

So, symmetric composition always generates a strongly bisimilar transition system compared to parallel composition. In addition, it is clear that the generated state space is strictly smaller than that obtained from parallel composition (except for $n = 1$, or if P is deadlocked or terminated). This raises the question whether this transition system is also *minimal* with respect to bisimilarity. As discussed in Section 2.3, a transition system is minimal if it possesses the least possible number of states necessary to represent a certain equivalence class of behaviours.

The experiences we sketched so far seem to suggest that symmetric composition has this desirable property. However, we are only able to show minimality for a rather restrictive class of chains, *linear* chains, except if no synchronisation among replicas is forced.

Definition 5.5.3. *A chain of $\mathsf{IMC}_{\mathsf{XXL}}$ is linear if it does at most involve recursion, (delay and action) prefix, termination and abstraction, but neither choice nor parallel (or symmetric) composition (and time constraints).*

Theorem 5.5.2. *If P is minimal with respect to strong bisimilarity, then $\{n \,!\, P\}A$ is minimal with respect to strong bisimilarity provided that P is linear or that A is empty.*

Proof. See Appendix B.7.

Example 5.5.7. The minimal representation of a one-place buffer, $E_1 := \mathsf{in}.\mathsf{out}.E_1$ (Figure 2.5) is linear. By Theorem 5.5.2, symmetric composition of hundred replicas of a one-place buffer with $\{100 \,!\, E_1\}\emptyset$ is minimal. The state space consists of 101 states. Note that no synchronisation among buffers is forced, as well.

One major reason for the above restriction to linear chains is the possibility to introduce deadlocks by means of synchronisation. An example is symmetric composition of $P := \mathsf{a}.\mathsf{b}.0 + \mathsf{a}.\mathsf{c}.0$. The chain $\{2 \,!\, P\}\{\mathsf{a}, \mathsf{b}, \mathsf{c}\}$ possesses a

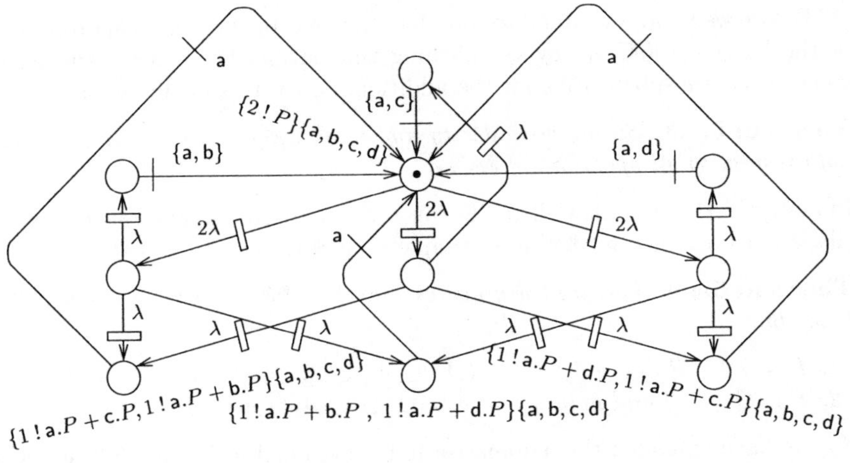

Figure 5.4. Non-minimal symmetric composition

deadlock $\{1 ! \mathsf{b}.0, 1 ! \mathsf{c}.0\}\{\mathsf{a}, \mathsf{b}, \mathsf{c}\}$ that trivially behaves in the same way as $\{2 ! 0\}\{\mathsf{a}, \mathsf{b}, \mathsf{c}\}$, which is also reachable. So, the semantics is not minimal. Analysing such examples seems to suggest that deterministic chains do not pose such problems. A chain is deterministic if for each choice, the initial actions are pairwise different. This property is closed under symmetric composition for actions appearing in the synchronisation set. Since the above problem arises in the context of synchronisation, determinism of all actions appearing in a synchronisation set appears to avoid this. But again, counterexamples exist, and it is even not enough to consider only chains where $\{n ! P\}A$ does not deadlock.

Example 5.5.8. An example is depicted in Figure 5.4. The process $\{2 ! P\}\{\mathsf{a}, \mathsf{b}, \mathsf{c}, \mathsf{d}\}$, *where*

$$P := \quad (\lambda).(\mathsf{a}.P + \mathsf{b}.P) \quad + \quad (\lambda).(\mathsf{a}.P + \mathsf{c}.P) \quad + \quad (\lambda).(\mathsf{a}.P + \mathsf{d}.P)$$

possesses three equivalent states, appearing in the last row of this figure. Therefore the semantics is not minimal. Note however, that Markovian transitions are maximally lumped, and that applying parallel composition would lead to 16 instead of 10 states.

The syntax of this counterexample possesses a specific pattern. Nevertheless it is not obvious how to characterise the absence of this pattern and similar patterns syntactically. A syntactic criterion, excluding such situations is a prerequisite to extend Theorem 5.5.2 beyond linear processes. This deficiency is contrasted by the observation that symmetric composition leads to minimal transition systems in most practical examples we investigated so far. This will be exemplified in Chapter 6, where we will illustrate the value of Interactive Markov Chains in practice, with particular emphasis on symmetric composition.

Before we do so, we intend to complete the picture of theoretical properties of the language $\mathsf{IMC_{XXL}}$, by establishing that strong bisimilarity and weak congruence are substitutive for the additional operators of this section.

Theorem 5.5.3. *Strong bisimilarity and weak congruence are congruences with respect to all operators of $\mathsf{IMC_{XXL}}$.*

In addition, the set of laws listed in this section are sound, and together with the laws established for IMC also complete for $\mathsf{IMC_{XXL}}$.

Theorem 5.5.4. *For arbitrary expressions $P \in \mathsf{IMC_{XXL}}$ and $Q \in \mathsf{IMC_{XXL}}$ it holds that*

1. $P \sim Q$ *if and only if* $(\mathcal{A}_\sim \cup \mathcal{A}_X \cup \mathcal{A}_\sharp \cup \mathcal{A}_!) \vdash P = Q$, *and*
2. $P \simeq Q$ *if and only if* $(\mathcal{A}_\simeq \cup \mathcal{A}_X \cup \mathcal{A}_\sharp \cup \mathcal{A}_!) \vdash P = Q$.

So, we have extended the axiomatisations developed in Section 5.3 to cover the four additional operators as well. Apart from providing a clear characterisation of both operators and congruences, these axiomatisations can be used to prove equality by means of purely syntactic transformations. For this purpose, it is worthwhile to derive further laws from the basic laws that abbreviate tedious syntactic transformations. Some examples are given below (In order to avoid superfluous brackets, 2. has already been used in Theorem 5.5.1 and Figure 5.2.).

Lemma 5.5.1. *Let ' $\vdash$ ' denote derivability by $\mathcal{A}_\sim$, respectively $\mathcal{A}_\simeq$, plus $(\mathcal{A}_X \cup \mathcal{A}_\sharp \cup \mathcal{A}_!)$. Then,*

1. $\vdash$ $P \overline{\overline{a_1 \ldots a_n}} \, Q = Q \overline{\overline{a_1 \ldots a_n}} \, P$,

2. $\vdash$ $(P \overline{\overline{a_1 \ldots a_n}} \, Q) \overline{\overline{a_1 \ldots a_n}} \, R = P \overline{\overline{a_1 \ldots a_n}} \, (Q \overline{\overline{a_1 \ldots a_n}} \, R)$,

3. $\vdash$ $\boxed{P\ \boxed{a}\ \boxed{b}} = \boxed{P\ \boxed{b}\ \boxed{a}}$,

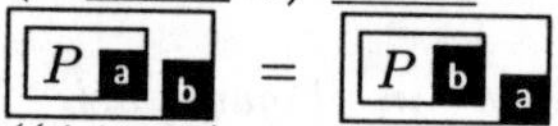

4. $\vdash$ $((\mathcal{M} \oplus P) \oplus Q)A = ((\mathcal{M} \oplus Q) \oplus P)A$,

5. $\vdash$ $P\ \begin{array}{l} \flat\ S_1 \\ \natural\ I_1 \\ \sharp\ D_1 \end{array}\ Q_1\ \begin{array}{l} \flat\ S_2 \\ \natural\ I_2 \\ \sharp\ D_2 \end{array}\ Q_2 \;=\; P\ \begin{array}{l} \flat\ S_2 \\ \natural\ I_2 \\ \sharp\ D_2 \end{array}\ Q_2\ \begin{array}{l} \flat\ S_1 \\ \natural\ I_1 \\ \sharp\ D_1 \end{array}\ Q_1$.

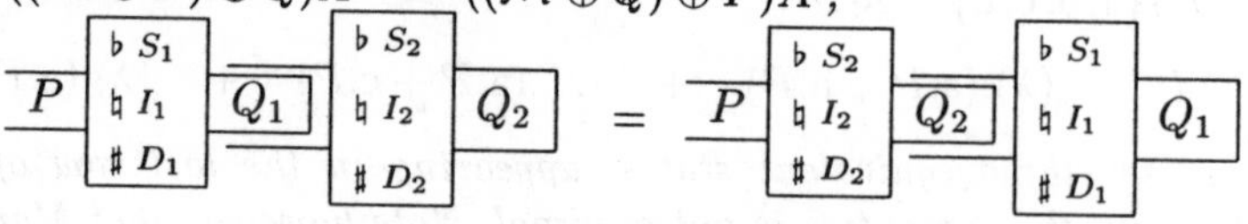

5.6 Discussion

In this chapter we have introduced a language to generate IMC and have built a rigid algebraic theory on top of this language. Starting from a basic calculus, we have added a set of composition operators for different purposes, while retaining a sound and complete axiomatisation (Theorem 5.5.4).

We like to point out that the axiomatisation established for the basic calculus IML is far more than a trivial exercise. On the contrary, a satisfactory equational treatment of bisimilarity is rare in the literature of process calculi with a notion of time and maximal progress. Apart from some work that restricts to finite behaviours (i.e., excludes recursive definitions at all), we are only aware of the work of Aceto and Jeffrey [1] who give an axiomatisation of strong bisimilarity for Wang's timed CCS [187]. Their equational theory is restricted to closed and weakly guarded expressions.

Furthermore, the issue of a sound and complete axiomatisation for *weak* congruence has been an open problem for any process calculus with maximal progress for a long time. The reason is that time-divergent expressions cannot be treated in the way pursued by Milner in [144]. In particular, Milner's law $(_{:=}7)$

$$\underline{_{X:=}\,\tau.X + E} = \underline{_{X:=}\,\tau.E}$$

is easily shown to contradict maximal progress, expressed by (cf. Lemma 5.3.1)

$$(\lambda).E + \tau.F = \tau.F.$$

The set $\mathcal{A}_{\simeq}$ solves this problem for the language IML, (including open and fully unguarded recursive expressions). The key idea is to introduce a specific symbol $\bot$ to indicate ill-definedness and to replace the above law $(_{:=}7)$ by a set of laws that equate time-divergence and ill-definedness. Together with a law that makes it possible to escape from ill-definedness (and hence time-divergence) by means of an internal step (law $(\bot 2)$), this treatment is sufficient to achieve completeness. The necessary proofs, given in detail in Appendix B, are non-standard and fairly involved, in particular the proofs of Theorem 5.3.3, Lemma 5.3.3 and Theorem 5.3.8. They are based on joint work with Markus Lohrey [105].

The notion of ill-definedness is borrowed from Walker [185]. Walker has studied divergence in the context of regular CCS and weak congruence. It is interesting to discuss $\simeq$ in the context of regular CCS that arises from IML by disallowing delay prefixing. We use $\mathsf{IML}_{\not\lambda}$ to denote this subset of IML. With the technical means of Section 5.1 the following result is easy to show.

Theorem 5.6.1. *For arbitrary expressions $E, F \in \mathsf{IML}_{\not\lambda}$ it holds that*

$$E \simeq F \text{ if and only if } \left(\{(\bot 2), (_{:=}5), (_{:=}8)\} \cup \left(\mathcal{A}_{\simeq}^{ccs} \setminus \{(_{:=}3), (_{:=}7)\}\right)\right) \vdash E = F.$$

Stated differently, we have obtained an equational theory for CCS modulo weak congruence. The theory (and hence the congruence itself) differs from other treatments of divergence in CCS: Our notion of weak congruence does neither coincide with Milner's divergence insensitive notion (denoted $\simeq_{\mathsf{Milner}}$) nor with Walker's divergence sensitive variant ($\simeq_{\mathsf{Walker}}$).[7] Roughly, the reason is that, different from Walker, it is possible to escape from *unstable* divergence. But, deviating from Milner, it is not possible to escape from *stable*

[7] Walker's basic notion is a preorder rather than an equivalence. The induced equivalence turns out to be incomparable with Milner's original notion of weak congruence.

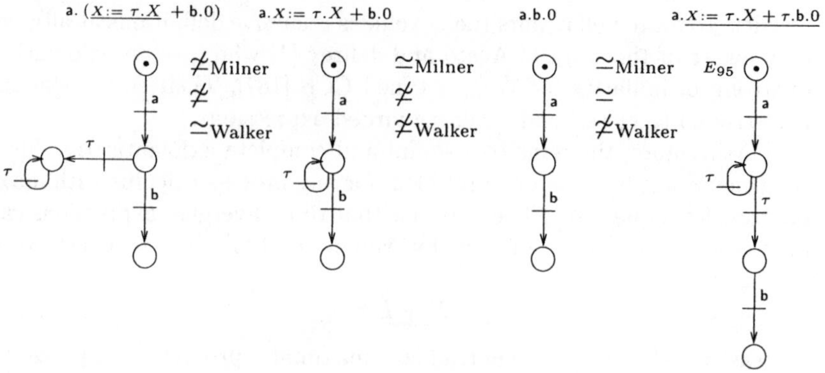

Figure 5.5. Weak congruence is finer than Milner's notion and incomparable with Walker's

divergence. As a whole, it can be shown that $\simeq$ is incomparable with Walker's notion (cf. the first and last pair in Figure 5.5). In contrast, $\simeq$ turns out to be strictly finer than Milner's observational congruence.

Theorem 5.6.2. *For arbitrary expressions $E, F \in \mathsf{IML}_{\mathcal{X}}$ it holds that $E \simeq F$ implies $E \simeq_{Milner} F$.*

The inclusion is strict, as testified by the middle pair in Figure 5.5. This compares favourably to the treatment of divergence in [185] that is incomparable with the original definition. Furthermore, our treatment of divergence can be adopted to establish (formerly unknown) sound and complete equational theories for a variety of process algebras. We briefly sketch some results, focusing on a calculus comparable with IML, i.e., including prefix, choice, termination and recursion.

- The set of laws $\{(C), (A), (I1), (I3), (N), (\perp 1), (:=1), (:=2), (:=4), (:=5)\}$ provides a sound and complete axiomatisation of strong equivalence of PEPA [114] (where '(a, _);' replaces '(_).' in law $(I1)$ and law $(\perp 1)$.).
- The same set of laws is sound and complete for Markovian bisimilarity of MTIPP [107] (giving an implicit proof that strong equivalence and Markovian bisimilarity agree on this common fragment).
- The set of laws $\{(\perp 1), (\perp 2), (:=1), (:=2), (:=4), (:=5)\}$ forms the basis of a sound and complete equational theory with respect to strong bisimilarity on the timed calculus CSA with a single clock [49]. This fragment of CSA agrees with Hennessy and Regan's TPL [93], but originally TPL is developed in a testing setting.
- The set of laws $\{(C), (A), (I), (N), (\perp\ 1), (\perp\ 2), (:=1), (:=2), (:=4), (:=5)\}$ (where '$\underline{a}$.' replaces '(λ).' in each of the laws) is sound and complete with respect to strong congruence on CCS^{prio}, the *prioritised* calculus of [148].
- For a simplified variant of weak congruence on CCS^{prio} the above set of laws without law $(:=4)$ but together

with $\{(\text{:=}4'), (\tau 1), (\tau 2), (\tau 3), (\tau 4), (\text{:=}6), (\text{:=}8)\}$ and a law $(\lambda).(E + \tau.F) + (\lambda).F = (\lambda).(E + \tau.F)$ is sound and complete (again replacing each '(λ).' by '$\underline{\mathsf{a}}$.'). The details are carried out in [104].

Indeed for prioritised calculi such as CCS^{prio} the problems are very similar to those faced in the context of timed calculi with maximal progress. The law that nicely reflects priority

$$\tau.E + \underline{\mathsf{a}}.F = \tau.E.$$

(where $\underline{\mathsf{a}}$ has a lower priority than τ) contradicts Milner's treatment of divergence, law $(\text{:=}7)$. As a consequence, no axiomatisation of weak congruence for process calculi including priority has appeared in the literature before this one, as it is the case for calculi with maximal progress. An alternative to our approach has recently been proposed by Bravetti and Gorrieri [35], who manage to retain $(\text{:=}7)$ in an axiomatisation of priority.

It seems also worth to include a small historical discussion. Law $(\text{:=}7)$ realises a notion of *fairness*, since it may be used to abstract from a computation that just consists of internal steps without actually progressing. Koomen [131] was the first to define a 'fair abstraction rule' (KFAR) similar to law $(\text{:=}7)$. Bergstra *et al.* have introduced a weaker version of fair abstraction (WFAR) that allows to escape from such an internal computation only if an internal alternative exists [23]. In our setting, a WFAR law can be formulated as

$$\underline{X{:=}\,\tau.X + \tau.E + F} = \underline{X{:=}\,\tau.(\tau.E + F)}\,.$$

Actually, this law can be derived by means of law $(\text{:=}8)$ and $(\bot 2)$, i.e.,

$$\{(\text{:=}8), (\bot 2)\} \vdash \underline{X{:=}\,\tau.X + \tau.E + F} = \underline{X{:=}\,\tau.(\bot {+} \tau.E + F)} = \underline{X{:=}\,\tau.(\tau.E + F)}\,.$$

So, one may view Theorem 5.6.1 (and Corollary 5.3.2) as a sound and complete equational theory on CCS (respectively IML) modulo weak congruence with WFAR instead of KFAR. This variant of weak bisimulation appears already in the *linear time – branching time spectrum II* of R. van Glabbeek [74], where it is called stable weak bisimulation. Recently, the equational theory has been extended to the entire lattice of divergence sensitive weak bisimulations of the spectrum [135]. Furthermore, Bravetti has very recently proposed an interesting revision of the IMC algebra, where KFAR remains valid [31].

In Section 4.7 we have compared our definition of weak bisimulation with the one of Rettelbach. We still owe the reader the answer why weak bisimulation (and weak congruence) a lá Rettelbach is not satisfactory in the context of an equational theory. The reason is, that the definition of [163] fails to be a congruence for recursion.

Example 5.6.1. Following Rettelbach's definition and applying the 'lifting' according to Definition 5.2.3, the two expression $\tau.X$ and $\tau.\tau.X$ are equivalent, while $\underline{X{:=}\,\tau.X}$ and $\underline{X{:=}\,\tau.\tau.X}$ are not. The first describes a loop of length one and is treated different from the latter, a loop of length two. This makes the congruence property fail for recursion.

It is hence not reasonable to base an equational theory for IML that accounts for recursion on the definition of Rettelbach.

Based on the subalgebra $\mathsf{IMC_L}$ we have introduced a language to generate Interactive Markov Chains. The language $\mathsf{IMC_{XXL}}$ contains further operators, two of which have appeared before, parallel composition and abstraction. We have shown that these operators can be encoded into the basic calculus by means of additional laws. The same is true for the two additional operators that we have introduced for specific modelling and analysis purposes, time constraints and symmetric composition. Time constraints are compositionally introduced by means of the elapse operator. In fact, this operator is an instantiation of a more general operator, the 'trap'-operator. This operator has been brought up by Garavel and Sighireanu as a means to express a general exception handling mechanism in LOTOS [68]. The second additional operator, symmetric composition, has been introduced in [95] and further studied in [109], as a joint work with Marina Ribaudo. The operator exploits ideas similar to those developed for SWN [46], for SAN [167], and also for PEPA [72]. The specific benefits of either of these two operators will be emphasised in the next chapter, by means of a few case studies.

6. Interactive Markov Chains in Practice

In this chapter we use Interactive Markov Chains to investigate some illustrative examples and case studies. We first study the effect of symmetric composition by means of a simple producer–consumer example. In particular, we compare the growth of the state space to other methods to generate an aggregated state space. In a second case study, we use IMC to model a real world application, namely an ordinary telephony system. The constraint oriented specification of time dependencies will be a central issue. In order to circumvent state space sizes of more than 10 millions of states we make excessive use of the theoretical properties and concepts achieved so far, and obtain a Markov chain of manageable size. These two case studies show how performance estimation can be based on a Markov chain obtained from a highly modular and hierarchical IMC specification. A third example will then be used to highlight some implicit limitations to this approach. In particular, the issue of *nondeterminism* will discussed.

6.1 State Space Aggregation by Example

In this section we study the benefits of symmetric composition by means of an example. In particular, we compare the growth of the state space with other methods to generate an aggregated state space. We investigate a simple producer–consumer example, parametric in the number of producers and consumers. Its structure is depicted in Figure 6.1. Each producer generates jobs and then delivers them to a buffer, common to all producers. We restrict the buffer size to five places. The buffer is specified as explained in Section 5.5, by (non-synchronised) symmetric composition of one-place buffers

$$\mathsf{Buffer}_5 := \{5\,!\,\mathsf{Buffer}_1\}\emptyset \qquad \text{where} \qquad \mathsf{Buffer}_1 := \mathsf{put.get.Buffer}_1.$$

Consumers simply take jobs out of the buffer and work on them (with a certain rate). One of them is given by

$$\mathsf{Consumer} := \mathsf{get.}(\lambda_{\mathsf{work}})\mathsf{.Consumer}.$$

The specification is parametric in the number m of consumers. Since consumers do not interact with each other, we may specify a multiset of m consumers as follows, using symmetric composition,

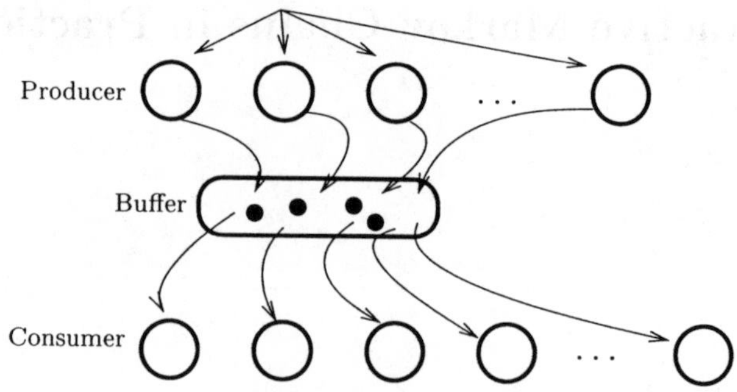

Figure 6.1. A simple producer–consumer example

$$\mathsf{Consumers}_m := \{m\,!\,\mathsf{Consumer}\}\emptyset.$$

Using ordinary parallel composition, the same situation can be specified by[1]

$$\mathsf{Consumers}_m := \underbrace{\boxed{\mathsf{Consumer} \quad \ldots \quad \mathsf{Consumer}}}_{m \text{ times}}.$$

We now turn our attention to the specification of producers. Each producer generates jobs. To make the example a bit more interesting, the individual rate for the generation of jobs is assumed to alternate between two different values. In a high load phase, the rate is smaller, i.e., jobs are generated more often than in a phase of low load. The change of phase is synchronised between all producers by a synchronising action c, i.e.,

$$\begin{aligned} \mathsf{Producer} \quad &:= (\lambda_{\mathsf{high}}).\mathsf{put}.\mathsf{Producer} + \mathsf{c}.\mathsf{Producer\_low} \\ \mathsf{Producer\_low} \quad &:= (\lambda_{\mathsf{low}}).\mathsf{put}.\mathsf{Producer} + \mathsf{c}.\mathsf{Producer}. \end{aligned}$$

As with consumers, a parametric specification of a set of producers can be achieved in two ways, using either symmetric or parallel composition, i.e., by means of

$$\begin{aligned} \mathsf{Producers}_n \quad &:= \quad \{n\,!\,\mathsf{Producer}\}\{\mathsf{c}\}, \quad \text{or} \\ \mathsf{Producers}_n \quad &:= \quad \underbrace{\mathsf{Producer} \mathrel{|[\mathsf{c}]|} \ldots \mathrel{|[\mathsf{c}]|} \mathsf{Producer}}_{n \text{ times}}. \end{aligned}$$

Note that all producers synchronise on action c. The synchronous change of phase between high and low load occurs periodically. For this purpose, we impose a time constraint between successive actions c. We assume an exponential distribution with a rather long mean duration, expressed by an absorbing Markov chain Delay as

---

[1] Be reminded that $\boxed{P \quad Q}$ abbreviates parallel composition with empty synchronisation set, and that Lemma 5.5.1.*2*. allows us to avoid bracketing.

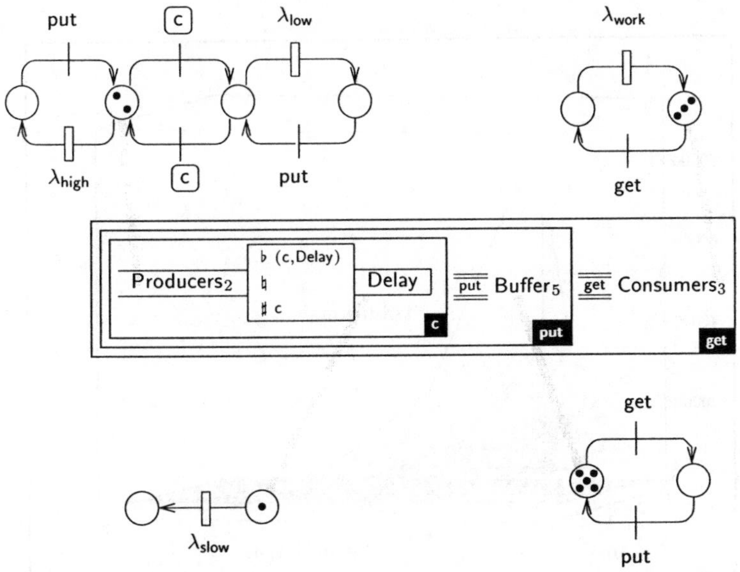

Figure 6.2. Hierarchical structure of the producer-consumer example

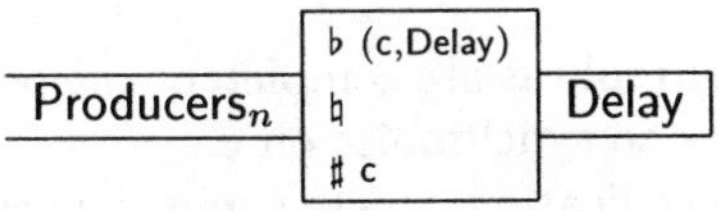

where $\mathsf{Delay} := (\lambda_{\mathsf{slow}}).0$

Note that the time constraint is specified to be active initially, thus interaction on action c is prohibited until Delay has evolved to 0. The overall structure of the specification is as follows:

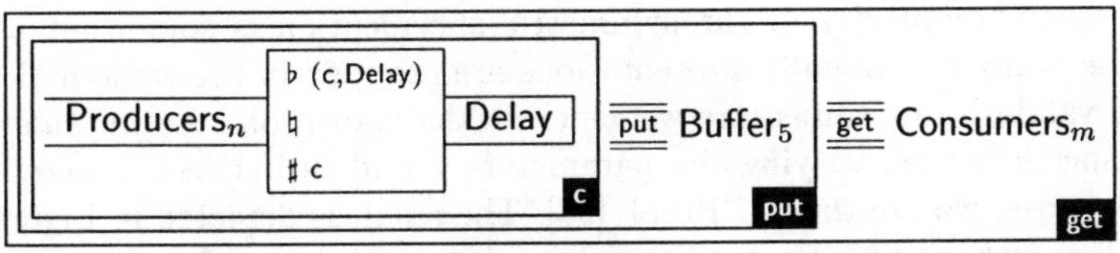

Abstraction is used to internalise actions (such as action c and **put**) as soon as (from a bottom-up perspective) they are irrelevant for external purposes, i.e., for further synchronisation.

In Figure 6.2 the specification is depicted for two producers and three consumers, using symmetric composition. As a graphical representation of symmetric composition $\{k\,!\,P\}\mathrm{A}$ of k identical replicas of a chain P we mark the states of chain P with the respective multiplicities (inspired by Petri net notation), i.e., the initial state is initially marked with multiplicity k. Transitions that have to be performed synchronously by all replicas (because the action label appears in A) are distinguished by changing the respective ac-

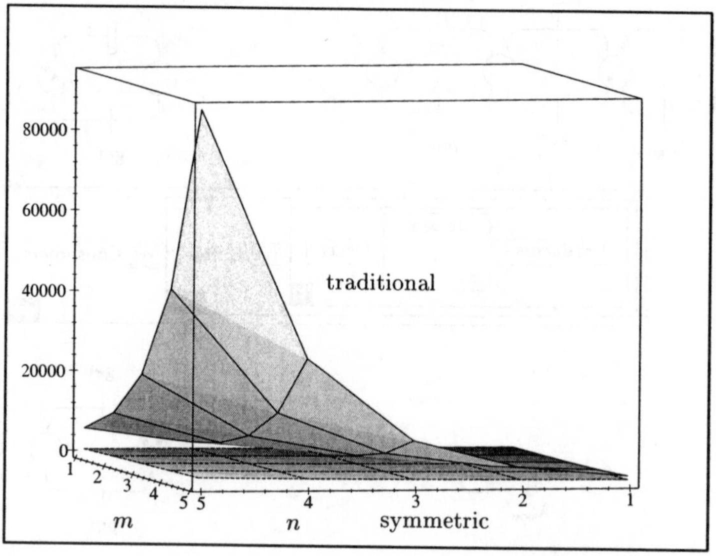

Figure 6.3. Size of the transition system for increasing n and m

tion label 'a' to '[a]'. Note that Consumer replicas are completely independent from each other, whereas Producers have to synchronise on c.

In the sequel, we will refer to this specification, where symmetric composition is applied to producers and consumers, as the *symmetric* specification. On the contrary we will call the specification where parallel composition is applied instead as the *traditional* specification. From Theorem 5.5.1 it is clear that both specifications are strongly bisimilar. However, we suppose that the state space complexity of the symmetric specification is somewhat better, because many equivalent representations are grouped in the same multisets.

To validate this conjecture we generate the transition system underlying both specifications, varying the parameters n and m between 1 and 5. For this purpose, we use the TIPPtool [98]. The result is depicted in Figure 6.3. Solid lines are used for the traditional specification, dotted for the symmetric specification. The traditional specification grows truly exponential in both parameters, from 72 states ($n = 1, m = 1$) to 93312 states ($n = 5, m = 5$). In contrast, symmetric composition leads to a quadratic growth of the state space, from 72 states to only 1512 states. Note that both transition systems are strongly and (also weakly) bisimilar.

Using the bisimulation algorithms described in Section 4.5 we can check whether the state space obtained by symmetric composition can be further aggregated. Indeed, for strong bisimilarity, no further aggregation can be achieved, hence the state space is minimal. This is not self-evident, since our specification is not covered by Theorem 5.5.2: The symmetric composi-

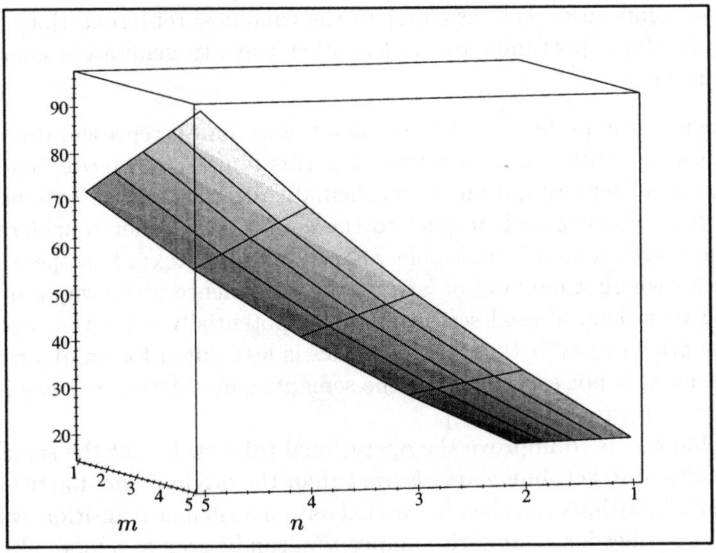

Figure 6.4. Size of the lumped Markov chain for increasing n and m

tion $\mathsf{Producers}_n$ is not necessarily minimal, since it involves synchronisation on action c, and because Producer is not a linear chain, it contains choice operators.

With respect to weak bisimilarity, further aggregation is possible. Since interactions are internalised, weak bisimilarity can be used to abstract from these internal computations (that will take place without delay, due to maximal progress). For fixed n and m the minimal transition system (obtained by partition refinement according to Table 4.3) does not possess any action transitions at all. The size of this minimal transition system is depicted in Figure 6.4, it grows quadratically from 16 to 96 states, again varying n and m between 1 and 5. Since it does not contain action transitions, it directly corresponds to a lumped Markov chain, and it can be seen as the *canonical representation* of the behaviour of the specification.

In general, there are many different ways to reach this canonical representation. Since Markov chain analysis is usually facing the infamous state space explosion problem, our pragmatic concern is simple. We are aiming to generate the canonical representation as directly as possible, in order to avoid intractably large intermediate state spaces.

Among the various ways that lead to the canonical representation, we have, so far, considered two variants of the same way, namely applying the operational semantics and subsequent partition refinement with respect to weak bisimilarity. The variants differ in the actual semantics rules used, since they rely on parallel, respectively symmetric composition. It turns out that symmetric composition gets us closer to the canonical representation.

We take the opportunity to discuss other ways to generate a canonical representation.

- In principle, the preferable way to reach the canonical representation is to define the semantics in such a way that this semantics *directly* generates the canonical representation. In mathematical terms, such a semantics is called *fully abstract* with respect to the equivalence under consideration. But this way is generally infeasible, at least in the context of an operational semantics where some kind of 'structural' congruence is anyway required. Moving to a denotational semantics could potentially solve the problem, but the price to pay is that the semantics is less suited for an algorithmic treatment. It is not easy to define the semantics in constructive way in the presence of recursion [170, 18].
- A feasible way is to improve the operational rules such that the semantics is not fully abstract, but *more abstract* than the original one. Partition refinement algorithms can then be carried out on a smaller transition system. In fact, the rules for symmetric composition can be seen as a more abstract semantics for parallel composition (for the specific case of symmetries). We will see another example below where improved operational rules lead to a more abstract semantics.
- Compositional aggregation also avoids large intermediate state spaces. It exploits substitutivity, by constructing transition systems along the hierarchical structure of the specification, interwoven with (partition refinement based) aggregation steps. As already exemplified in various examples during this book, compositional aggregation with respect to weak bisimilarity allows one to effectively exploit abstraction from internal details.
- An entirely different way to construct the canonical representation is based on *term rewriting.* The idea is to transform the specification into a *syntactic canonical form* in a preprocessing step. Applying the operational semantics to this canonical form will then directly generate the canonical representation. The preprocessing step is based on the equational theory established in Chapter 5. Since the set of laws developed therein is complete, they allow to transform an arbitrary specification into a canonical form.

In order to get more insight into the differences among these approaches, we will discuss them in the context of our producer–consumer example.

Improved Operational Rules. In order to evade an unnecessary growth of the state space, we want to avoid to generate states that are irrelevant with respect to our notion of equivalence. In other words, we are aiming to encode our equivalence notion *partly* into the operational semantics. A complete encoding is impossible, as discussed above in the context of 'fully abstract' operational semantics.

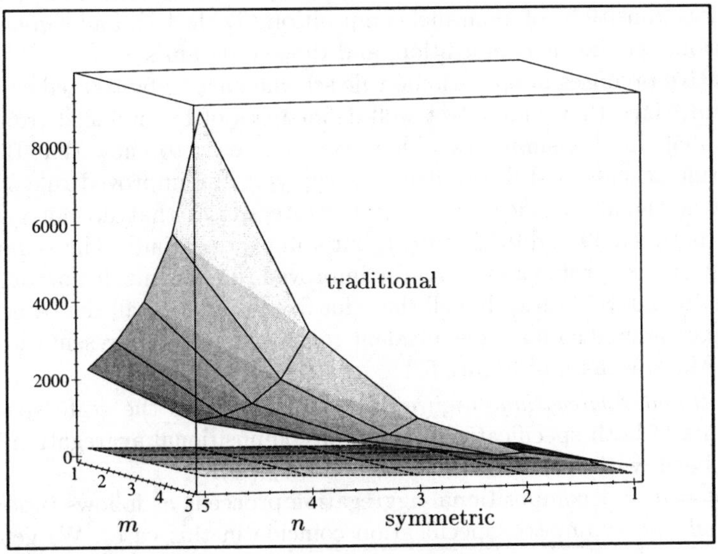

Figure 6.5. Size of the transition system with improved operational rules

Figure 6.5 shows the state spaces of the producer–consumer example when a slight improvement is incorporated into the operational semantics. The improvement encodes *maximal progress*, an important aspect of our notion of (strong and weak) bisimilarity. The fact that internal actions will preempt the passage of time has been formally established in Lemma 5.3.1 by

$$(\lambda).P + \tau.Q = \tau.Q.$$

This property ensures that it is not necessary to generate the state space of $(\lambda).P$, because bisimulation algorithms will anyway ignore this branch in the transition system. Justified by Lemma 5.3.1 we may therefore equip the operational rules for Markovian transitions of the choice operator '+' (Table 5.1) with additional premises without affecting the behaviour, but significantly reducing the size of the state space:

$$\frac{P \overset{\lambda}{\dashrightarrow} P' \qquad Q \overset{\tau}{\not\longrightarrow}}{P + Q \overset{\lambda}{\dashrightarrow} P'} \qquad\qquad \frac{Q \overset{\lambda}{\dashrightarrow} Q' \qquad P \overset{\tau}{\not\longrightarrow}}{P + Q \overset{\lambda}{\dashrightarrow} Q'}$$

The additional premise $Q \overset{\tau}{\not\longrightarrow}$ inhibits the generation of Markovian transition in the presence of maximal progress. Be reminded that $Q \overset{\tau}{\not\longrightarrow}$ denotes that Q does not possess the possibility to internally change to another behaviour. As a consequence of the expansion law (X) (Table 5.6) we are also allowed to include such premises in the operational rules that generate Markovian transitions of a parallel composition (Table 4.1), and similar with abstraction, symmetric composition, and time constraints.

Negative premises in operational rule scheme have to be treated carefully in general, since they may affect well-definedness of the induced transition relation [85]. In this simple case, however, it is easy to show that the improved rule scheme is still well-defined. Applying the improved rule scheme to our traditional specification, leads to state spaces that do not grow so rapidly (between 49 and 9728 states), but still exponentially. The symmetric specification also profits a lot from the improved rule scheme, it now occupies between 49 and 317 states. Recall that (for fixed n and m) all the transitions systems considered so far are equivalent to the canonical representation, the lumped Markov chain of Figure 6.4.

Compositional Aggregation. Figure 6.6 again compares the state space requirements of both specifications, but now compositional aggregation is applied to both, with respect to weak bisimilarity ($\approx$).

For $n = m = 1$ compositional aggregation proceeds as follows (note that traditional and symmetric specification coincide in this case). We generate the state space of $\boxed{\text{Producer} \parallel \text{Delay}}$ that consists of 6 states. Aggregation with respect to $\approx$ leads to an equivalent behaviour, described by a chain P with 4 states, that can be used instead of the former. In the next step we generate the state space of $\boxed{\text{P} \overline{\text{put}} \text{ Buffer}_5}$ (24 states) and afterwards aggregate it to, say PB, containing 14 states. Finally, we generate $\boxed{\text{PB} \overline{\text{get}} \text{ Consumer}}$ (28 states) and aggregate this transition system once again. The resulting transition system contains 16 states but faithfully describes the behaviour of the whole system. It is the canonical representation.

We have performed compositional aggregation for both types of our system specification, again varying n and m between 1 and 5. With compositional aggregation, it makes no sense to compare the result of the whole aggregation procedure (16, in the above case), since the result of the very last aggregation step is always canonical. In other words, compositional aggregation for traditional as well as symmetric specification produces the results depicted in Figure 6.4. A proper comparison is based on the maximum number of (intermediate) states that have to be stored during compositional generation and aggregation (28, in the above case). This is what is depicted in Figure 6.6. In this comparison, the symmetric specification (28 to 216) still behaves slightly better than the traditional specification (28 to 275). Interestingly, an increase of m does not always affect the maximum usage of state space. This is particularly true for the symmetry exploiting specification, but also for the traditional description (for small m). This phenomenon is caused

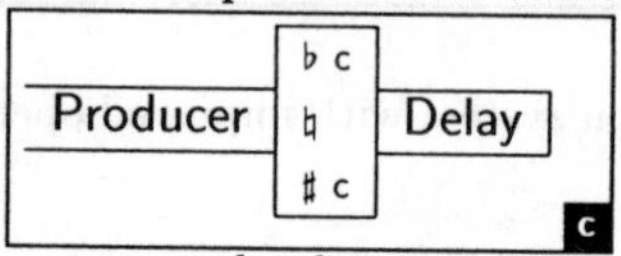

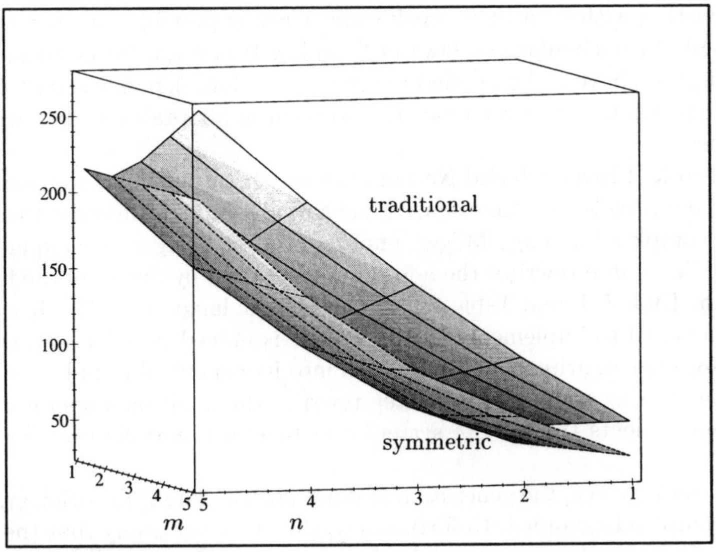

Figure 6.6. Maximal size of transition systems produced during compositional aggregation

by the fact that m comes into play in the very last composition step(s), but former composition steps have already led to larger intermediate state spaces. Compared to Figure 6.3 the state space (to be stored in memory) is reduced by several orders of magnitudes due to compositional aggregation.

Term Rewriting. In order to generate the canonical representation of a specification without generating unnecessary large state spaces, it is possible to exploit the equational theory developed in Chapter 5 in a preprocessing step. The idea is to transform the specification into a *syntactic canonical form.* The canonical *representation* directly corresponds to this canonical form and is obtained by applying the operational semantics. The necessary syntactic transformations are based on a set of *rewrite rules* that are used for *term rewriting* of the syntax of a specification. Roughly speaking, each rewrite rule has the form $E \to F$ and is obtained from a law $E = F$ of the equational theory. The direction of $\to$ indicates that E may be replaced by F, and it is chosen such that each application of a rewrite rule gets us closer to the canonical form. Under certain conditions, it is assured that whenever multiple rewrite rules can be applied to a term, they can be applied in any order.

This explanation gives a rough idea of the general strategy of term rewriting. However, in order to apply term rewriting to IMC$_{\text{XXL}}$, many details have to be addressed. In particular, we have to clarify how the canonical form looks like, and how to direct the laws in Table 5.5. For regular CCS (Table 5.4), the problem has been discussed by Inverardi and Nesi in [121]. They show that rewriting with respect to weak congruence is possible, but not easy to implement. In particular the laws ($\tau 2$) and ($\tau 3$) can not be given just one direction, they have to be applied in either direction, dependent on the context. Their solution to overcome this problem is adaptable to IML without difficulties.

The work of Inverardi and Nesi only solves the problem for specifications without any parallelism, time constraints and abstraction. In order to extend it to the complete language $\mathsf{IMC}_{\mathsf{XXL}}$, another preprocessing step is required beforehand. This step rewrites the additional operators by means of the laws in Table 5.6, Table 5.9, and Table 5.11 into the basic language, $\mathsf{IMC}_{\mathsf{L}}$. It is quite straightforward to implement. In this way, it is indeed possible to mechanically transform an arbitrary specification into its canonical form by means of two preprocessing steps. The first step rewrites the additional operators, the second implements the rewrite system a lá Inverardi and Nesi on the basic calculus.

As a consequence, the generation of unnecessarily large transition systems can be completely avoided. Unfortunately, this *does not* imply that the state space explosion problem is absent in this approach. The explosion problem is just turned into a 'syntax' explosion problem, now caused by the first preprocessing step. If the specification contains parallelism, this step expands the syntactic specification drastically by producing all the different possibilities of future behaviours syntactically. The usual term 'expansion law' for law (X) is quite illustrative.

A possible way out is to interweave both preprocessing steps. This means that a specification is iteratively expanded and then rewritten according to the laws of Table 5.5. The general strategy of interweaving both steps works quite well, and has been applied manually for instance in [99] and [108]. However, its mechanisation has not been tackled so far. A comparison of such a technique with the ones discussed here is anyway difficult. First, syntactic rewriting does not produce a state space. Instead, it shrinks and expands the syntactic specification by applying rewrite rules. It is therefore difficult to compare storage requirements. In addition, the efficiency of rewriting depends on the inclusion of derived laws, such as Lemma 5.3.1 and Lemma 5.5.1. Such laws greatly simplify the effort needed by term rewriting, because they abbreviate space and time consuming rewriting steps. On the other hand, the more rules are applicable to a specification the more it is required to implement a kind of heuristics that governs the application of rewrite rules, in order to reduce storage requirements, since different (orderings of) rewrite rule applications can lead to drastically different storage requirements.

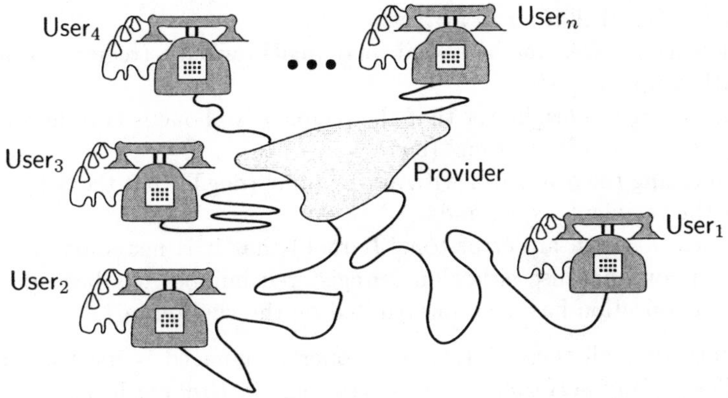

Figure 6.7. The telephony system

6.2 An Application Study: Ernberg's Ordinary Telephony System

In this section we use Interactive Markov Chains to investigate an ordinary telephony system. The constraint-oriented specification of timing information will be a central issue. In order to obtain a Markov chain representation of manageable size we make excessive use of the theoretical properties and concepts achieved so far. The core of this case study originates from the Swedish Institute of Computer Science where it has been studied by P. Ernberg.

The system structure is simple, as depicted in Figure 6.7. Several users are connected to a telephony network and they can thus phone each other using the service of the network. In textual terms, the system may be described as follows:

$$\boxed{\mathsf{User}_1 \;\ldots\; \mathsf{User}_n} \;\overline{\underline{A_1,\ldots,A_n}}\; \mathsf{Provider}$$

where A_i is the list of possible interactions between User_i and the controlling unit of the service provider, Provider. We introduce the details of the specification in four steps. First, we address details of the provider specification, then we turn our attention to the specification of the users. The next step explains how time constraints govern the stochastic behaviour of the system. To complete the specification, we introduce the actual numerical values that will be used. Finally, the specification is analysed in order to obtain interesting results characterising the stochastic behaviour of the system.

Provider Specification. The main complexity of this system lies in the controlling unit Provider that is realised by a single IMC. It is responsible for

- checking whether the originator and the recipient of a call are registered users,
- establishing a connection between originator and recipient,
- handling of various signals:

- ringing the bell if a call arrives,
- indicating a free line by a 'dial tone' (dialT) when a (registered) user has picked up the phone,
- indicating the originator that the recipient's phone is currently off hook by means of a 'busy tone' (busyT),
- indicating the originator with a 'ring tone' (ringT) that the bell is ringing at the recipient's side, and
- indicating with an 'error tone' (errorT) that it is necessary to hang up the phone because either an unregistered number has been dialled, or the connection has been interrupted by the counterpart.

To ensure that all these duties are properly managed is itself a challenge, regardless of any performance considerations. P. Ernberg has given a specification of Provider consisting of more than thousand lines of Full LOTOS. This specification has been used as a common example during the EUCALYPTUS project, an European/Canadian project that focused on the elaboration of a toolset for LOTOS [64]. During this project, 17 requirements have been formulated to be fulfilled by Ernberg's specification. The project participants have shown that these requirements are indeed satisfied by the specification, using different techniques such as equivalence checking and model checking, that are implemented in the *Caesar/Aldebaran Development Package* (CADP) [61, 66].

For our purposes, it is sufficient to know that the Full LOTOS specification has been extensively verified and that the semantics of Full LOTOS is defined in terms of transition systems, or more precise in terms of interactive processes. Since Interactive Markov Chains are a superset of interactive processes we can directly adopt the transition system generated by Ernberg's specification as an IMC Provider of our system.[2]

User Specification. We focus on the basic model of user interaction, given by User_i. It is depicted in Figure 6.8 (the index i is omitted to enhance readability), in graphical as well as in textual notation. It restricts the possibilities of the user to a subset of what is actually possible. For instance, in Ernberg's specification, a user has the possibility to dial an arbitrary number at an arbitrary system state, even if the phone is on hook. This possibility reflects a definite property of any real-world telephony system. However, in order to study performance aspects, we decide to restrict ourselves to those behavioural patterns that are most likely to have an impact on performance. In other words, we exclude the possibility to dial a number while the phone is on hook, since this is negligible with respect to performance issues. Similarly, we exclude some interaction possibilities that are very unlikely to happen, such as striking the hook during a running conversation. The reason for such

[2] It is a tedious exercise to extend the language in Chapter 5 such that it contains Full LOTOS. Instead we decide to link to Full LOTOS on the level of the transition system, where the situation is obvious, as expressed by Theorem 4.2.1, since the semantics of Full LOTOS maps onto interactive processes.

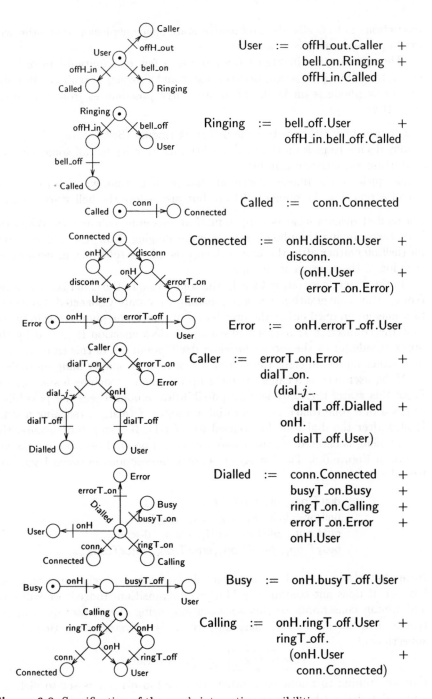

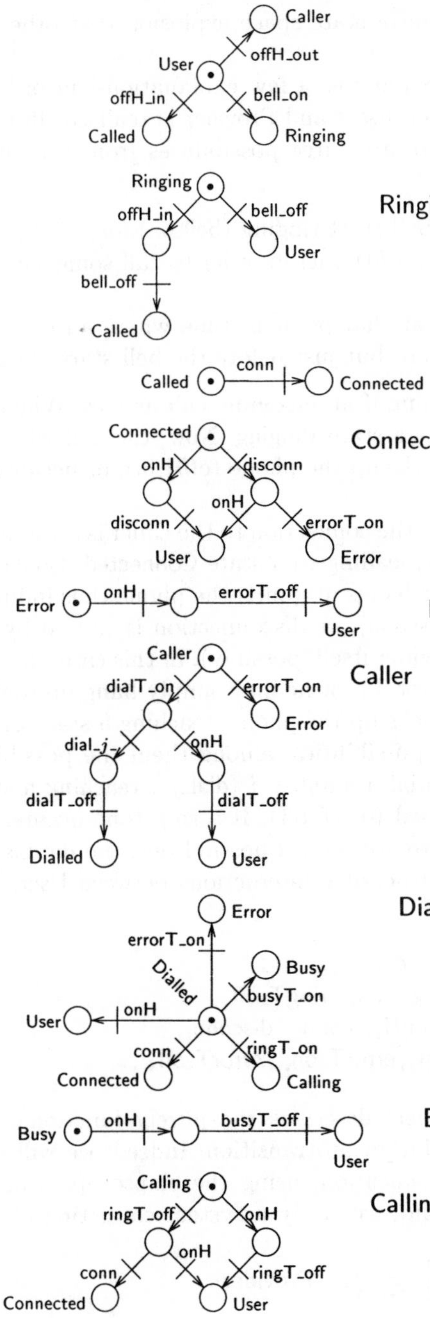

$$
\begin{aligned}
\mathsf{User} &:= \mathsf{offH\_out.Caller} + \mathsf{bell\_on.Ringing} + \mathsf{offH\_in.Called}\\
\mathsf{Ringing} &:= \mathsf{bell\_off.User} + \mathsf{offH\_in.bell\_off.Called}\\
\mathsf{Called} &:= \mathsf{conn.Connected}\\
\mathsf{Connected} &:= \mathsf{onH.disconn.User} + \mathsf{disconn.}(\mathsf{onH.User} + \mathsf{errorT\_on.Error})\\
\mathsf{Error} &:= \mathsf{onH.errorT\_off.User}\\
\mathsf{Caller} &:= \mathsf{errorT\_on.Error} + \mathsf{dialT\_on.}(\mathsf{dial\_}j\mathsf{\_.dialT\_off.Dialled} + \mathsf{onH.dialT\_off.User})\\
\mathsf{Dialled} &:= \mathsf{conn.Connected} + \mathsf{busyT\_on.Busy} + \mathsf{ringT\_on.Calling} + \mathsf{errorT\_on.Error} + \mathsf{onH.User}\\
\mathsf{Busy} &:= \mathsf{onH.busyT\_off.User}\\
\mathsf{Calling} &:= \mathsf{onH.ringT\_off.User} + \mathsf{ringT\_off.}(\mathsf{onH.User} + \mathsf{conn.Connected})
\end{aligned}
$$

Figure 6.8. Specification of the user's interaction possibilities

restrictions is basically the problem of state space explosion that otherwise would prevent analysis at all.

The specification in Figure 6.8 requires a few explanations, in order to understand how interaction between users and Provider is realised. Initially, the user's phone is on hook. There are three possibilities from this initial state User.

- The user may either notice the bell start ringing (bell_on), or
- may decide to pick up the phone (offH_out) in order to call someone, thus starting an outgoing call, or
- may pick up the phone exactly at that point in time where an incoming call is reaching the phone (offH_in), but just before the bell starts ringing.

Let us first discuss what is happening if an incoming call arrives. When the bell starts ringing the phone enters a state Ringing. Now, the bell will turn off (bell_off) either after the user picks up the phone (offH_in), or because no reaction of the user occurs in time.

In the former case (state Called), the connection to the caller is established (conn), thus conversation can start, leading to a state Connected. Conversation may go on until either the user hangs up (onH) the phone thus inducing that the connection is released (disconn), or disconnection is caused by the opposite side (or by the service provider itself, possibly). In this case, the user may either notice an error tone (error_T), or he may simply hang up (onH).

If the user acts as a caller, he picks up the phone, reaching a state Caller. From this state, there are several possibilities, among them the possibility to get a dial tone (dialT_on), and dial a number j (dial_j_) reaching a state Dialled after the dial tone has ceased (dialT_off). If a ring tone occurs, the state Calling is entered. Again, there are several possibilities, the details are shown in Figure 6.8. The list A_i of possible interactions between User_i and Provider is

$$\begin{array}{l}\mathsf{dialT\_on}_i,\ \mathsf{dialT\_off}_i,\ \mathsf{dial\_}j_{\text{-}i},\\ \mathsf{bell\_on}_i,\ \mathsf{bell\_off}_i, \mathsf{ringT\_on}_i,\ \mathsf{ringT\_off}_i,\\ \mathsf{offH\_out}_i,\ \mathsf{offH\_in}_i,\ \mathsf{onH}_i,\ \mathsf{conn}_i,\ \mathsf{disconn}_i,\\ \mathsf{busyT\_on}_i,\ \mathsf{busyT\_off}_i, \mathsf{errorT\_on}_i,\ \mathsf{errorT\_off}_i\ .\end{array}$$

Note that User_i (as well as Provider) describes the purely functional behaviour, it does not contain any Markovian transition. Indeed, we will now add timing constraints to this specification, using the elapse operator introduced in Section 5.5. To that end, we study a system consisting of two subscribers,

$$\boxed{\mathsf{User}_1\ \ \mathsf{User}_2}\ \ \underset{\overline{}}{\overline{\overline{A_1,A_2}}}\ \ \mathsf{Provider}\ .$$

that are aiming to phone each other, thus action $\mathsf{dial\_}j_{\text{-}1}$ is set to $\mathsf{dial\_2}_{\text{-}1}$, and action $\mathsf{dial\_}j_{\text{-}2}$ becomes $\mathsf{dial\_1}_{\text{-}2}$.

Time Constraints. Now, we want to specify several delays that have to elapse between certain interactions. There is, of course, a variety of different choices where to introduce timing constraints. We consider a specific scenario here, that turns out to be quite interesting. In total, we add 14 time constraints to the specification, summarised in Figure 6.9. In this pictorial representation we have used associativity of the elapse operator, technically justified by Lemma 5.5.1.*5*.

- Whenever an existing connection is released, the error tone is raised after a while, except if the respective user hangs up the phone in the meanwhile. The delay is governed by ErrorDly, and is imposed for User_1, as well as User_2.
- After a user has picked up the phone in order to make a call, it requires some time (to check whether the user is registered, for instance) before some tone is raised. The corresponding time constraint is governed by InitDly, for both User_1 and User_2.
- Establishing a connection (or detecting that no connection can be established) requires some time. Thus we impose a delay DialDly between dialling a number and getting a reaction from the providers side, symmetrically for both users.
- When noticing a ring tone, User_1 waits for a connection. He may hang up the phone after a while, except if a connection has been established in the meanwhile. The time until hanging up the phone is determined by $\mathsf{WaitDly}_1$. For User_2, a time constraint governed by $\mathsf{WaitDly}_2$ is imposed in the same manner.
- Between two phone-calls User_1 takes a rest, but if the phone rings, he decides to go and pick up the phone. We let $\mathsf{IdleDly}_1$ govern the time that elapses between two phone-calls originated by User_1. For User_2, the situation is symmetric, but the delay is given by $\mathsf{IdleDly}_2$. Note that we specify an active time constraint, i.e., initially both users have to wait before they may pick up the phone.
- If the phone bell starts ringing, User_1 needs a certain time to reach the phone, given by $\mathsf{PickupDly}_1$. For User_2, the corresponding constraint is governed by $\mathsf{PickupDly}_2$.
- Once a connection is established, User_1 contributes to the conversation a certain amount of time, but a disconnection may occur anyway in the meanwhile. The corresponding delay is imposed by a time constraint governed by $\mathsf{SpeakDly}_1$ for User_1, and similar for User_2.

This summarises the major timing constraints we assume to exist. Note that the time constraints that are due to the provider (InitDly, DialDly, and ErrorDly) are identical to both users. So, the provider is assumed to be fair and not to prefer either of the users. On the other hand, those time constraints that are caused by the users ($\mathsf{SpeakDly}_i$, $\mathsf{IdleDly}_i$, $\mathsf{PickupDly}_i$, and $\mathsf{WaitDly}_i$) are parametric in the identity of the user. This makes it possible to incorporate different user profiles.

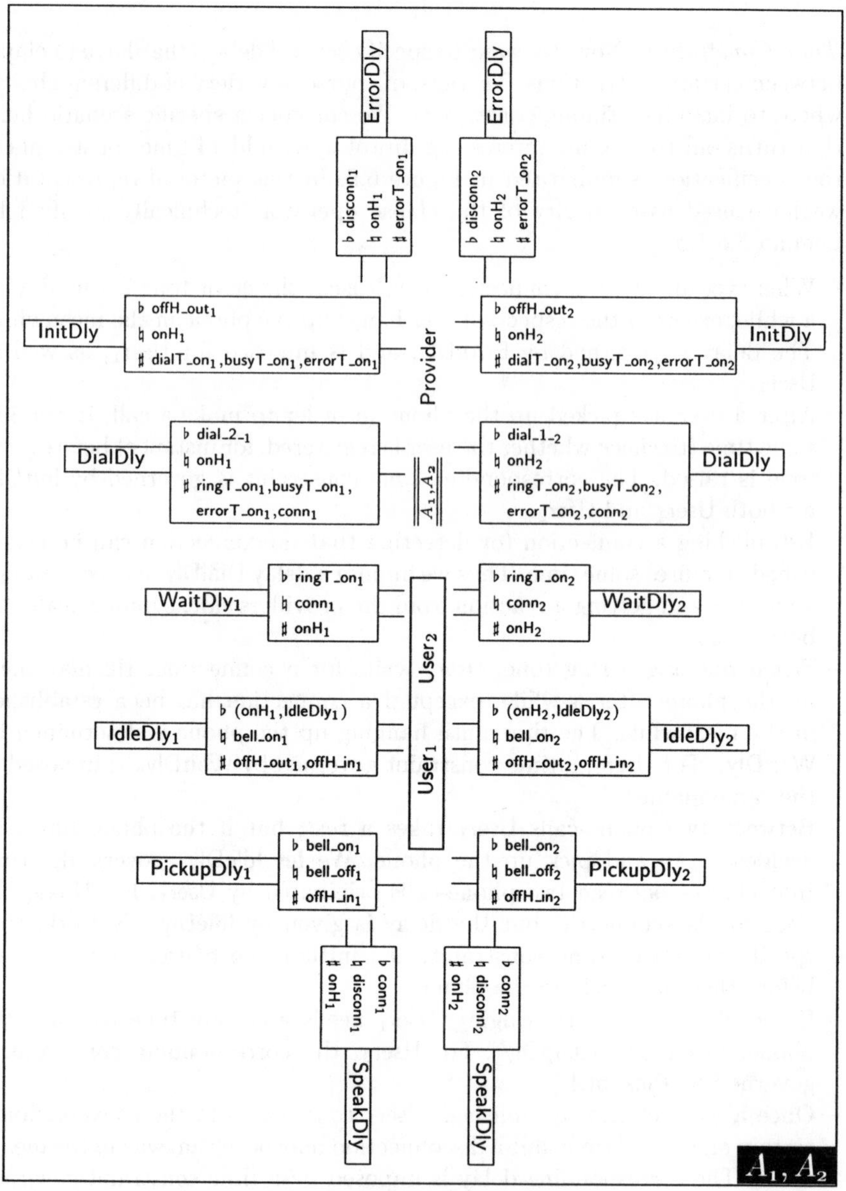

Figure 6.9. The time constrained telephony system specification

Time Values. Table 6.1 shows the time values we have used for numerical analysis of the telephony system specification, listing the mean durations of each delay. In order to estimate the nature of the distributions involved, we have included the parameter CV, the *coefficient of variation.* This parameter is a measure for the variance exhibited by a distribution. Intuitively, the

Table 6.1. Time values for the telephony system specification

	User_1	User_2	CV	type of distribution
ErrorDly	15 sec		1	exponential
InitDly	2 sec		1	exponential
DialDly	6 sec		0.408	Erlang_6
WaitDly	60 sec	30 sec	1	exponential
SpeakDly	15 min	5 min	1.5	branching Erlang
IdleDly	60 min	75 min	0.316	Erlang_{10}
PickupDly	40 sec	10 sec	1	exponential

higher the value of CV the more the sample values drawn from the distribution differ from each other. For exponential distributions, $CV = 1$, while for deterministic distributions, describing a delay of a fixed length, we have $CV = 0$.

The profiles of User_1 and User_2 differ with respect to their time values in a particular manner. We assume that User_2 is generally reacting quicker than User_1, while User_1 likes to phone more often and longer than User_2. The majority of delays simply consist of a single exponential phases. DialDly describes an Erlang distribution, because we assume that the time required to establish a connection should not exhibit too much variance. On the other hand, the time of conversation my vary a lot, the variance of the respective distribution, given by $\mathsf{SpeakDly}_i$, should therefore be rather high. We use a superposition of exponential distributions for this purpose, giving rise to a so-called branching Erlang distribution [129]. The delays that are imposed between two successive phone-calls, $\mathsf{IdleDly}_i$, are assumed to be governed by an Erlang distribution with fairly low variances, consisting of ten exponential phases each.

This concludes the description of the details of the telephony system specification. We may now apply the operational rules in order to analyse the specification.

Analysing the Specification. The specification we want to analyse is quite complex. Therefore, the state space, obtained by means of the operational semantics, is very large. It consists of more than 10 million states. But it is weakly bisimilar to an IMC that consists of 720 states. However, this aggregation is beyond what is computable with contemporary computer assistance. Indeed we have not even been able to generate the complete state space of the specification at all. The tools we are employing, TIPPtool [98] and Caesar [67] ran out of memory after generating 0.3, respectively 10.2 million states. Caesar is one of the ingredients of the CADP toolset.

So, we are going to use compositional aggregation, as exemplified in previous chapters to tackle this model (and to formally establish the connection to the aggregated model). Instead of considering the complete specification, we apply the operational semantics rules to the non-constrained specification,

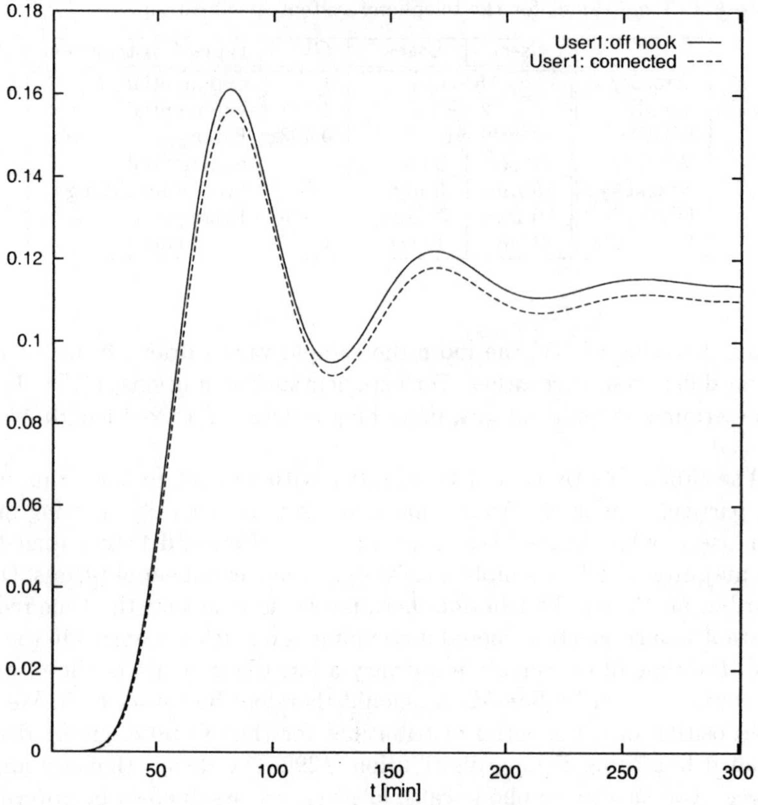

Figure 6.10. Probability of being off hook and of being connected for User_1

$$\boxed{\mathsf{User}_1\ \mathsf{User}_2}\ \underline{\underline{\overline{\overline{A_1,A_2}}}}\ \mathsf{Provider}\ .$$

Indeed, the state space generated from this specification turns out to be smaller by one order of magnitude, compared to the original state space. It consists of 1040529 states and 2527346 transitions. This is still very large but it can be aggregated, using strong bisimulation, to a state space with 327 states only. To perform this aggregation, we can rely on the algorithm of Table 2.2, since neither User_i, nor Provider give rise to any Markovian transitions. This is indeed a consequence of the fact that we used a constraint-oriented style, where time delays are specified separately. (For the aggregation we used the implementation of Aldebaran [62], another component of the CADP toolset.)

The reason for the enormous reduction of state space complexity essentially originates from the fact that Ernberg's LOTOS specification of Provider uses a variety of data variables to keep track of the current status of each individual subscriber. Now, the operational semantics of LOTOS produces differ-

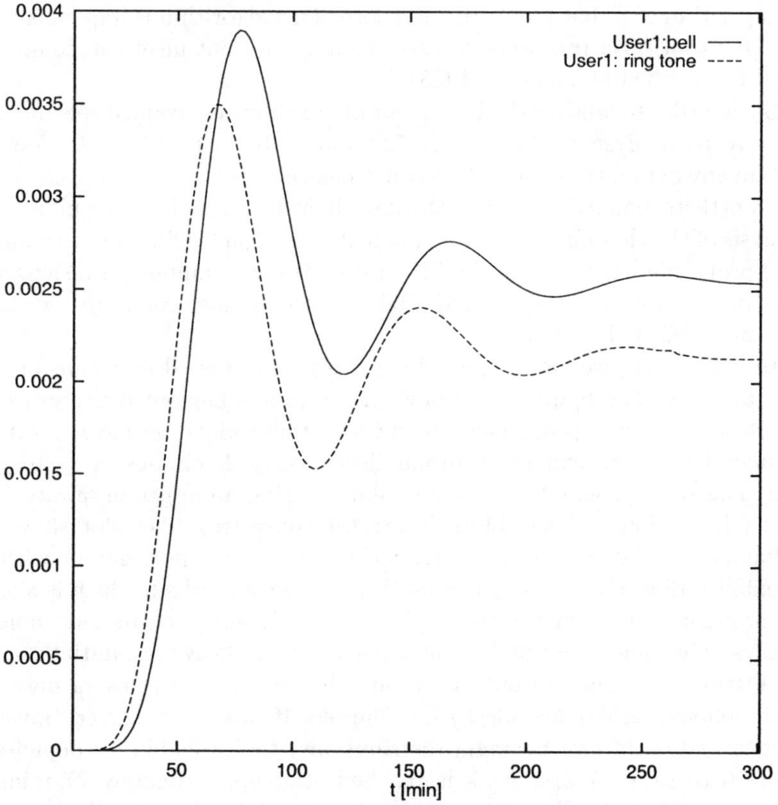

Figure 6.11. Probability of ringing bell and ring tone at User_1

ent states for every (reachable) combination of actual variable assignments, as it is the case in a real implementation. However, from a behavioural point of view, many states are equivalent, since they exhibit the same behaviour. They are therefore equated by strong bisimilarity.

If we let UUP denote the aggregated transition system, we obtain that

$$\boxed{\mathsf{User}_1 \ \mathsf{User}_2} \,\overline{\overline{\underline{\underline{\ A_1,A_2\ }}}}\, \mathsf{Provider} \sim \mathsf{UUP}\,.$$

As a consequence of Theorem 5.5.3, our time constrained specification of Figure 6.9 is bisimilar to the specification obtained by imposing the same time constraints on UUP and subsequent abstraction of A_1, A_2. Hence, we may investigate the latter specification instead. The resulting state space consists of 10421 transitions and 3215 states, instead of more than 10 million states, as it was the case for the original specification.

This state space, in turn, can be aggregated once again. It is weakly bisimilar to a minimal representation consisting of 720 states. Since this aggregation involves both action transitions and Markovian transitions we use the algorithm of Table 4.3 for this purpose. This algorithm is implemented in the TIPPtool (in a prototypical version that does not meet the complexity result established in Theorem 4.5.3).

So, in order to analyse the behaviour of our time-constrained specification, we have to analyse an IMC with 720 states. In fact, this IMC does not contain any action transition. As a consequence, it is a lumped Markov chain. It is worth to remark that this Markov chain has a highly irregular shape. Analysis of the chain proceeds as explained in Section 3.4, in order to compute state probabilities of each of the 720 states. We can compute transient state probabilities, or since the Markov chain is ergodic, also compute the steady state probability distribution.

In Figure 6.10 we have depicted some results obtained by means of transient analysis. The figure shows how the system behaviour converges to an equilibrium as time progresses. To achieve these plots we have iteratively calculated the transient state probabilities every 2 minutes of system life time. The state probabilities are cumulated, using appropriate reward functions (cf. Section 3.4) to obtain interesting measures. The plot shows the probability that User_1 has currently picked up the phone, together with the probability that a speech connection is actually established, which is slightly less probable. The shape of these plots is mostly governed by our choice of IdleDlys. The time constraints controlled by these delays are initially active (cf. Figure 6.9). This initially prohibits that either of the users may pick up the phone, until some IdleDly has elapsed. If this has occurred (given by a superposition of two Erlang_{10} distributions) the probabilities depicted in Figure 6.10 raise. A first peak is reached after approximately 77 minutes. The probabilities oscillate for a while, because idle phases and connection phases alternate. An equilibrium is reached, due to the stochastic perturbation caused by the distributions, especially the branching Erlang distributed SpeakDlys.

Figure 6.11 shows the probability that User_1 is noticing a ring tone, respectively a ringing bell. The probabilities are fairly small compared to the ones in Figure 6.10. The probability of a ringing bell increases later than that of a ring tone. This is a consequence of the fact that User_1 (in the mean) is calling User_2 earlier than the other way round. The first attempt to call User_2 occurs after roughly 60 minutes. The opposite direction is delayed for about 75 minutes.

In summary, the influence of the time that elapses between two phone calls is decisive for the transient behaviour of the system. If we change the distribution of IdleDly such that the variance is increased, the oscillation is flattened. On the other hand, equilibrium probabilities are not affected significantly. To study the effect of different distribution is quite easy, due to the constraint oriented specification style used for time constraints. Of course, employing different distributions for a time constraint requires that the state spaces have to be generated again (starting from UUP), since different dis-

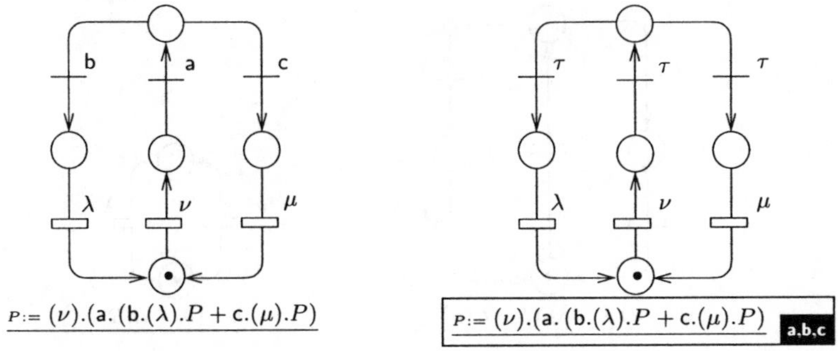

Figure 6.12. External and Internal Nondeterminism

tributions lead to different state spaces. For instance, changing both IdleDlys to single exponential phases leads to a lumped Markov chain of 351 states.

6.3 Nondeterminism and Underspecification

In the preceding sections of this chapter, as well as in Section 4.6 we have shown how to use Interactive Markov Chains as compositional means to generate small Markov chains from complex specifications. It seems necessary to recapitulate the decisive steps used.

First, it is worth to recall that Interactive Markov Chains are a strict superset of Markov chains. There are infinitely many IMC that do not belong to the class of Markov chains. The reason is that one of our core concepts, the potential of interaction with an environment, is not expressible with Markov chains.

Example 6.3.1. Consider the IMC $P = (\nu).(\mathsf{a}.(\mathsf{b}.(\lambda).P + \mathsf{c}.(\mu).P)$, *depicted in Figure 6.12. It is not possible to view this system as a Markov Chain, since it depends on interaction with the environment when action* a, b *or* c *will actually occur.*

Therefore, in order to associate a Markov chain with some IMC specification, it is always required that the specification does not possess any potential of interaction with an environment. For this purpose, we rely on *abstraction.* We can completely internalise the interaction potential of an IMC by means of the abstraction operator. This is a prerequisite in order to determine a Markov chain that represents the behaviour of the specification. The reader is invited to verify that whenever we have generated a Markov chain somewhere in this book, we have used abstraction of all relevant actions as the outermost operator. This is necessary, but unfortunately not sufficient to determine a Markov chain.

Example 6.3.2. Consider the above chain P. *Using abstraction as the outermost operator we obtain a chain* $\boxed{P\ \mathsf{a,b,c}}$. *It is depicted in Figure 6.12.*

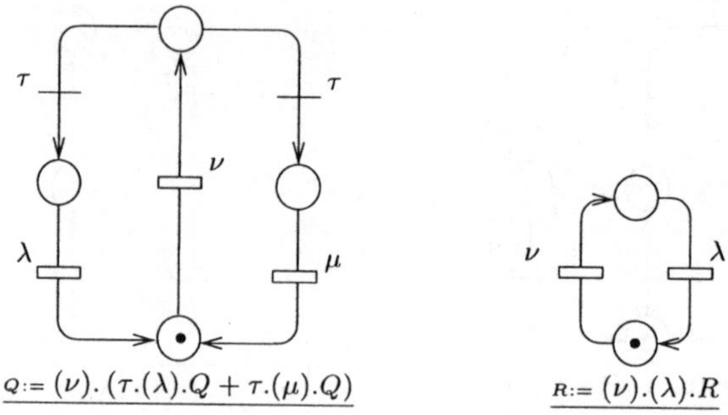

Figure 6.13. Weak bisimilarity may, or may not eliminate nondeterminism

Now, all actions are internalised. Anyhow, from a stochastic point of view, the behaviour is not completely determined: the decision between the left and the right branch of the choice is nondeterministic.

In the examples and case studies we have investigated so far, it has always been the case that a final application of weak bisimilarity led to an IMC without any action transition, and hence it was a Markov chain. For the above example, this property cannot be guaranteed.

Example 6.3.3. Aggregation with respect to weak bisimilarity does not *eliminate all (internal) action transitions from* P a.b.c*. Instead, it results in the IMC Q depicted in Figure 6.13. So, due to nondeterminism, the stochastic process to be associated with this IMC is* underspecified. *This evaluation implicitly assumes that the rates* λ *and* μ *are distinct values. If, on the other hand, we have* $\lambda = \mu$ *the situation is different, since the states reached by means of internal steps turn out to be bisimilar. Thus the nondeterministic choice is not a decisive one, since any outcome of this choice leads into the same class of behaviours. Hence, aggregating* P a.b.c *with respect to weak bisimilarity removes action transitions and nondeterminism. The resulting Markov chain R is depicted in Figure 6.13.*

This example provides some insight into the subtle influence of nondeterminism. In fact, the presence of nondeterminism may hamper a stochastic analysis of many Interactive Markov Chains. On the other hand, a wide range of IMC can be transformed into a Markov chain by applying weak bisimilarity after abstraction of all interactions. From a performance analysis perspective, the issue of nondeterminism may seem as the price to pay for the nice properties enjoyed by IMC. Indeed, the presence of nondeterminism is a direct consequence of our decision to separate delays from actions, taken in Section 4.1. As explained therein, the notion of nondeterminism is essential for process algebra in general. We have classified different important concepts expressible by means of nondeterminism: the possibility to express implementation freedom, scheduling freedom, and influence of an external

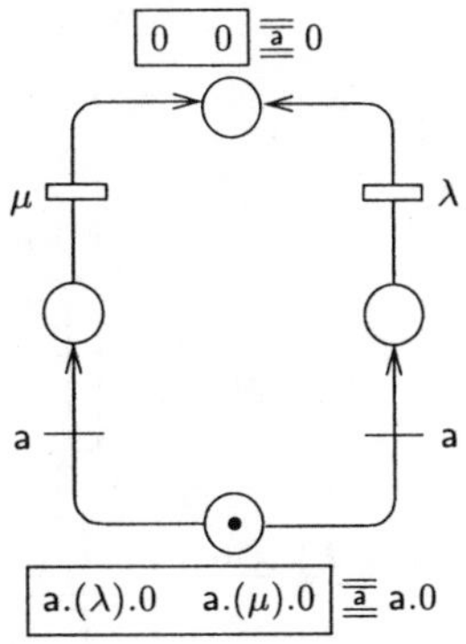

Figure 6.14. Internal nondeterminism due to auto-concurrency

environment. In particular, the use of nondeterministic interleaving to represent scheduling freedom avoids the shortcomings observed in other Markovian process algebras' parallel composition.

So, nondeterminism is a substantial concept for Interactive Markov Chains. However, the question arises how to associate a stochastic process (a Markov chain) to an IMC that contains nondeterminism. A stochastic process does not exhibit nondeterministic behaviour, since each possible behaviour at any point in time has a certain, unique, probability. Indeed, weak bisimilarity may eliminate nondeterminism, but it does not necessarily eliminate all nondeterminism.[3]

Let us fix some terminology. For a given transition system, we say that a state is a *nondeterministic state* whenever several action transitions emanate this state. The action labels of these transition may, or may not be different from each other. If at least one action label differs from all the others, we call this situation *external* nondeterminism. Otherwise, if each action label occurs at least twice, we call it *internal* nondeterminism.

Example 6.3.4. Figure 2.1 already contains examples for internal and external nondeterminism, in E_2. *The IMC P in Figure 6.12 contains external nondeterminism, while* $\boxed{P}$ a,b,c *has an internally nondeterministic state. Another example where internal nondeterminism is present is* $\boxed{a.(\lambda).0 \quad a.(\mu).0} \overset{a}{\equiv} a.0$, *depicted in Figure 6.14. Here, nondeterminism does not arise from a choice operator, but from a phenomenon usually called auto-concurrency. Two actions compete for synchronisation, and hence there is an implicit, internal nondeterministic choice about which process will succeed in the competition.*

Note that internal nondeterminism is not bound to internal actions. The important difference between external and internal nondeterminism is that

[3] Due to the presence of nondeterminism, IMC can be viewed as a variant of continuous-time Markov decision processes [118, 161].

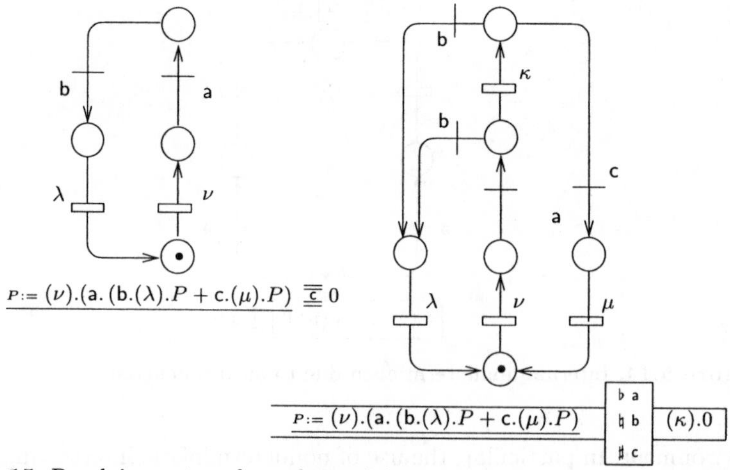

Figure 6.15. Resolving external nondeterminism

external nondeterminism can be resolved by means of some external influence. On the contrary, internal nondeterminism is effectively out of control of the environment, there is always a decision between branches with the same action label to be taken locally, by the IMC under consideration.

Example 6.3.5. Figure 6.15 shows the result of parallel composition $P \mathbin{\|_{\mathsf{c}}} 0$. *The external nondeterminism in P is influenced by requiring a synchronisation on action* c *with the terminated chain* 0 *that will prevent occurrence of action* c. *As a consequence, the nondeterminism in P has been resolved by some (rather destructive) external influence. As another example, one may resolve nondeterminism by adding a time constraint on action* c, *such as* $P \; [\flat\,\mathsf{a} \;\; \natural\,\mathsf{b} \;\; \sharp\,\mathsf{c}] \; (\kappa).0$. *(cf. Figure 6.15). In both cases, we obtain the Markov chain R of Figure 6.13 after abstraction and factoring with respect to weak bisimilarity. For the time-constrained example, this is a consequence of maximal progress.*

Internal nondeterminism remains unresolved by such external influences, since the latter affects transitions with the same label in the same way. Thus whenever internal nondeterminism is faced, the only thing that we can hope is that (after abstraction) the subsequent behaviours are equivalent, because then weak bisimilarity will eliminate the nondeterministic state. Otherwise, the specification is underspecified from a stochastic point of view.

In order to avoid nondeterminism in general, one may resort to a rigorous solution. The simplest way to get rid of nondeterminism is to rule it out by imposing syntactic constraints. To rule out *any* nondeterminism requires severe restrictions [139]: The choice operator has to be excluded and parallel composition has to proceed completely synchronously (with respect to action transitions). Clearly, this restrict the applicability of the approach.

Since internal nondeterminism is more difficult to cope with, compared to external nondeterminism, one may whish to rule out just internal nondeterminism. This requires weaker syntactic restrictions: The choice operator can be used with some restrictions that avoid expressions like $\mathsf{a}.P + \mathsf{a}.Q$, and parallel composition has to be changed to what is known as CSP-style synchronisation [116], in order to rule out auto-concurrency (cf. Figure 6.12). In CSP-style synchronisation, parallel components are forced to interact on all their common actions. Unfortunately, ***abstraction*** does not preserve the absence of internal nondeterminism, as exemplified in Figure 6.12. Hence, apart from trivial cases, our general approach to generate a Markov chain is not applicable, because it substantially relies on the use of abstraction (and subsequent application of weak bisimilarity).

As a consequence, we do not believe that it is wise to impose syntactic restrictions on $\mathsf{IMC}_{\mathsf{XXL}}$ in order to ensure that each expressible IMC possesses a directly corresponding Markov chain. We take the view that whenever the IMC generated from a specification contains nondeterminism, then this truly reflects what the specification is supposed to express, namely an underspecified stochastic process. Based on this viewpoint there is still a way to cope with underspecified stochastic processes. Inspired by Vardi, Hansson, and others [183, 87, 118, 161, 19] we may employ the notion of a ***scheduler*** in order to determinate an underspecified stochastic process. This scheduler resolves nondeterminism ***a posteriori***. The stochastic process associated to an IMC is thus parametrised with a specific scheduler. Technically, a scheduler is a function that for any state determines the next transition to be executed, taking into account the history of the system. As e.g. in Segala's work, a scheduler may decide probabilistically [171], scheduling certain transitions with a certain probabilities. Performance and dependability results are then either conditioned on a particular scheduler, or they are obtained as the minimal and/or maximal measures derivable for a set of allowed scheduler. We believe that the concept of schedulers is worth to be studied in depth in the context of IMC.

6.4 Discussion

In this chapter we have discussed how Interactive Markov Chains can be employed in practice. We have focused on the question how to generate a (small) Markov chain from a (complex) compositional IMC specification. We only gave a partial answer to the question, because IMC are more expressive than Markov chains. This additional expressiveness comes with the ability to express the potential of interaction, and to express nondeterminism among different interactions. Due to the presence of nondeterminism, IMC can be viewed as a variant of continuous-time Markov decision processes [118, 161].

Apart from the issue of nondeterminism, we have tried to show that IMC can be successfully used to specify and analyse systems of high complexity. The ordinary telephony system is one such example. The lumped Markov chain associated with this system is of a modest size (720 states), but it has a highly irregular shape, and we doubt that it is possible to generate this Markov chain by hand – or by any other formalism. The specification (taken from the Eucalyptus project) is quite involved, and a nice example why process algebra in general is advantageous when dealing with complex scenarios. Furthermore, the benefits of a constraint-oriented specification of timing dependencies have become obvious. The state space of the time-constrained specification is beyond what our computer assistance is able to manage. Anyhow, we have been able to generate the lumped Markov chain by means of compositional aggregation. Besides our own tool prototype, the TIPPtool [98], we used Caesar [67] for generating some of the intermediate state spaces, and Aldebaran [62] for the initial aggregation step. The study of the telephony system is joint work with Joost-Pieter Katoen [101].

The TIPPtool has also been used to study the producer–consumer example. This example illustrated how an increasing number of parallel components leads to an exponential blow-up of the state space. If symmetric composition is used instead of parallel composition, this exponential blow-up is reduced to a quadratic growth. We have also compared other approaches to evade the state space explosion problem by means of this example.

- We have shown that a slight improvement of the operational rules drastically reduces the state space complexity. This improvement basically encodes maximal progress into the rules.
- Once again, the benefits of compositional aggregation, have become obvious. Compositional aggregation can alleviate the state space explosion considerably.
- We have briefly sketched a term rewriting approach in order to avoid state space explosion. This approach has already been used in [99] and [108]. However, it is due to our completeness result (Theorem 5.5.4) that we can assure to be able to rewrite arbitrary expressions successfully.

The practical considerations in this chapter have been complemented with a discussion of nondeterminism in the context of IMC. We have seen that (internal) nondeterminism is a phenomenon that hampers the association of a Markov chain with IMC in general. IMC that 'suffer' from nondeterminism do not possess a unique stochastic interpretation, the Markov chain is underspecified. We have proposed different possible ways to overcome this problem.

7. Conclusion

This chapter gives a retrospective view on the main contributions of this book. It summarises the major building blocks of compositional performance and dependability estimation with Interactive Markov Chains, and addresses the general question whether the *challenge of compositional performance and dependability estimation* has been met. In addition, we point out relevant directions for further work.

7.1 Major Achievements

In this book we have addressed the issue of compositional specification and analysis of continuous-time Markov chains. In Chapter 2 and Chapter 3 we have introduced the basics of process algebra, respectively Markov chains. Chapter 4 integrates these two antipodes in a single formalism, Interactive Markov Chains. The ideas behind IMC are substantially different from other existing approaches to compositional Markov chain generation. Due to a deliberate decision to separate delays from actions we avoid their shortcomings in the definition of synchronisation. The interrelation of delays and actions is governed by the notion of maximal progress: Internal actions are executed without letting time pass, while external actions are not.

IMC is more than 'yet another' formalism to describe Markov chains. This claim is substantiated by the results achieved in Chapter 5 and Chapter 6. A sound and complete axiomatisation for strong bisimilarity and weak congruence has been given for the core calculus IMC, as well as for the specification language $\mathsf{IMC}_{\mathsf{XXL}}$. As highlighted in Section 5.6, our axiomatic treatment of maximal progress solves an open problem for timed process calculi in general, and can be adapted to solve similar problems for process calculi with priorities.

These theoretical results have been complemented by investigations about the practicability and benefits of the approach. For this purpose, we have described efficient partition refinement based algorithms to compute strong and weak bisimilarity on IMC. We have shown that the algorithmic complexity of these relations does not increase, when moving from non-stochastic process algebra to IMC (cf. Section 4.5). It is important to note that non-stochastic process algebra forms a proper sub-algebra of IMC, obtained by omitting delay prefix. On the other hand, by disallowing action prefix we obtain a sub-algebra of continuous-time Markov chains. On this sub-algebra, strong and weak bisimilarity, and also weak congruence, coincide with the notion of lumpability. As a side result, we have presented a fast algorithm to compute the best possible lumping of a given Markov chain (cf. Section 3.7).

The above algorithms have been successfully applied to aggregate IMC, by factoring the state space with respect to strong and weak bisimilarity. Due to the substitutivity property, we have been able to establish a compositional aggregation strategy. By means of several examples, we have seen that compositional aggregation can alleviate the infamous state space explosion problem substantially. Among others, we have applied compositional aggregation to attack state spaces of several millions of states, as in the ordinary telephony system case study (cf. Section 6.2).

In addition, we have discussed further means to avoid large state spaces and to enhance specification convenience. Section 6.1 has shown how symmetric composition reduces an exponential blow-up to a quadratic growth of the state space, if identical components are contained in a specification. In order to support a compositional specification style, we have introduced the elapse operator. With this operator, time constraints can be specified in a constraint-oriented style. Each time constraint can be governed by some arbitrary phase-type distributed delay.

7.2 Has the Challenge Been Met?

Interactive Markov Chains can be used to specify complex dependencies compositionally, and to analyse their properties by transformation into Markov chains. From a performance engineering perspective, the most important question is whether complex IMC specifications can be transformed efficiently into small Markov chains. The examples we investigated suggest a positive answer to this questions. However, we have pointed out in Section 6.3 that this result can only be a partial answer, due to the fact that Markov chains are less expressive than IMC. The additional expressiveness of IMC comes with the ability to express the potential of interaction, and to express nondeterminism among different interactions.

In order to eliminate nondeterminism (and to aggregate the state space) we have developed a general recipe leading from an IMC specification to a Markov chain. We summarise the main steps:

1. The first step consists of developing a specification of the system under investigation, using the composition operators provided by $\mathsf{IMC}_{\mathsf{XXL}}$. One may start from an existing interactive process (obtained, for instance, from a Statechart, SDL, Promela, or LOTOS specification), and enrich the specification by incorporating time constraints. The elapse operator is convenient to use for this purpose.

2. In order to transform the system into a Markov chain, a second step is mandatory, namely abstraction of all actions.
3. After generating the state space, weak bisimilarity is applied in order to aggregate the state space, to eliminate action transitions, and to remove nondeterminism. This aggregation step is preferably done compositionally, in order to circumvent the state space explosion problem. Of course, other strategies can be applied as well, such as the term rewriting approach sketched in Section 6.1.
4. If the aggregated, minimal transition system does not contain action transitions, it corresponds to a lumped Markov chain. Analysis can start as briefly outlined in Section 3.4.
5. If the minimal transition system still contains action transitions the stochastic process is underspecified. Assuming that this result is not intended, it indicates that the specification contains an error. Hence, one has to return to the specification step (1.). If underspecification is caused by external nondeterminism, the specification can be changed by incorporating an external influence that resolves nondeterminism. The interpretation of the resulting Markov chain should then be conditioned on this external influence. Internal nondeterminism is a more severe problem, it requires a detailed analysis of the specification, cf. Section 6.3.

Nondeterminism is one of the vital ingredients of IMC, though it appears as an additional hurdle when it comes to performance estimation. As already pointed out in Chapter 1 (and illustrated in Figure 1.1), this hurdle is not an artifact, it reflects that many system designs behave nondeterministically at a certain level of abstraction, unless all relevant information is specified. Since $\mathsf{IMC}_{\mathsf{XXL}}$ can be straightforwardly extended to a full specification language, the means to specify the necessary information are at hand.

It is worth to mention that the above recipe has become a substantial part of the TIPPtool [98]. Recently, also the CADP toolset [61, 66] has been augmented with efficient algorithms for state space generation, aggregation and Markov chain analysis, in order to support performance and dependability estimation with IMC [65]. This extension substantially improves the effectiveness of the IMC approach in practice. CADP is an industrial strength verification environment, originally focused on functional modelling and verification[1].

7.3 The Challenge Continues

While the main focus of this work is on compositional generation and aggregation of Markov chains, it has fallen somewhat short with respect to exploiting compositionality in the numerical analysis process. In the meanwhile, substantial work in this direction has been undertaken, for instance,

[1] `http://www.inrialpes.fr/vasy/cadp/`.

by Mertsiotakis [139] and by Bohnenkamp[28], in contexts that are readily applicable within the IMC framework, and more inspiring contributions can be foreseen in this area in the future.

Our approach to a compositional generation of Markov chains is driven by experiences with the first generation of stochastic process algebras, in particular TIPP [79], PEPA [114], and EMPA [25]. While the IMC approach has exploited the lessons learned thereof, the next generation of stochastic process algebras has already appeared in the literature. They aim to break the limitations that are due to the restriction to exponential distributions [111, 160, 56, 34, 36]. Our approach also contributes to these efforts, since time constraints can be arbitrarily phase-type distributed. More foundational approaches broaden the class of stochastic processes to be generated beyond Markov chains, for instance to the class of Generalised Semi Markov Processes (GSMPs) [174].

In this context, we are particularly enthusiastic about the algebra $\spadesuit$ of D'Argenio [54, 56] and IGSMP of Bravetti [32, 36], not least because these approaches follow a spirit very similar to IMC. Though conceptually different, both algebras separate delays and actions. They even go a step further, in order to achieve support for non-Markovian distributions: Instead of a two-way split, both are based on a three-way split where delays are again separated into explicit initialisation and termination. In this way, concurrently running delays can be interleaved. As long as the Markov property holds, this does not provide benefits. But when moving to non-Markovian distributions a three-way split turns into a key for an elegant interleaving semantics. The work of Bravetti is of independent interest since it contains a valuable extension of IMC with weighted probabilistic choice, aside the main work on IGSMP [32].

It should however be mentioned that numerical analysis of GSMPs is computationally expensive. Thus one usually resorts to discrete-event simulation. On the other hand, numerical analysis is still feasible if some syntactical restrictions are imposed on the stochastic process algebra allowing for non-Markovian distributions. Herzog has outlined this approach in [111] providing a semantics that maps onto aperiodic stochastic graph models. In the absence of nondeterminism, performance estimates can be obtained efficiently by means of discretisation of distributions [90, 165]. The problems faced in the presence of nondeterminism are akin to those occuring in the context of IMC (and $\spadesuit$ as well as IGSMP). Possibly, some of the solutions developed within this book are valuable. Of course, the potential benefits are of a more speculative nature, since stochastic graph models are non-interleaving models, closely resembling stochastic event structure models [126].

Finally, we remark that this book has focused on the specification of models, while it has mostly neglected an important related issue, namely the specification of requirements the model should satisfy, such as particular stochastic timed properties. A usable methodology should however cover both aspects (at least we think so). In the setting described here, properties of a specification are evaluated via the calculation of state probabilities. The interpretation of these probabilities by means of reward functions (cf. Section 3.4) is not easy, because the behavioural view is lost on the level of the Markov chain. We do prefer a model checking approach to this problem, where requirements are expressed in terms of a temporal logics [48]. For CTMCs, such an approach has recently been developed [9, 17, 15, 103], and is implemented in tools such as $\mathsf{E} \vdash \mathsf{MC}^2$ [102]. However, to make model checking usable together with IMC in its full generality requires to address nondeterminism in the model checking algorithm. In this context the work of de Alfaro [5, 6], is particularly promising. It allows, at least in principle, model checking of IMC-like models with respect to long-run performance properties.

A. Proofs for Chapter 3 and Chapter 4

A.1 Theorem 3.6.1

The algorithm of Table 3.1 computes Markovian bisimilarity on S. It can be implemented with a time complexity of $\mathcal{O}(m_M \log n)$ where m_M is the number of Markovian transitions and n is the number of states. The space complexity of this implementation is $\mathcal{O}(m_M)$.

Sketch of Proof. Correctness: For any partitioning *Part*, $Refine(Part, \mathsf{a}, C)$ is finer than *Part*. In addition, $Refine(Part, \mathsf{a}, C)$ is coarser than $S/\sim$ if *Part* is coarser than $S/\sim$. If the set *Spl* is empty and *Part* is coarser than $S/\sim$, then *Part* coincides with $S/\sim$. The algorithm is correct because the initial partition is coarser than $S/\sim$. It terminates because no splitter is processed twice and the number of possible splitters is bounded.

Complexity: As mentioned in Section 2.3, Paige and Tarjan [154] describe an implementation of the general algorithm to compute strong bisimilarity on interactive processes (Table 2.2). The time complexity is $\mathcal{O}(m_I \log n)$ and space complexity is $\mathcal{O}(m_I)$.

The central idea of the algorithm is an adaption of the 'process the smaller half' algorithm [2]. The basic observation is that, if some class is split into (at least) two new classes, one of them has at most half the size of the original class. In further refinement steps, only the smaller half is used as a splitter, and refinement with respect to the larger one is done implicitly. In this way, each state P appears in at most $\log n + 1$ different splitters used during refinement.

This strategy can be adopted to our scenario because of the following reasons. Assume that $Part'$ is the result of refining a partitioning *Part* with respect to a splitter C. In other words, $Part' = M\_Refine(Part, C)$. Then any $Part''$ that is finer than $Part'$ will not be refined further when applying $M\_Refine(Part'', C)$. Assume now that class C is for some reason split into C_1 and C_2. Now, we would have to refine $Part''$ with respect to both, C_1 and C_2.

However, if we remember the values of $\gamma_M(P, C)$ for each P (that we have indeed computed earlier, during computation of $M\_Refine(Part, C)$), we can convince ourselves that refining with respect to one of the classes is

sufficient. The reason is that $\gamma_M(P, C) = \gamma_M(P, C_1) + \gamma_M(P, C_2)$. So, if we split a class X according to the values $\gamma_M(\_, C_2)$, this is the same as splitting with respect to $\gamma_M(\_, C_1)$, since we have ensured before that all values $\gamma_M(\_, C)$ are the same for states contained in class X.
Thus, only refinement with respect to the smaller class is required, refinement with respect to the larger will not lead to further refinement. If C is refined into more classes than just C_1 and C_2, the same reasoning applies in order to avoid to refine with respect to one of the largest classes. In this way we can assure that each state P appears in at most $\log n + 1$ different splitters used during refinement.
The overall time bound of Paige and Tarjan, $\mathcal{O}(m_I \log n)$, is obtained by summing over all splitters and all states in these splitters, together with an estimation of the time consumption of a refinement step. They show that a refinement step with respect to a splitter $(\mathtt{a}, C)$ can be computed in $\mathcal{O}(|C| + |pre_{\mathtt{a}}(C)|)$ time, where $pre_{\mathtt{a}}(C)$ is the set of a-transitions that lead into the class C, i.e., $\{P \xrightarrow{\mathtt{a}} P' | P' \in C\}$.[1]
To get the same complexity result is not straightforward, since we fail to show that in our setting a refinement step with respect to a splitter C requires the corresponding $\mathcal{O}(|C| + |pre(C)|)$ time, where $pre(C) = \{P \xrightarrow{\lambda} P' \mid P' \in C\}$. The crucial detail is that we need to sort the results of the real valued function γ_M in order to group states with the same value. Using some form of binary search tree, this extra sorting gives an extra logarithmic overhead, resulting in $\mathcal{O}(|C| + |pre(C)| \log(|pre(C)|))$ time requirement to split C. This approach thus leads to an overall time complexity bound of $\mathcal{O}(m_M \log^2 n)$.
The bound can however be improved to $\mathcal{O}(m_M \log n)$ by resorting to statically optimal trees, e.g. to *splay trees* [179]. The crucial step is to show that the cumulated sorting effort that is needed in the entire refinement is of order $\mathcal{O}(n \log n)$. To establish this property, assume that during the entire refinement algorithm a state s is involved in j refinements, i.e., it is moved from larger partitions into smaller partitions j times, and assume $m_0 \geq m_1 \geq \ldots \geq m_j$ being the size of these partitions. By breaking the static optimality theorem [179] over the individual accesses to states, we obtain that when s is moved into a smaller partition the i-th time, it takes $\mathcal{O}(\log m_{i-1} - \log m_i)$ amortised time to determine the partition – plus some linear overhead that in total costs $\mathcal{O}(m_M \log n)$ time. Excluding the overhead and taking the sum over all moves needed for state s we obtain a bound $\mathcal{O}(\log m_0 - \log m_j)$ for the effort related to state s, which is bounded by $\mathcal{O}(\log n)$. Summing over all n states, we obtain $\mathcal{O}(n \log n)$ for the overall sorting, from which the overall time complexity $\mathcal{O}(m_M \log n)$ follows. We refer to [59] for a detailed proof.

[1] To be precise, Paige and Tarjan consider unlabelled transition relations, thus the label a is not required.

A.2 Theorem 4.3.1

Strong bisimilarity is substitutive with respect to parallel composition and abstraction, i.e.,

$$P_1 \sim P_2 \quad implies \quad P_1 \overline{\underline{a_1 \ldots a_n}} P_3 \sim P_2 \overline{\underline{a_1 \ldots a_n}} P_3,$$

$$P_1 \sim P_2 \quad implies \quad P_3 \overline{\underline{a_1 \ldots a_n}} P_1 \sim P_3 \overline{\underline{a_1 \ldots a_n}} P_2,$$

$$P_1 \sim P_2 \quad implies \quad \boxed{P_1\ \blacksquare_{a_1, \ldots, a_n}} \sim \boxed{P_2\ \blacksquare_{a_1 \ldots a_n}}.$$

Sketch of Proof. The proof is not difficult. We include a sketch of it in order to highlight a general proof strategy that will re-appear in various later proofs. It exploits the coinductive definition of bisimilarity. In order to show the first two implications, let $a_1 \ldots a_n$ be some actions and assume $P_1 \sim P_2$, i.e., there is some strong bisimulation $\mathcal{B}$ containing the pair (P_1, P_2). For an arbitrary IMC P_3 we have to show that $P_1 \overline{\underline{a_1 \ldots a_n}} P_3 \sim P_2 \overline{\underline{a_1 \ldots a_n}} P_3$. According to Definition 4.3.1 it is sufficient to show that a strong bisimulation $\mathcal{B}'$ exists containing the pair $(P_1 \overline{\underline{a_1 \ldots a_n}} P_3, P_2 \overline{\underline{a_1 \ldots a_n}} P_3)$. The relation

$$\mathcal{B}' := \left\{ \left(P' \overline{\underline{a_1 \ldots a_n}} P''', P'' \overline{\underline{a_1 \ldots a_n}} P''' \right) \mid P' \mathcal{B} P'' \wedge P''' \in \mathcal{S}^{all} \right\} \cup \mathit{Id}_{\mathcal{S}^{all}}$$

contains this pair. $\mathcal{B}'$ is an equivalence relation, because it is reflexive and because $\mathcal{B}$ is transitive and symmetric. In addition, by means of a detailed case analysis, it can be proven that $\mathcal{B}'$ satisfies *(i)* and *(ii)* of Definition 4.3.1. Verifying the second implication of this theorem is completely analogous, using a relation

$$\mathcal{B}' := \left\{ \left(P''' \overline{\underline{a_1 \ldots a_n}} P', P''' \overline{\underline{a_1 \ldots a_n}} P'' \right) \mid P' \mathcal{B} P'' \wedge P''' \in \mathcal{S}^{all} \right\} \cup \mathit{Id}_{\mathcal{S}^{all}},$$

since parallel composition (Definition 4.2.2) is symmetric. The proof of the third implication proceeds similarly, but uses a relation

$$\mathcal{B}' := \left\{ \left(\boxed{P'\ \blacksquare_{a_1 \ldots a_n}}, \boxed{P''\ \blacksquare_{a_1 \ldots a_n}} \right) \mid P' \mathcal{B} P'' \right\} \cup \mathit{Id}_{\mathcal{S}^{all}}.$$

A.3 Theorem 4.3.2

Two interactive processes are strongly bisimilar according to Definition 4.3.1 if and only if they are strongly bisimilar according to Definition 2.2.2.

Two Markovian Chains are strongly bisimilar according to Definition 4.3.1 if and only if they are Markovian bisimilar according to Definition 3.5.1.

Proof. For interactive processes, $\longrightarrow\!\!\!\!\!\square\!\!\!\rightarrow$ is empty. Then the second clause of Definition 4.3.1 follows from the first. The first clause in turn is equivalent to what is expressed in Lemma 2.2.2 and hence to Definition 2.2.2.

For Markovian chains, $\longrightarrow\!\!\!\!\!\rightarrow$ is empty. In this case, the first clause is irrelevant and, since all states are stable, the second clause directly reduces to the constraints on γ_{M} required in Definition 3.5.1.

A.4 Lemma 4.4.2

An equivalence relation $\mathcal{E}$ on $\mathcal{S}^{\mathrm{all}}$ is a weak bisimulation iff $P\ \mathcal{E}\ Q$ implies for all $\mathsf{a} \in Act$ and for all equivalence classes C of $\mathcal{E}$,

1. $P \xrightarrow{\mathsf{a}} P'$ *implies* $Q \stackrel{\mathsf{a}}{\Longrightarrow} Q'$ *for some* Q' *with* $P'\ \mathcal{E}\ Q'$*,*
2. $P \not\xrightarrow{\tau}$ *implies* $\gamma_{\mathrm{M}}(P, C) = \gamma_{\mathrm{M}}(Q'', C)$ *for some* $Q'' \not\xrightarrow{\tau}$ *such that* $Q \stackrel{\tau}{\Longrightarrow} Q''$ *and* $P\mathcal{E}Q''$*.*

Proof. We have to show that an equivalence relation $\mathcal{E}$ satisfying this lemma also satisfies Definition 4.4.1, and vice versa.

Take an arbitrary pair $(P, Q) \in \mathcal{E}$. To begin with, the first condition of Definition 4.4.1 directly implies the first clause of Lemma 4.4.2. The converse direction is shown by tracing the weak transition $P \stackrel{\mathsf{a}}{\Longrightarrow} P'$ stepwise. (The equivalence of these two clauses is indeed a well known property for non-stochastic weak bisimulation).

In order to show that the respective second clauses are interchangeable, we first point out that C^τ is a (disjoint) union of equivalence classes of $\mathcal{E}$ if C is a class of $\mathcal{E}$.

We then proceed and verify that the second clause of Definition 4.4.1 follows from Lemma 4.4.2. Choose an equivalence class C and assume $P \stackrel{\tau}{\Longrightarrow} P'$ and $P' \not\xrightarrow{\tau}$. By means of the first clause of Lemma 4.4.2 (applied iteratively, see above) we obtain that $Q \stackrel{\tau}{\Longrightarrow} Q'$ such that $P'\mathcal{E}Q'$. We deduce from the second clause of Lemma 4.4.2 (since $P' \not\xrightarrow{\tau}$) that $Q' \stackrel{\tau}{\Longrightarrow} Q''$ for some $Q'' \not\xrightarrow{\tau}$ and that $P'\mathcal{E}Q''$. As a whole, we have $Q \stackrel{\tau}{\Longrightarrow} Q''$.

Thus, in order to prove the implication, it remains to be shown that $\gamma_{\mathrm{M}}(P', C^\tau) = \gamma_{\mathrm{M}}(Q'', C^\tau)$ for an arbitrary class C of $\mathcal{E}$. Since C^τ is a union of equivalence classes of $\mathcal{E}$, it is sufficient to show that for all classes D of $\mathcal{E}$, $\gamma_{\mathrm{M}}(P', D) = \gamma_{\mathrm{M}}(Q'', D)$. This is indeed the case, because of the following property: We know that $P'\mathcal{E}Q''$ and that both P' and Q'' are stable. Thus Lemma 4.4.2 says that for all equivalence classes D of $\mathcal{E}$ there is a stable state Q''' reachable from Q'' by means of internal transitions such that $\gamma_{\mathrm{M}}(P', D) = \gamma_{\mathrm{M}}(Q''', D)$. But since $Q'' \not\xrightarrow{\tau}$, Q''' coincides with Q'' independent of the actual class D.

We continue the proof by verifying that the second clause of Lemma 4.4.2 is a consequence of Definition 4.4.1. Assume that $P \not\xrightarrow{\tau}$. This implies (by

Definition 4.4.1) that there is some stable state Q'' such that $Q \stackrel{\tau}{\longrightarrow\!\!\!\!\!\rightarrow} Q''$ and for an arbitrary class $\mathcal{E}$ it holds that $\gamma_{\mathrm{M}}(P', C^{\tau}) = \gamma_{\mathrm{M}}(Q'', C^{\tau})$. In order to complete the proof, we will first show that $Q''\mathcal{E}P$ and then derive that $\gamma_{\mathrm{M}}(P, D) = \gamma_{\mathrm{M}}(Q'', D)$ holds for an arbitrary class D. (The latter property is slightly stronger than required because the universal quantifier over D appears inside the existential quantifier over Q'', while Lemma 4.4.2 requires merely the reversed ordering of quantifiers).

In order to derive $Q''\mathcal{E}P$ note that $Q \stackrel{\tau}{\Longrightarrow} Q''$ implies that there is some P'' such that $P \stackrel{\tau}{\Longrightarrow} P''$ and $P''\mathcal{E}Q''$, because $\mathcal{E}$ satisfies Definition 4.4.1 and $P\mathcal{E}Q$ by assumption. But since we know $P \stackrel{\tau}{\not\longrightarrow\!\!\!\!\!\rightarrow}$, P'' coincides with P and hence $Q''\mathcal{E}P$.

The only remaining requirement to complete the proof is to show $\gamma_{\mathrm{M}}(P, D) = \gamma_{\mathrm{M}}(Q'', D)$ for an arbitrary class D of $\mathcal{E}$. We show the required result by an induction on the number of equivalence classes subsumed by D^{τ}. Let n denote this number, where $n = 0$ is impossible, since D^{τ} at least includes D. If $n = 1$, $D^{\tau} = D$ and the result is immediate. Now assume that $\gamma_{\mathrm{M}}(P, D') = \gamma_{\mathrm{M}}(Q'', D')$ holds for all those D' where D'^{τ} subsumes at most n equivalence classes of $\mathcal{E}$. Assume that D^{τ} subsumes $n+1$ equivalence classes, but that $\gamma_{\mathrm{M}}(P, D) = \gamma_{\mathrm{M}}(Q'', D)$ does *not* hold (heading for a contradiction). Lemma 4.4.2 implies $\gamma_{\mathrm{M}}(P, D^{\tau}) = \gamma_{\mathrm{M}}(Q'', D^{\tau})$. We may split the computation of γ_{M} into a sum over disjoint sets. Specifically,

$$\gamma_{\mathrm{M}}(P, D^{\tau}) = \sum_{\substack{D' \subseteq D^{\tau} \\ D' \in \mathcal{S}^{\mathrm{all}}/\mathcal{E}}} \gamma_{\mathrm{M}}(P, D') = \sum_{\substack{D' \subseteq D^{\tau} \\ D' \in \mathcal{S}^{\mathrm{all}}/\mathcal{E}}} \gamma_{\mathrm{M}}(Q'', D') = \gamma_{\mathrm{M}}(Q'', D^{\tau}).$$

Since we have assumed $\gamma_{\mathrm{M}}(P, D) \neq \gamma_{\mathrm{M}}(Q'', D)$ we deduce

$$\sum_{\substack{D' \subseteq D^{\tau} \\ D' \in \mathcal{S}^{\mathrm{all}}/\mathcal{E} \\ D' \neq D}} \gamma_{\mathrm{M}}(P, D') \neq \sum_{\substack{D' \subseteq D^{\tau} \\ D' \in \mathcal{S}^{\mathrm{all}}/\mathcal{E} \\ D' \neq D}} \gamma_{\mathrm{M}}(Q'', D')$$

Be reminded that the induction hypothesis has been that whenever D'^{τ} subsumes less than $n+1$ equivalence classes we already know that $\gamma_{\mathrm{M}}(P, D') = \gamma_{\mathrm{M}}(Q'', D')$.

This means that there has to be at least one class D'' subsumed by D where D''^{τ} subsumes at least $n+1$ equivalence classes. Since, by definition, $D''^{\tau} \subseteq D^{\tau}$ and $D \subseteq D^{\tau}$ but $D \not\subseteq D''^{\tau}$, D has to subsume at least $n+2$ equivalence classes. This completes the proof of Lemma 4.4.2 with a contradiction, since we have assumed that D subsumes $n+1$ classes.

A.5 Theorem 4.4.1

Weak bisimilarity is substitutive with respect to parallel composition and abstraction, i.e.,

$$P_1 \sim P_2 \quad \textit{implies} \quad P_1 \;\|_{a_1\ldots a_n}\; P_3 \sim P_2 \;\|_{a_1\ldots a_n}\; P_3,$$

$$P_1 \sim P_2 \quad \textit{implies} \quad P_3 \;\|_{a_1\ldots a_n}\; P_1 \sim P_3 \;\|_{a_1\ldots a_n}\; P_2,$$

$$P_1 \sim P_2 \quad \textit{implies} \quad \boxed{P_1 \;\; a_1,\ldots,a_n} \sim \boxed{P_2 \;\; a_1\ldots a_n}.$$

Sketch of Proof. The proof follows the lines of the strong bisimulation proof (Theorem 4.3.1). In each of the respective three cases, relation $\mathcal{B}'$ appearing therein can be shown to satisfy Lemma 4.4.2 if $\mathcal{B}$ is a weak rather than a strong bisimulation.

A.6 Theorem 4.4.2

Two interactive processes are weakly bisimilar according to Definition 4.4.1 if and only if they are weakly bisimilar according to Definition 2.2.5.

Two Markovian Chains are weakly bisimilar according to Definition 4.4.1 if and only if they are Markovian bisimilar according to Definition 3.5.1.

Sketch of Proof. For interactive processes, $\dashrightarrow$ is empty and therefore the second clause of Definition 4.4.1 trivially follows from the first. This clause, in turn, is equivalent to Definition 2.2.5 apart from the fact that this definition does not presuppose that a weak bisimulation is an equivalence relation. As in the strong bisimulation case (Lemma 2.2.2) it can be easily shown that this does not affect the union of all weak bisimulations, weak bisimilarity.

For Markovian chains, $\twoheadrightarrow$ is empty. In this case, the first clause is irrelevant and, since all states are stable, the second clause directly reduces to the constraints on γ_{M} required in Definition 3.5.1.

A.7 Theorem 4.5.1

Let P be an IMC with finite (reachable) state space S. Strong bisimilarity on S is the unique fixed-point of

- $\smile_0 = S \times S$.
- $\smile_{k+1} = \smile_k \cap E_k$
  where $(P', Q') \in E_k$ *iff* $(\forall \mathsf{a} \in Act)\ (\forall C \in S/_{\smile_k})$

(*i*) $\gamma_O(P', \mathsf{a}, C) = \gamma_O(Q', \mathsf{a}, C)$,
(*ii*) *if* $P' \not\xrightarrow{\tau}$ *and* $Q' \not\xrightarrow{\tau}$, *then* $\gamma_M(P', C) = \gamma_M(Q', C)$.

Proof. We have to show

- that each $\smile_k$ is an equivalence relation,
- that a unique fixed-point exists, and
- that this fixed-point coincides with strong bisimilarity.

We first show, by induction, that each $\smile_k$ is an equivalence relation. The initial relation $\smile_0$ is a trivial equivalence relation. Now assume that $\smile_k$ also is. Then $\smile_{k+1}$ contains the identity relation on S, since for each $P' \in S$ the pair (P', P') (trivially) satisfies condition (*i*) and (*ii*). $\smile_{k+1}$ is symmetric, since $\smile_k$ is and because by construction E_k is symmetric. So, only transitivity remains to be verified. Assume that $(P', Q') \in E_k$ and $(Q', R') \in E_k$. We have to check whether also $(P', R') \in E_k$. From transitivity of equality on boolean values it follows that $\gamma_O(P', \mathsf{a}, C) = \gamma_O(R', \mathsf{a}, C)$, for each $\mathsf{a} \in Act$ and $C \in S/_{\smile_k}$. Hence, only (*ii*) remains to be checked. If all three states are stable, $\gamma_M(P', C) = \gamma_M(R', C)$ follows (again) from transitivity of equality (on real values). If otherwise P' or R' are unstable, nothing has to be checked. So the only remaining case is that P' and R' are stable while Q' is not. But this case is impossible since $(P', Q') \in E_k$, which (due to (*i*)) implies that $\gamma_O(P', \tau, S) = \gamma_O(Q', \tau, S)$. Hence either both, P' and Q', or neither of them is stable. This completes the induction, each $\smile_k$ is an equivalence indeed.

A fixed-point exists, because $S \times S$ is finite, $|\smile_k| \geq |\smile_{k+1}|$, and the chain of relations is bounded by the identity relation on S. The fixed point is unique by construction. Hence, there is some n such that $\smile_n = \smile_{n+k}$ for all $k > 0$.

To show that $\smile_n$ coincides with strong bisimilarity on S, we first verify $\smile_n \subseteq \sim_S$, where $\sim_S$ denotes $\sim \cap (S \times S)$. For this purpose, it is sufficient to show that $\smile_{n+1}$ ($\smile_{n+1} \cup \mathcal{I}d_{S^{\text{all}}}$, to be pedantic) is a strong bisimulation according to Definition 4.3.1.

Take an arbitrary pair $(P', Q') \in \smile_{n+1}$, an action a and an equivalence class C of $\smile_{n+1} = \smile_n$. The first condition of Definition 4.3.1 follows directly from $\gamma_O(P', \mathsf{a}, C) = \gamma_O(Q', \mathsf{a}, C)$. In order to verify the second clause, assume $P' \not\xrightarrow{\tau}$. Hence, for all classes D we have that $\gamma_O(P', \tau, D) = \texttt{false}$. This implies $\gamma_O(Q', \tau, D) = \texttt{false}$ and, ranging over all classes, $Q' \not\xrightarrow{\tau}$. We can therefore use the second clause of Theorem 4.5.1 to obtain $\gamma_M(P', C) = \gamma_M(Q', C)$, as required.

Thus only $\smile_n \supseteq \sim_S$ remains to be verified. We show by induction that $\smile_k \supseteq \sim_S$ for all $k \geq 0$. The base case is clear. Assume that $\smile_k \supseteq \sim_S$, but $\smile_{k+1} \not\supseteq \sim_S$. Hence, there is a pair (P', Q') contained in $\sim_S$ and $\smile_k$ but not in $\smile_{k+1}$. Note that each equivalence class of $\smile_k$ is a (disjoint) union of equivalence classes of $\sim_S$, because $\smile_k$ is coarser than $\sim_S$. Due to

Definition 4.3.1, it is therefore impossible that for any $a \in Act$ and $C \in S/\smile_k$ condition (i) or (ii) are violated by the pair (P', Q').

A.8 Theorem 4.5.2

The algorithm of Table 4.2 computes strong bisimilarity on S. It can be implemented with a time complexity of $\mathcal{O}((m_I + m_M)\log n)$ where m_I is the number of interactive transitions, m_M the number of Markovian transitions and n is the number of states. The space complexity of this implementation is $\mathcal{O}(m_I + m_M)$.

Sketch of Proof. Correctness: The main loop of the algorithm computes the partitioning

$$\Big(\mathit{M\_Refine}\big(\mathit{Refine}(\mathit{S\_Part}, a, C), C\big) \cup \mathit{Refine}(\mathit{U\_Part}, a, C)\Big)$$

from the current partitioning ($\mathit{S\_Part} \cup \mathit{U\_Part}$) and a chosen splitter (a, C). To prove correctness, observe that for each partitioning ($\mathit{S\_Part} \cup \mathit{U\_Part}$), the above is finer than ($\mathit{S\_Part} \cup \mathit{U\_Part}$). In addition, it is coarser than $S/\sim$ if ($\mathit{S\_Part} \cup \mathit{U\_Part}$) is. If the set *Spl* is empty and $\mathit{S\_Part} \cup \mathit{U\_Part}$ is coarser than $S/\sim$, then it indeed coincides with $S/\sim$. The algorithm is correct because the initial partition is coarser than $S/\sim$. It terminates because no splitter is processed twice and the number of possible splitters is bounded.

Complexity: As mentioned in Section 2.3, Paige and Tarjan's algorithm computes strong bisimilarity on interactive processes (Table 2.2) with a time complexity of $\mathcal{O}(m_I \log n)$ and space complexity $\mathcal{O}(m_I)$. The data structures and the management of splitters can be directly adopted to compute strong bisimilarity on IMC.

Compared to Table 2.2 there are two additional computations required in Table 4.2. During initialisation, the state space is split into *S_Part* and *U_Part* according to the abilities of performing internal transitions. This can be implemented in $\mathcal{O}(m_I)$, hence the initialisation does not worsen the time complexity.

In the main loop of the algorithm, we have added, again compared to Table 2.2, an additional refinement *M_Refine* (apart from the fact that *Refine* is split into two separate invocations, which does not contribute any additional computational complexity).

In order to establish the time complexity for the algorithm in Table 4.2, we separate the considerations w.r.t. the two transition relations. Since $(m_I + m_M)\log n = (m_I \log n) + (m_M \log n)$, where the first summand of the right hand side will be consumed by *Refine*, we need to show that the invocations of *M_Refine* in total need $\mathcal{O}(m_M \log n)$ time. For this we can resort to the arguments used in Appendix A.1 in the proof of Theorem 3.6.1.

A.9 Lemma 4.5.1

An equivalence relation $\mathcal{E}$ on $\mathcal{S}^{\text{all}}$ is a weak bisimulation iff $P\ \mathcal{E}\ Q$ implies for all $\mathsf{a} \in Act$ and all equivalence classes C of $\mathcal{E}$

1. $\gamma_{\mathrm{O}}(P, \mathsf{a}, C) = \gamma_{\mathrm{O}}(Q, \mathsf{a}, C)$,
2. $P \overset{\tau}{\searrow} P'$ *and* $P\mathcal{E}P'$ *imply* $Q \overset{\tau}{\searrow} Q'$ *for some* Q' *such that* $Q\mathcal{E}Q'$ *and* $\gamma_{\mathrm{M}}(P', C) = \gamma_{\mathrm{M}}(Q', C)$, *and*
3. $P \overset{\tau}{\searrow}$ *iff* $Q \overset{\tau}{\searrow}$.

Proof. It is clear that an equivalence $\mathcal{E}$ satisfying this lemma also satisfies Lemma 4.4.2. So let assume the converse, let $\mathcal{E}$ satisfy Lemma 4.4.2. The first clauses are identical. In order to verify the second clause of Lemma 4.5.1 assume $P\ \mathcal{E}\ Q$, $P \overset{\tau}{\searrow} P'$ and $P\mathcal{E}P'$. Thus $P' \not\xrightarrow{\tau}$ can be used in the third clause of Lemma 4.4.2 to derive the required constraints on Q. To prove the third clause we use an argument that already appeared in the proof of Lemma 4.4.2. Assume $P\ \mathcal{E}\ Q$ and $P \overset{\tau}{\searrow} P'$. By the first clause of Lemma 4.4.2 we have that there is some Q' such that $Q \overset{\tau}{\Longrightarrow} Q'$ and $P'\mathcal{E}Q'$. If this Q' is stable we are done. If it is not stable, we apply Lemma 4.4.2 to the pair $(P', Q') \in \mathcal{E}$, in particular the second clause. Since P' is stable, there is some stable Q'' such that $Q' \overset{\tau}{\searrow} Q''$. So, there is indeed is some Q'' such that $Q \overset{\tau}{\searrow} Q''$ (remind that $Q \overset{\tau}{\Longrightarrow} Q'$). Thus we have proven Lemma 4.5.1.

A.10 Theorem 4.5.3

The algorithm of Table 4.3 computes weak bisimilarity on S. The initialisation phase can be computed in $\mathcal{O}(n^{2.376})$ time. The body of the algorithm can be implemented such that it requires $\mathcal{O}(n\ (m_I'' + m_M))$ where n is the number of states, m_M is the number of Markovian transitions and m_I'' is the number of interactive transitions after transitive closure of internal transitions. The space complexity of this implementation is $\mathcal{O}(m_I'' + m_M)$.

Proof. Correctness: We show that

1. the initial partitioning is coarser than $S/\approx$.
2. each iteration preserves this property, i.e., if *TC_Part* $\cup$ *TD_Part* is coarser than $S/\approx$ and (a, C) is an element of *Spl*, then

$$\Big(\mathit{IM\_Refine}\big(\mathit{IRefine}(\mathit{TC\_Part}, \mathsf{a}, C), C\big) \cup \mathit{IRefine}(\mathit{TD\_Part}, \mathsf{a}, C)\Big)$$

is still coarser than $S/\approx$.

If these two requirements are fulfilled, it is easy to verify that, once *Spl* is empty, *TC_Part* $\cup$ *TDPart* coincides with $S/\approx$. The algorithm terminates, since no splitter will be processed twice and the size of splitters added to *Spl* becomes successively smaller, bounded by splitters of one element each.

The first requirement is an immediate consequence of Lemma 4.5.1: The third clause of this lemma implies that for all pairs (P, Q) contained in some bisimulation, both P and Q belong either to *TD_Part* or to *TC_Part*. Since weak bisimilarity is itself a weak bisimulation (Lemma 4.4.1) this gives us the first proof obligation.

We proceed to the second proof obligation. Assume that *TC_Part* $\cup$ *TD_Part* is coarser than $S/\approx$. Due to the definition of weak bisimilarity, Theorem 4.4.2, and the discussion of the algorithm for weak bisimilarity on interactive processes (Table 2.3) we do not discuss *IRefine* formally. Instead we simply claim that, under the above assumptions,

$$\mathit{IRefine}(\mathit{TC\_Part}, \mathsf{a}, C) \cup \mathit{IRefine}(\mathit{TD\_Part}, \mathsf{a}, C)$$

is still coarser than $S/\approx$. So we only have to address the question why *IM_Refine*(*TC_Part'*) is coarser than $S'/\approx$ if *TC_Part'* is, where S' denotes the subset of time convergent states of S (the initially computed *TC_Part* indeed). Assume that there is a pair (P, Q) contained in $\approx$ and thus also in the same partition X of *TC_Part'* but not contained in the same partition of *IM_Refine*(*TC_Part'*, C). From these assumption we will derive a contradiction. We distinguish four cases, dependent on the stability of P and Q.

- If $P \overset{\tau}{\not\twoheadrightarrow}$ and $Q \overset{\tau}{\not\twoheadrightarrow}$, then P and Q must be separated because $\gamma_{\mathrm{M}}(P, C) \neq \gamma_{\mathrm{M}}(Q, C)$. On the other hand, we have that $\gamma_{\mathrm{M}}(P, D) = \gamma_{\mathrm{M}}(Q, D)$ for all equivalence classes D of $\approx$. But C is the disjoint union of some equivalence classes of $\mathcal{E}$. A contradiction arises since the sum over some equivalence classes has to result in the same γ_{M} values for both P and Q.
- If $P \overset{\tau}{\not\twoheadrightarrow}$ and $Q \overset{\tau}{\longrightarrow\!\!\!\!\twoheadrightarrow}$, then, by assumption (Lemma 4.5.1), there is some $Q' \overset{\tau}{\not\twoheadrightarrow}$ such that $Q \approx Q'$, $Q \overset{\tau}{\longrightarrow\!\!\!\!\twoheadrightarrow} Q'$ and $\gamma_{\mathrm{M}}(P, D) = \gamma_{\mathrm{M}}(Q', D)$ for all equivalence classes D of $\approx$. In particular, Q and Q' are contained in the same class X of *TC_Part'* and we have that $\gamma_{\mathrm{M}}(P, C) = \gamma_{\mathrm{M}}(Q', C)$. So, P and Q' will be in the same class of *IM_Refine*(*TC_Part'*, C). Anyhow, Q is not contained in this class. Since it may internally evolve to Q', the only reason for not being in this class is (cf. Definition of *M_Spread*) that it may also evolve to a different stable state Q'' in class X that satisfies $\gamma_{\mathrm{M}}(Q', C) \neq \gamma_{\mathrm{M}}(Q'', C)$. In this case, Q will be inserted into *Rest*(*TC_Part'*, C). But in this case, we have, by assumption, also that $\gamma_{\mathrm{M}}(P, D) = \gamma_{\mathrm{M}}(Q'', D)$ for all equivalence classes D of $\approx$. This leads to a contradiction, because we can conclude $\gamma_{\mathrm{M}}(Q'', C) = \gamma_{\mathrm{M}}(P, C) = \gamma_{\mathrm{M}}(Q', C)$.
- If $P \overset{\tau}{\longrightarrow\!\!\!\!\twoheadrightarrow}$ and $Q \overset{\tau}{\not\twoheadrightarrow}$ the proof is symmetric to the above case.
- If $P \overset{\tau}{\longrightarrow\!\!\!\!\twoheadrightarrow}$ and $Q \overset{\tau}{\longrightarrow\!\!\!\!\twoheadrightarrow}$, we can assume that either P or Q is not contained in *Rest*(*TC_Part'*, C) (otherwise P and Q would be contained

in the same class of $\mathit{IM\_Refine}(\mathit{TC\_Part}', C)$). So, w.l.o.g. assume that P can only internally evolve to stable states that possess all the same value $\gamma_{\mathrm{M}}(\_, C)$. Function *Enrich* will then add P to this class of stable states. However, Q is not added to this class, by assumption. Since we have also assumed that $P \approx Q$ holds, we can derive (see the above cases) that also Q will be able to internally evolve to a stable state Q' that has the same $\gamma_{\mathrm{M}}(\_, C)$ value. So, it seems that Q will be inserted into $\mathit{Rest}(\mathit{TC\_Part}', C)$ because it may also evolve to some other stable state Q'' with a $\gamma_{\mathrm{M}}(Q'', C)$ value different from $\gamma_{\mathrm{M}}(Q', C)$. However, since $P \approx Q$, also P must have a similar possibility. This means that P must be inserted into $\mathit{Rest}(\mathit{TC\_Part}', C)$ as well, contradicting our assumption.

Complexity: In [86] Groote and Vaandrager show how to compute branching bisimilarity with an implementation that requires $\mathcal{O}(n\ m_I'')$ time and $\mathcal{O}(m_I'')$ space – apart from the initialisation phase requiring $\mathcal{O}(n^{2.376})$ time when using the algorithm of [52] – and computes the transitive closure $\xrightarrow{\tau}{}^{+}$ of internal transitions. Bouali adopted this machinery to also compute weak bisimilarity [30] with the same complexity. Since our definition combines weak bisimilarity for interactive transitions with branching bisimilarity for Markovian transitions, we will combine both implementations.

The main property of both algorithms is that they decide in $\mathcal{O}(m_I'')$ time whether a refinement of the current partitioning is required, and they compute this refinement in $\mathcal{O}(m_I'')$ time. Since there are at most n refinements required in the worst case (each partition contains a single state) this yields the total time complexity $\mathcal{O}(n\ m_I'')$.

Our method to compute weak bisimilarity proceeds in the same way. During initialisation, we compute $\xrightarrow{\tau}{}^{+}$. We then proceed and split the state space into *TC_Part* and *TD_Part*. This can be done in $\mathcal{O}(m_I'')$ time as follows. Scan all transitions. During this scan mark those states that have an outgoing $\xrightarrow{\tau}{}^{+}$ transition as potentially belonging to *TD_Part*. During another scan, unmark all states having an outgoing transition $\xrightarrow{\tau}{}^{+}$ to an unmarked state. After these two scans, all marked states are collected in *TD_Part*, all unmarked states in *TC_Part*.

During the rest of the algorithm *TD_Part* is refined according to weak bisimilarity exactly as in [30]. For the elements of *TC_Part* we remove all cycles of τ-transitions, by collapsing all states in such a cycle (they are weakly bisimilar). For this purpose we use an $\mathcal{O}(|\xrightarrow{\tau}{}^{+}|)$ algorithm from [2] solving the problem of finding strongly connected components in a graph. We are thus dealing with a cycle-free relation $\xrightarrow{\tau}{}^{+}$. Then, still during initialisation, we sort unstable states in a topological order according to their τ-transitions. This requires a depth first search with time and space complexity $\mathcal{O}(m_I'')$ [2]. All states are decorated with a special flag indicating absence or presence of stability. Markovian transitions pointing to a state are held in a separate list, but only if the origin is stable. Otherwise they are omitted.

The complexity of the whole initialisation phase is dominated by the transitive closure operation, its time complexity is $\mathcal{O}(n^{2.376})$ using the algorithm of [52] and it requires $\mathcal{O}(m_I'' + m_M)$ space (which is bounded by $\mathcal{O}(n^2)$ since *Act* is fixed).

Now, we describe how to decide in $\mathcal{O}(m_I''+m_M)$ time whether a refinement of the current partitioning is required. We only have to consider *TC_Part*, as indicated above. To check whether a refinement is needed because interactive transitions violate the first clause of Lemma 4.5.1 proceeds as in [30], as well as the refinement itself (if necessary).

The third clause of Lemma 4.5.1 is assured by construction. So, how can we check whether the second clause induces a refinement of *TC_Part* for a given partitioning (*TC_Part* $\cup$ *TD_Part*) in time $\mathcal{O}(m_I''+m_M)$? For each class C of (*TC_Part* $\cup$ *TD_Part*) proceed as follows. Compute $\gamma_M(P, C)$ for all stable states P that may reach C. This can be done in time proportional to $pre(C)$. During this computation, construct a list of classes that contain any such stable state. Afterwards process this list class by class. For a class C' in this list, check whether all stable states possess the same value $\gamma_M(\_, C)$. If this is the case, proceed to the next class in the list. If the list is empty, proceed to the computation of $\gamma_M(\_, C)$ for the next class C. If, on the other hand, two stable states in a class C' have different $\gamma_M(\_, C)$ values, then refinement is required. Ranging over all classes C the complexity to decide this is $\mathcal{O}(m_M)$, because m_M is the sum of $|pre(C)|$ over all classes C.

If a class C' contains multiple stable states with different γ_M values, then refinement is required, and thus the stable states in C' have to be sorted according to their γ_M values. Using a similar method as in Appendix A.1 in the proof of Theorem 3.6.1, this sorting effort is $\mathcal{O}(n \log n)$, for the entire refinement algorithm.

We now discuss how to achieve the $\mathcal{O}(n\ (m_I'' + m_M))$ bound for the steps of actual refinement, excluding the sorting effort. Since there are at most n refinements required, it is again enough to discuss how to compute a single refinement in $\mathcal{O}(m_I'' + m_M)$ time. Assuming that we found out that C' has to be split, we move each stable state P with nonzero γ_M value from its old class C' to a new class, determined by its γ_M value. A new class is created if it does not exist before. Unstable states of C' are treated afterwards. By construction, a class C' points to a list of unstable states and this list satisfies a topological ordering, as initially achieved for *TC_Part*. (The refinement with respect to interactive transitions also ensures this, as in [30].). Scan the list stepwise and insert an unstable state into a new class if all outgoing $\overset{\tau}{\Longrightarrow}^{+}$ lead into this class. Unstable states leading to more than one new class are collected in another separate class

(For this class we have to provide a list of 'bottom-states' in order match the requirements of weak bisimilarity computation of Bouali's implementation). New classes are inserted into the data structures as described in [86].
One can easily check that in $\mathcal{O}(m_I'' + m_M)$ time we have refined the current partitioning. The space complexity is $\mathcal{O}(m_I'' + m_M)$ completing the proof.

B. Proofs for Chapter 5

B.1 Theorem 5.2.2

Weak congruence is a congruence with respect to all operators of IML*: If* E_1*,* E_2 *and* E_3 *are expressions of* IML *and* $\mathsf{a} \in Act$, $\lambda \in \mathbb{R}^+$, $X \in \mathcal{V}$*, then*

- $E_1 \simeq E_2$ *implies* $\mathsf{a}.E_1 \simeq \mathsf{a}.E_2$,
- $E_1 \simeq E_2$ *implies* $(\lambda).E_1 \simeq (\lambda).E_2$,
- $E_1 \simeq E_2$ *implies* $E_1 + E_3 \simeq E_2 + E_3$ *and* $E_3 + E_1 \simeq E_3 + E_2$,
- $E_1 \simeq E_2$ *implies* $\underline{X{:=}\,E_1} \simeq \underline{X{:=}\,E_2}$.

Congruence of $\simeq$ with respect to the operators of IML follows the lines of [108], except for recursion. In order to prove congruence with respect to recursion we use a notion of 'bisimulation up to $\approx$' [169]. We only consider closed expressions $P \in$ IMC. Once we have shown for closed $\underline{X{:=}\,E}$ and $\underline{X{:=}\,F}$ that $E \simeq F$ implies $\underline{X{:=}\,E} \simeq \underline{X{:=}\,F}$, Definition 5.2.3 assures that this implication also holds for arbitrary expressions. This result will be established as Corollary B.1.1.

Let $\mathcal{S}$ be a binary relation on IMC. A sequence $(R_1, \ldots, R_n)$ with $n \geq 1$ and $R_i \in$ IMC for $i \in \{1, \ldots, n\}$ is a $(\mathcal{S} \cup \approx)$-path (or a path, for short) if

- $R_i\ (\mathcal{S} \cup \approx)\ R_{i+1}$ for all $i \in \{1, \ldots, n-1\}$ and
- $R_i {\approx} R_{i+1}$ implies $R_{i+1} \not\approx R_{i+2})$ for $i \in \{1, \ldots, n-2\}$.

It is worth to remark that the second requirement can always be guaranteed, due to the transitivity of $\approx$. In the sequel we will use $\mathcal{P}$, $\mathcal{P}'$, $\mathcal{Q}$, $\mathcal{Q}' \ldots$ to denote paths. If $\mathcal{P} = (R_1, \ldots, R_n)$ is a path, let

- $\mathcal{P}_i \equiv R_i$ and $\mathcal{P}^{(i)} = (R_i, R_{i+1}, \ldots, R_n)$ for $1 \leq i \leq n$,
- $|\mathcal{P}| = n$, and $|\mathcal{P}|_{\not\approx} = |\{i \mid 1 \leq i \leq n-1 \wedge R_i \not\approx R_{i+1}\}|$.

For two paths, $\mathcal{P} = (P_1, \ldots, P_n)$ and $\mathcal{Q} = (Q_1, \ldots, Q_m)$ with $P_n \equiv Q_1$, we let

$$\mathcal{PQ} = \begin{cases} (P_1, \ldots, P_n, Q_2, \ldots, Q_m) & \text{if } P_{n-1} \not\approx P_n \text{ or } Q_1 \not\approx Q_2 \\ (P_1, \ldots, P_{n-1}, Q_2, \ldots, Q_m) & \text{else} \end{cases}$$

In the sequel, we use $\stackrel{\mathsf{a}}{\Longrightarrow\!\!\!\!\!\!+}$ to abbreviate $\stackrel{\tau}{\longrightarrow\!\!\!\!\!+}{}^{*} \stackrel{\mathsf{a}}{\longrightarrow\!\!\!\!\!+} \stackrel{\tau}{\longrightarrow\!\!\!\!\!+}{}^{*}$. Note that $\stackrel{\mathsf{a}}{\Longrightarrow\!\!\!\!\!\!+}$ agrees with $\stackrel{\mathsf{a}}{\Longrightarrow}$, except if $\mathsf{a} = \tau$. In this case the latter includes the reflexive closure (Definition 2.2.4), while the former does not.

Definition B.1.1. *Let $\mathcal{S}$ be a symmetric relation on* IMC *and let $\mathcal{T} = (\mathcal{S} \cup \approx)^*$ be the induced equivalence relation on* IMC. *$\mathcal{S}$ is a weak congruence up to $\approx$ iff $P\ \mathcal{S}\ Q$ implies for all* $\mathsf{a} \in Act$, *$C \in \mathsf{IMC}/\mathcal{T}$ and $P' \in \mathsf{IMC}$ that*

1. *$P \xrightarrow{\mathsf{a}} P'$ implies $Q \xRightarrow{\hat{\mathsf{a}}} Q'$ and $P'\ \mathcal{S}\ R \approx Q'$ for some $R, Q' \in \mathsf{IMC}$,*
2. *$P\surd\!\downarrow$ (or $Q\surd\!\downarrow$) implies $\gamma_{\mathrm{M}}(P, C) = \gamma_{\mathrm{M}}(Q, C)$,*
3. *$P\surd\!\downarrow$ iff $Q\surd\!\downarrow$.*

We are aiming to show that $P \simeq Q$ holds already if there is a weak congruence up to $\approx$ between P and Q. This will be expressed in Lemma B.1.7. We need a few lemmas beforehand, the first just being a restatement of Lemma 4.4.2.

Lemma B.1.1. *An equivalence relation $\mathcal{E}$ on* IMC *is a weak bisimulation iff $P\ \mathcal{E}\ Q$ implies for all $\mathsf{a} \in Act$ and all $C \in \mathsf{IMC}/\mathcal{E}$:*

1. *$P \xrightarrow{\mathsf{a}} P'$ implies $Q \xRightarrow{\mathsf{a}} Q'$ for some Q' with $P'\ \mathcal{E}\ Q'$,*
2. *$P\surd\!\downarrow$ implies $\gamma_{\mathrm{M}}(P, C) = \gamma_{\mathrm{M}}(Q'', C)$ for some $Q''\surd\!\downarrow$ such that $Q \xRightarrow{\tau} Q''$ and $P\ \mathcal{E}\ Q''$.*

Proof. Rework of the proof of Lemma 4.4.2, following the lines of Appendix A.4.

Lemma B.1.2. *Let $\mathcal{B}_1$, $\mathcal{B}_2$ be equivalence relations on* IML. *$\mathcal{B}_1 \subseteq \mathcal{B}_2$ implies that*

$$\begin{array}{ll} \textit{if} & (\forall C_1 \in \mathsf{IML}/\mathcal{B}_1)\ \gamma_{\mathrm{M}}(E, C_1) = \gamma_{\mathrm{M}}(F, C_1), \\ \textit{then} & (\forall C_2 \in \mathsf{IML}/\mathcal{B}_2)\ \gamma_{\mathrm{M}}(E, C_2) = \gamma_{\mathrm{M}}(F, C_2). \end{array}$$

Proof. Follows from the fact that each $C \in \mathsf{IML}/\mathcal{B}_2$ is the disjoint union of classes from $\mathsf{IML}/\mathcal{B}_1$.

Lemma B.1.3. *Let $\mathcal{B} \subseteq \mathsf{IMC} \times \mathsf{IMC}$ be a symmetric relation such that $P\ \mathcal{B}\ Q$ implies for all $\mathsf{a} \in Act$ and all $C \in \mathsf{IMC}/\mathcal{B}^*$:*

1. *$P \xrightarrow{\mathsf{a}} P'$ implies $Q \xRightarrow{\mathsf{a}} Q'$ for some $Q' \in \mathsf{IMC}$ with $P'\ \mathcal{B}\ Q'$,*
2. *$P\surd\!\downarrow$ implies $\gamma_{\mathrm{M}}(P, C) = \gamma_{\mathrm{M}}(Q'', C)$ for some $Q''\surd\!\downarrow$ such that $Q \xRightarrow{\tau} Q''$ and $P\ \mathcal{B}\ Q''$.*

Then the transitive and reflexive closure $\mathcal{B}^*$ of $\mathcal{B}$ is a weak bisimulation.

Proof. If $P\ \mathcal{B}^*\ Q$, then $P\ \mathcal{B}^n\ Q$ for some $n \geq 0$. Now, (1) and (2) of Lemma B.1.1 can be shown by an induction on n, using the fact that $P\ \mathcal{B}\ Q$ and $P \xRightarrow{\mathsf{a}} P'$ implies $Q \xRightarrow{\mathsf{a}} Q'\ \mathcal{B}\ P'$ for some $Q' \in \mathsf{IMC}$ (which follows from the first assumption of Lemma B.1.3).

Lemma B.1.4. *For $P, Q \in \mathsf{IMC}$, $P \approx Q$ holds iff for every $\mathsf{a} \in Act$ and $C \in \mathsf{IMC}/\approx$:*

1. $P \overset{a}{-\!\!\!+\!\!\!\!\rightarrow} P'$ *implies* $Q \overset{a}{\Longrightarrow} Q'$ *and* $P' \approx Q'$ *for some* $Q' \in \mathsf{IMC}$,
2. $Q \overset{a}{-\!\!\!+\!\!\!\!\rightarrow} Q'$ *implies* $P \overset{a}{\Longrightarrow} P'$ *and* $P' \approx Q'$ *for some* $P' \in \mathsf{IMC}$,
3. $P \surd\!\!\downarrow$ *implies* $\gamma_M(P, C) = \gamma_M(Q'', C)$ *for some* $Q'' \surd\!\!\downarrow$ *such that* $Q \overset{\tau}{\Longrightarrow} Q''$ *and* $P \; \mathcal{E} \; Q''$,
4. $Q \surd\!\!\downarrow$ *implies* $\gamma_M(P'', C) = \gamma_M(Q, C)$ *for some* $P'' \surd\!\!\downarrow$ *such that* $P \overset{\tau}{\Longrightarrow} P''$ *and* $P'' \; \mathcal{E} \; Q$.

Proof. The direction saying that $P \approx Q$ implies the four conditions of lemma follows from the fact that $\approx$ is weak bisimulation. The converse direction can be proven by verifying that $\{(P, Q), (Q, P)\} \cup \approx)$ satisfies the conditions of Lemma B.1.3 if P and Q satisfy the four conditions of the lemma.

Lemma B.1.5. *If* $\mathcal{S}$ *is a weak congruence up to* $\approx$, *then* $P \; \mathcal{S} \; Q$ *and* $P \overset{\hat{a}}{=\!\!\!+\!\!\!\!\Rightarrow} P'$ *imply* $Q \overset{\hat{a}}{=\!\!\!+\!\!\!\!\Rightarrow} Q'$ *and* $P' \; \mathcal{S} \; R \approx Q'$ *for some* $Q', R \in \mathsf{IMC}$.

Proof. It is sufficient to show that

$$\text{if} \quad P' \; \mathcal{S} \; R' \approx Q' \; \wedge \; P' \overset{\hat{a}}{=\!\!\!+\!\!\!\!\Rightarrow} P'',$$
$$\text{then} \quad \exists Q'', R'' \in \mathsf{IMC} \; (Q' \overset{a}{\Longrightarrow} Q'' \; \wedge \; P'' \; \mathcal{S} \; R'' \approx Q'')$$

because this implies that we can trace the chain $P \overset{a}{=\!\!\!+\!\!\!\!\Rightarrow} P'$ transition by transition choosing $R' \equiv Q$ for the first step. Since the first step, according to (1) of Definition B.1.1 and $P \; \mathcal{S} \; Q$ implies that we will perform at least one transition from Q we can deduce that $Q \overset{a}{=\!\!\!+\!\!\!\!\Rightarrow} Q'$ instead of the weaker $Q \overset{a}{\Longrightarrow} Q'$. So, let $P' \; \mathcal{S} \; R' \approx Q' \wedge P' \overset{a}{-\!\!\!+\!\!\!\!\rightarrow} P''$. From $P' \overset{a}{-\!\!\!+\!\!\!\!\rightarrow} P''$, $P' \; \mathcal{S} \; R'$ we derive using property (1) of Definition B.1.1 $R' \overset{\hat{a}}{=\!\!\!+\!\!\!\!\Rightarrow} R'''$ and $P'' \; \mathcal{S} \; R'' \approx R'''$ for some $R'', R''' \in \mathsf{IMC}$. Now, $R' \overset{\hat{a}}{=\!\!\!+\!\!\!\!\Rightarrow} R'''$ and $R' \approx Q'$ imply $Q' \overset{a}{\Longrightarrow} Q''$ and $R''' \approx Q''$ for some $Q'' \in \mathsf{IMC}$. As a whole, we obtain $P'' \; \mathcal{S} \; R'' \approx R''' \approx Q''$. The above reasoning is illustrated (for a specific example) by the following diagram.

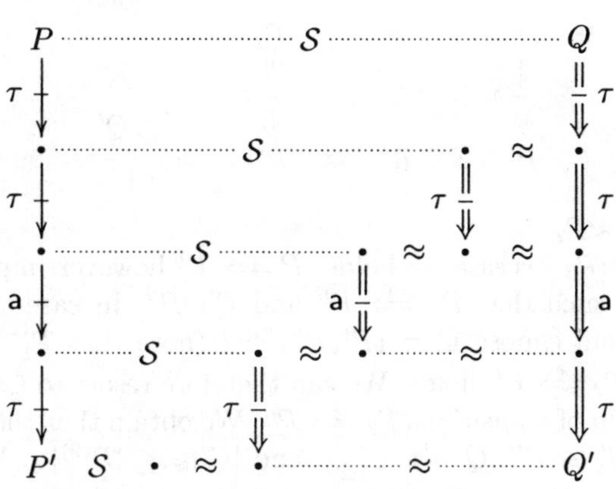

Lemma B.1.6. *If* $\mathcal{S}$ *is a weak congruence up to* $\approx$, *then* $\mathcal{T} = (\mathcal{S} \cup \approx)^*$ *is a weak bisimulation.*

Proof. Since $\mathcal{S}$ is symmetric, $\mathcal{T}$ is an equivalence relation. In order to show condition (1) of Lemma B.1.1, assume $P\ \mathcal{T}\ Q$. Thus, there is a $(\mathcal{S} \cup \approx)$–path $\mathcal{P}$ with $\mathcal{P}_1 \equiv P$ and $\mathcal{P}_{|\mathcal{P}|} \equiv Q$. In order to simplify the proof of condition (2) of Lemma B.1.1 it is helpful to show the following stronger condition (1') instead of (1).

> Let $\mathcal{P}$ be a path with $P \equiv \mathcal{P}_1$ and $Q \equiv \mathcal{P}_{|\mathcal{P}|}$ and let $P \stackrel{\hat{a}}{\Longrightarrow} P'$. Then there is a path $\mathcal{Q}$ such that $Q \stackrel{a}{\Longrightarrow} \mathcal{Q}_{|\mathcal{Q}|}$, $\mathcal{Q}_1 \equiv P'$ and
> – If $P \equiv \mathcal{P}_1 \approx \mathcal{P}_2$, then $|\mathcal{Q}|_{\not\approx} \leq |\mathcal{P}|_{\not\approx}$
> – If $P \equiv \mathcal{P}_1 \not\approx \mathcal{P}_2$, then $((\mathcal{Q}_1 \not\approx \mathcal{Q}_2 \wedge |\mathcal{Q}|_{\not\approx} \leq |\mathcal{P}|_{\not\approx}) \vee |\mathcal{Q}|_{\not\approx} < |\mathcal{P}|_{\not\approx})$

The proof of (1') is by induction on $|\mathcal{P}|_{\not\approx}$. So, let $\mathcal{P}$ be a path with $P \equiv \mathcal{P}_1$ and $Q \equiv \mathcal{P}_{|\mathcal{P}|}$ and let $P \stackrel{\hat{a}}{\Longrightarrow} P'$.

First, assume $|\mathcal{P}|_{\not\approx} = 0$, i.e., $P \approx Q$. Then $P \stackrel{\hat{a}}{\Longrightarrow} P'$ implies, as expected, that there is some Q' such that $Q \stackrel{a}{\Longrightarrow} Q'$ and $P' \approx Q'$. We therefore choose $\mathcal{Q} = (P', Q')$. Now, let $|\mathcal{P}|_{\not\approx} > 0$. We distinguish the following two cases.

Case 1: $P \equiv \mathcal{P}_1 \not\approx \mathcal{P}_2$
This implies $\mathcal{P}_1\ \mathcal{S}\ \mathcal{P}_2$. From $P \stackrel{\hat{a}}{\Longrightarrow} P'$ and Lemma B.1.5 we derive that there are some $P'', R' \in \mathrm{IMC}$ such that $\mathcal{P}_2 \stackrel{\hat{a}}{\Longrightarrow} P''$ and $P'\ \mathcal{S}\ R' \approx P''$. Since we also know that $|\mathcal{P}^{(2)}|_{\not\approx} < |\mathcal{P}|_{\not\approx}$ holds we make use of the induction hypotheses for the path $\mathcal{P}^{(2)}$ and the chain of transitions $\mathcal{P}_2 \stackrel{\hat{a}}{\Longrightarrow} P''$ in order to deduce that there is a path $\mathcal{Q}'$ satisfying $\mathcal{Q}'_1 \equiv P''$, $Q \stackrel{a}{\Longrightarrow} \mathcal{Q}'_{|\mathcal{Q}'|}$, and $|\mathcal{Q}'|_{\not\approx} \leq |\mathcal{P}^{(2)}|_{\not\approx}$. If $P' \not\approx R'$ define a path $\mathcal{Q}$ as $\mathcal{Q} = (P', R', P'')\mathcal{Q}'$. Then, $\mathcal{Q}_1 \equiv P' \not\approx R' \equiv \mathcal{Q}_2$ and $|\mathcal{Q}|_{\not\approx} = |\mathcal{Q}'|_{\not\approx} + 1 \leq |\mathcal{P}^{(2)}|_{\not\approx} + 1 = |\mathcal{P}|_{\not\approx}$. Hence $\mathcal{Q}$ satisfies all requirements. If, otherwise $P' \approx R'$, then $R' \approx P''$ implies $P' \approx P''$. Define $\mathcal{Q} = (P', P'')\mathcal{Q}'$. Then $|\mathcal{Q}|_{\not\approx} = |\mathcal{Q}'|_{\not\approx} \leq |\mathcal{P}^{(2)}|_{\not\approx} < |\mathcal{P}|_{\not\approx}$ holds as required. A diagram illustrates the proof of Case 1:

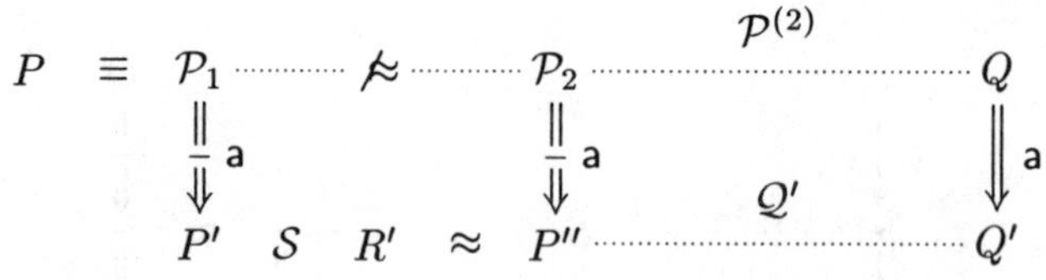

Case 2: $P \equiv \mathcal{P}_1 \approx \mathcal{P}_2$
Thus, $\mathcal{P}_2 \not\approx \mathcal{P}_3$ necessarily holds. $P \stackrel{\hat{a}}{\Longrightarrow} P'$ however implies that there is some P'' such that $\mathcal{P}_2 \stackrel{a}{\Longrightarrow} P''$ and $P' \approx P''$. In case that $\mathsf{a} = \tau$ and $\mathcal{P}_2 \equiv P''$ hold choose $\mathcal{Q} = (P', P'')\mathcal{P}^{(2)}$ (note that $\mathcal{P}^{(2)}_1 \equiv \mathcal{P}_2 \equiv P''$). Otherwise $\mathcal{P}_2 \stackrel{\hat{a}}{\Longrightarrow} P''$ holds. We can therefore resort to Case 1 with $\mathcal{P}^{(2)}$ and the chain of transitions $\mathcal{P}_2 \stackrel{\hat{a}}{\Longrightarrow} P''$. We obtain that there is a path $\mathcal{Q}'$ satisfying $\mathcal{Q}'_1 \equiv P''$, $Q \stackrel{a}{\Longrightarrow} \mathcal{Q}'_{|\mathcal{Q}'|}$, and $|\mathcal{Q}'|_{\not\approx} \leq |\mathcal{P}^{(2)}|_{\not\approx}$. We now choose $\mathcal{Q} = (P', P'')\mathcal{Q}'$ and obtain $|\mathcal{Q}|_{\not\approx} = |\mathcal{Q}'|_{\not\approx} \leq |\mathcal{P}^{(2)}|_{\not\approx} = |\mathcal{P}|_{\not\approx}$. Again a diagram is helpful to illustrate the proof of (1') for Case 2.

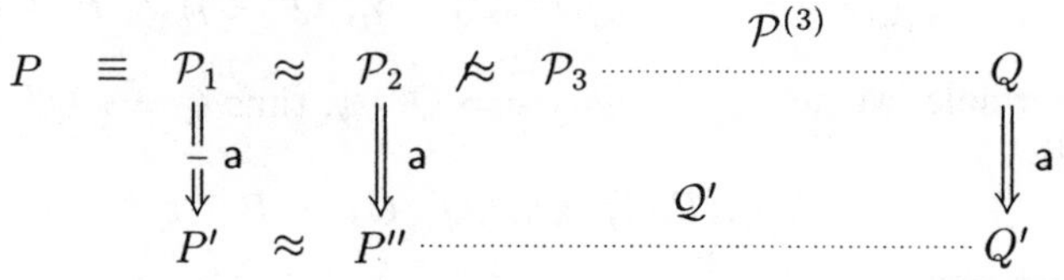

We are now ready to verify condition (2) of Lemma B.1.1. Let $P \mathcal{T} Q$, $C \in \mathsf{IMC}/\mathcal{T}$, and $P\surd\!\downarrow$. Furthermore let $\mathcal{P}$ be a path satisfying $\mathcal{P}_1 \equiv P$ and $\mathcal{P}_{|\mathcal{P}|} \equiv Q$. We show the following property by induction on $|\mathcal{P}|_{\not\approx}$.

$$\exists Q''\surd\!\downarrow : (Q \stackrel{\tau}{\Longrightarrow} Q'' \ \wedge \gamma_{\mathrm{M}}(P,C) = \gamma_{\mathrm{M}}(Q'',C) \wedge P \mathcal{T} Q''). \qquad \text{(B.1)}$$

The base case, $|\mathcal{P}|_{\not\approx} = 0$, (i.e., $P{\approx}Q$) follows directly from property (3) of Lemma B.1.4 and Lemma B.1.2 because $\approx \subseteq \mathcal{T}$. Now let $|\mathcal{P}|_{\not\approx} > 0$. Again we distinguish two cases.

Case 1: $P \equiv \mathcal{P}_1 \not\approx \mathcal{P}_2$

So, $\mathcal{P}_1 \ \mathcal{S} \ \mathcal{P}_2$ holds. Since $P \equiv \mathcal{P}_1\surd\!\downarrow$ we obtain (using condition (2) and (3) of Definition B.1.1) that $\mathcal{P}_2\surd\!\downarrow$ and for all $C \in \mathsf{IMC}/\mathcal{T}$:

$$\gamma_{\mathrm{M}}(P,C) = \gamma_{\mathrm{M}}(\mathcal{P}_2,C). \qquad \text{(B.2)}$$

Since $|\mathcal{P}^{(2)}|_{\not\approx} < |\mathcal{P}|_{\not\approx}$ and $\mathcal{P}_2\surd\!\downarrow$ we make use of our induction hypotheses for the path $\mathcal{P}^{(2)}$. We obtain that there is some Q'' such that $Q \stackrel{\tau}{\Longrightarrow} Q''$,

$$\gamma_{\mathrm{M}}(\mathcal{P}_2,C) = \gamma_{\mathrm{M}}(Q'',C), \qquad \text{(B.3)}$$

and $\mathcal{P}_2 \ \mathcal{T} \ Q''$. This gives $P \ \mathcal{T} \ Q''$, using $P \ \mathcal{S} \ \mathcal{P}_2$. In addition, (B.2) and (B.3) imply (B.1).

Case 2: $P \equiv \mathcal{P}_1{\approx}\mathcal{P}_2$

Thus, necessarily $\mathcal{P}_2 \not\approx \mathcal{P}_3$. From $P\surd\!\downarrow$ and $P{\approx}\mathcal{P}_2$ we deduce, using $\approx \subseteq \mathcal{T}$ and Lemma B.1.2, the existence of some $P'\surd\!\downarrow$ such that

$$\mathcal{P}_2 \stackrel{\tau}{\Longrightarrow} P' \wedge \gamma_{\mathrm{M}}(P,C) = \gamma_{\mathrm{M}}(\mathcal{P}_2,C) = \gamma_{\mathrm{M}}(P',C) \wedge P \ \mathcal{T} \ \mathcal{P}_2 \ \mathcal{T} \ P'. \qquad \text{(B.4)}$$

If $\mathcal{P}_2 \equiv P'$, we resort to Case 1 with the path $\mathcal{P}^{(2)}$ taking into account that $\mathcal{P}^{(2)}_1 \equiv P'\surd\!\downarrow$ holds. We obtain that there is some Q'' such that $Q \stackrel{\tau}{\Longrightarrow} Q''$, $\gamma_{\mathrm{M}}(P',C) = \gamma_{\mathrm{M}}(Q''C)$, and $P' \ \mathcal{T} \ Q''$. Together with (B.4) this implies (B.1). Therefore we can assume that $\mathcal{P}_2 \stackrel{\tau}{\not\Longrightarrow} P'$. We can now make use of property (1') for the path $\mathcal{P}^{(2)}$ and the chain $\mathcal{P}^{(2)}_1 \equiv \mathcal{P}_2 \stackrel{\tau}{\not\Longrightarrow} P'$ taking into account that $\mathcal{P}^{(2)}_1 \equiv \mathcal{P}_2 \not\approx \mathcal{P}_3 \equiv \mathcal{P}^{(2)}_2$. We obtain that there is a path $\mathcal{Q}$ such that $Q \stackrel{\tau}{\Longrightarrow} \mathcal{Q}_{|\mathcal{Q}|}$, $\mathcal{Q}_1 \equiv P'$, and

$$(\mathcal{Q}_1 \not\approx \mathcal{Q}_2 \ \wedge \ |\mathcal{Q}|_{\not\approx} \leq |\mathcal{P}^{(2)}|_{\not\approx} = |\mathcal{P}|_{\not\approx}) \ \vee \ |\mathcal{Q}|_{\not\approx} < |\mathcal{P}^{(2)}|_{\not\approx} = |\mathcal{P}|_{\not\approx}.$$

If $|\mathcal{Q}|_{\not\approx} < |\mathcal{P}^{(2)}|_{\not\approx} = |\mathcal{P}|_{\not\approx}$ holds, we can directly use our induction hypotheses for the path $\mathcal{Q}$ taking $\mathcal{Q}_1 \equiv P'\surd\!\downarrow$ into account. We obtain that there is some $Q''\surd\!\downarrow$ such that

$$\mathcal{Q}_{|\mathcal{Q}|} \overset{\tau}{\Longrightarrow} Q'' \;\wedge\; \gamma_{\mathrm{M}}(P',C) = \gamma_{\mathrm{M}}(Q'',C)) \;\wedge\; P' \,\mathcal{T}\, Q''.$$

As a whole, we get $Q \overset{\tau}{\Longrightarrow} \mathcal{Q}_{|\mathcal{Q}|} \overset{\tau}{\Longrightarrow} Q''\surd\!\downarrow$, thus $Q \overset{\tau}{\Longrightarrow} Q''\surd\!\downarrow$ and, using (B.4),

$$\gamma_{\mathrm{M}}(P,C) = \gamma_{\mathrm{M}}(Q'',C) \;\wedge\; P \,\mathcal{T}\, Q''.$$

as required.
If, otherwise, $\mathcal{Q}_1 \not\approx \mathcal{Q}_2$ and $|\mathcal{Q}|_{\not\approx} \leq |\mathcal{P}^{(2)}|_{\not\approx} = |\mathcal{P}|_{\not\approx}$, we can resort to Case 1 with path $\mathcal{Q}$ taking (again) $\mathcal{Q}_1 \equiv P'\surd\!\downarrow$ into account. Once again, we obtain that there is some $Q''\surd\!\downarrow$ satisfying

$$\mathcal{Q}_{|\mathcal{Q}|} \overset{\tau}{\Longrightarrow} Q'' \;\wedge\; \gamma_{\mathrm{M}}(P',C) = \gamma_{\mathrm{M}}(Q'',C) \;\wedge\; P' \,\mathcal{T}\, Q''.$$

As above, this results in (B.1).

We have hence completed the proof of Lemma B.1.6.

Lemma B.1.7. *If $\mathcal{S}$ is a weak congruence up to $\approx$, then $\mathcal{S} \subseteq \simeq$.*

Proof. Lemma B.1.6 says that $\mathcal{T} = (\mathcal{S} \cup \approx)^*$ is a weak bisimulation, implying $\mathcal{S} \subseteq (\mathcal{S} \cup \approx)^* \subseteq \approx$. We can therefore deduce the following properties from $P\,\mathcal{S}\,Q$:

- Let $P \overset{a}{\longrightarrow} P'$. This implies (using (1) of Definition B.1.1) $Q \overset{a}{\Longrightarrow} Q'$ and $P'\,\mathcal{S}\,R \approx Q'$ for some $Q', R \in \mathrm{IMC}$. $P'\,\mathcal{S}\,R$ now implies (using $\mathcal{S} \subseteq \approx$) that $P' \approx R$ and thus $P' \approx Q'$.
- Symmetrically, $Q \overset{a}{\longrightarrow} Q'$ implies that there is some $P' \in \mathrm{IMC}$ such that $P \overset{a}{\Longrightarrow} P'$ and $P' \approx Q'$.
- Let $P\surd\!\downarrow$ and $Q\surd\!\downarrow$. Condition (2) of Definition B.1.1 implies $\gamma_{\mathrm{M}}(P,C) = \gamma_{\mathrm{M}}(Q,C)$ for each class $C \in \mathrm{IMC}/\mathcal{T}$. Since $\mathcal{T} \subseteq \approx$, Lemma B.1.2 implies $\gamma_{\mathrm{M}}(P,C) = \gamma_{\mathrm{M}}(Q,C)$ for each class $C \in \mathrm{IMC}/\approx$.
- Condition (3) of Definition B.1.1 gives rise to $P\surd\!\downarrow$ iff $Q\surd\!\downarrow$.

These four properties are equivalent to $P \simeq Q$ (Definition 5.2.4).

We are still collecting the necessary means to prove the congruence property for recursion. In order to prove $\underline{X{:=}E} \simeq \underline{X{:=}F}$ if X is the only free variable in E and F it is sufficient to construct a weak congruence up to $\approx$ containing the pair $(\underline{X{:=}E}\,, \underline{X{:=}F}\,)$. We will construct such a relation $\mathcal{S}$ in Lemma B.1.13. First, we state some simple lemmas. In all of them we assume that $G, H, E \in \mathrm{IML}$ and $X \in \mathcal{V}$. Lemma B.1.8 can be proven by structural induction on $\mathrm{IML}_{\downarrow}$, whereas Lemma B.1.9 to Lemma B.1.12 require induction on the heights of the proof trees of the involved transitions.

Lemma B.1.8. *$G\{E/X\}\!\downarrow$ iff $G\!\downarrow$ and ($E\!\downarrow$ or X weakly guarded in G).*

Lemma B.1.9. *If $G \overset{a}{\longrightarrow} H$, then $G\{E/X\} \overset{a}{\longrightarrow} H\{E/X\}$.*

Lemma B.1.10. *If $E \overset{a}{\longrightarrow} E'$ and X is fully unguarded in G, then $G\{E/X\} \overset{a}{\longrightarrow} E'$.*

Lemma B.1.11. *Let $G\{E/X\} \xrightarrow{a} F$. This implies that X is fully unguarded in G and $E \xrightarrow{a} F$ or that $G \xrightarrow{a} H$ and $F \equiv H\{E/X\}$. Furthermore, if X is strongly guarded in G and $a = \tau$, then X is strongly guarded in H.*

From the last four lemmas it follows immediately that

$$G\{E/X\}\surd\!\!\downarrow \;\Leftrightarrow\; (G\surd\!\!\downarrow \wedge (X \text{ weakly guarded in } G \vee E\surd\!\!\downarrow)). \qquad \text{(B.5)}$$

Lemma B.1.12. *Assume that $G\downarrow$, $G \overset{\lambda_i}{\dashrightarrow} G_i$ for $1 \le i \le n$ and $E \overset{\mu_j}{\dashrightarrow} E_j$ for $1 \le j \le m$ and furthermore assume that these are all $\dashrightarrow$-transitions emanating G and E, respectively.*

Then $G\{E/X\} \overset{\lambda_i}{\dashrightarrow} G_i\{E/X\}$ for $1 \le i \le n$. Furthermore, if k is the number of fully unguarded occurrences of X in E, then, for $1 \le j \le m$, the transition $G\{E/X\} \overset{\mu_j}{\dashrightarrow} E_j$ is contained in the multi-relation $\dashrightarrow$ k times. No other $\dashrightarrow$-transitions emanate $G\{E/X\}$.

Lemma B.1.13. *Let $\mathcal{V}(E) \cup \mathcal{V}(F) \subseteq \{X\}$ and $E \simeq F$. Furthermore let $\mathcal{R} = \{(G\{\underline{x{:=}E}/X\}, G\{\underline{x{:=}F}/X\}) \mid \mathcal{V}(G) \subseteq \{X\}\}$. Then, $\mathcal{S} = (\mathcal{R} \cup \mathcal{R}^{-1})$ is a weak congruence up to $\approx$.*

Sketch of Proof. We have to check condition (1)-(3) of Definition B.1.1, which is only necessary for pairs in $\mathcal{R}$ due to symmetry. Each condition involves a case analysis concerning the outermost operator of G. Condition (1) is shown by an induction on the height of the proof tree for $G\{\underline{x{:=}E}/X\} \xrightarrow{a} P'$. To show (2) we perform an induction on the sum of the heights of all proof trees for $\dashrightarrow$ transitions emanating $G\{\underline{x{:=}E}/X\}$. Structural induction on the definition of $\mathsf{IML}_\downarrow$ is needed to show condition (3). All cases except of $G \equiv X$ are straightforward. Thus, we only check condition (1)-(3) of Definition B.1.1 for the pair $(\underline{x{:=}E}, \underline{x{:=}F})$.

Assume $\underline{x{:=}E} \xrightarrow{a} P$. Thus $E\{\underline{x{:=}E}/X\} \xrightarrow{a} P$, which can be derived by a smaller proof tree. Thus, the induction hypothesis implies $E\{\underline{x{:=}F}/X\} \overset{a}{\Longrightarrow} Q'$ and $P\;\mathcal{S}\;R{\approx}Q'$ for some Q'. Since $E \simeq F$, i.e., (using Definition 5.2.3) $E\{\underline{x{:=}F}/X\} \simeq F\{\underline{x{:=}F}/X\}$ this implies $F\{\underline{x{:=}F}/X\} \overset{a}{\Longrightarrow} Q$ and $Q'{\approx}Q$ for some Q and thus finally $\underline{x{:=}F} \overset{a}{\Longrightarrow} Q$ and $P\;\mathcal{S}\;R{\approx}Q'{\approx}Q$.

In order to verify condition (2) of Definition B.1.1 assume $\underline{x{:=}E}\surd\!\!\downarrow$ and $\underline{x{:=}F}\surd\!\!\downarrow$. Thus, $E\{\underline{x{:=}E}/X\}\surd\!\!\downarrow$, $F\{\underline{x{:=}F}/X\}\surd\!\!\downarrow$. Furthermore, $\underline{x{:=}E}\downarrow$ implies that X is weakly guarded in E. (B.5) on page 181 implies $E\{\underline{x{:=}F}/X\}\surd\!\!\downarrow$. Therefore, for every $C \in \mathsf{IMC}/(\mathcal{S} \cup \approx)^*$

$$\begin{aligned}
\gamma_{\mathrm{M}}(\underline{x{:=}E}, C) &= \quad \text{(recursion rule of Table 5.1)}\\
\gamma_{\mathrm{M}}(E\{\underline{x{:=}E}/X\}, C) &= \quad \text{(induction hypothesis)}\\
\gamma_{\mathrm{M}}(E\{\underline{x{:=}F}/X\}, C) &= \quad (E\{\underline{x{:=}F}/X\} \simeq F\{\underline{x{:=}F}/X\}, \text{Lemma B.1.2})
\end{aligned}$$

$$\gamma_M(F\{\underline{X := F}/X\}, C) = \text{(recursion rule of Table 5.1)}$$
$$\gamma_M(\underline{X := F}, C).$$

Finally condition (3) of Definition B.1.1 can be shown as follows. Assume $\underline{X := E}\surd\!\downarrow$. As in the proof of condition (2) it follows $E\{\underline{X := F}/X\}\surd\!\downarrow$. Since $E \simeq F$, i.e., $E\{\underline{X := F}/X\} \simeq F\{\underline{X := F}/X\}$, this implies $F\{\underline{X := F}/X\}\surd\!\downarrow$, i.e., $\underline{X := F}\surd\!\downarrow$.

Eventually, we have all the means to derive that $\simeq$ is a congruence with respect to recursion.

Corollary B.1.1. *If $E, F \in$ IML, then $E \simeq F$ implies $\underline{X := E} \simeq \underline{X := F}$.*

Proof. Because of Definition 5.2.3 it is sufficient to consider only those $E, F \in$ IML where $\mathcal{V}(E) \cup \mathcal{V}(F) \subseteq \{X\}$. Assume that $E \simeq F$ holds. Then the relation $\mathcal{S}$ appearing in Lemma B.1.13 is a weak congruence up to $\approx$. Choosing $G \equiv X$ implies $\underline{X := E}\ \mathcal{S}\ \underline{X := F}$. Lemma B.1.7 now implies $\underline{X := E} \simeq \underline{X := F}$.

B.2 Theorem 5.3.2

For each weakly guarded expression E there is some weakly guarded SES in the free variables of E that E $\mathcal{A}_\sim$-provably satisfies.

Proof. The proof is by induction on the structure of E, it exhibits no differences to the proof of [142, Theorem 5.9]. We include some details here, since we will make use of them later, in the weak congruence case.

First consider the base cases: If E is 0, $\perp$ or Y choose an SES S with a single equation $X_1 := E$. This is a weakly (even strongly) guarded SES, and $\mathcal{A}_\sim \vdash E = E\{E/X_1\}$ is trivially satisfied.

Now, assume $E \equiv \mathsf{a}.E$. By induction there is a weakly guarded SES $S' = \{\boldsymbol{X}' := \boldsymbol{F}'\}$ $\mathcal{A}_\sim$-provably satisfied by expression E. Define a new SES S by a leading equation $X_1 := F_1$, where $F_1 \equiv \mathsf{a}.F_1'$, and appending S', i.e., fix $X_2 \equiv X_1'$ and $F_2 \equiv F_2'$, $X_3 \equiv X_2'$ and $F_3 \equiv F_2'$, and so on. This SES is weakly guarded and it is $\mathcal{A}_\sim$-provably satisfied by E. If $E \equiv (\lambda).E$ use the same construction, but let $F_1 \equiv (\lambda).F_1'$.

If $E \equiv E' + E''$, then we have, by assumption, two weakly guarded SES, say S' and S'', $\mathcal{A}_\sim$-provably satisfied by E', respectively E''. We can assume that the formal variables of S' and S'' are distinct. Define an ES S by setting $X_1 := F_1$, where $F_1 := X_1' + X_1''$ and appending S' and S''. This ES is a weakly guarded SES since S' and S'' are. In addition, E $\mathcal{A}_\sim$-provably satisfies S.

The only remaining case (and indeed the only nontrivial one) is that $E \equiv \underline{Y := E'}$, where Y has to be weakly guarded in E' (otherwise E would not be weakly guarded). The free variables of E' are those of E possibly extended with variable Y. By assumption a weakly guarded SES $S' = \{\boldsymbol{X} := \boldsymbol{F}'\}$ in the free variables of E' exists that is $\mathcal{A}_\sim$-provably satisfied by E'. In other words, there is some vector of expressions $\boldsymbol{E}' = (E_1, \ldots, E'_n)$ such that $\mathcal{A}_\sim \vdash E'_i = F'_i\{\boldsymbol{E}'/\boldsymbol{X}\}$ and $\mathcal{A}_\sim \vdash E' = E'_1$.

Define an ES S by setting $X_i := F_i$ where $F_i \equiv F'_i\{F'_1/Y\}$. This implies $F_1 \equiv F'_1$, i.e leaves the leading equation unchanged, since Y is weakly guarded in E' and thus cannot be a summand of F'_1. For the same reason, it is clear that S is a weakly guarded standard ES and that it has free variables in those of E. It remains to be shown that E $\mathcal{A}_\sim$-provably satisfies S. To verify this, we have to construct a vector $\boldsymbol{E} = (E_1, \ldots . E_n)$ such that $\mathcal{A}_\sim \vdash E = E_1$ and $\mathcal{A}_\sim \vdash E_i = F_i\{\boldsymbol{E}/\boldsymbol{X}\}$.

In order to achieve these two properties, let $\boldsymbol{E} \equiv \boldsymbol{E}'\{E'_1/Y\}$. Apply (:=2) and $\mathcal{A}_\sim \vdash E' = E'_1$ in order to obtain the first of the required properties:

$$\mathcal{A}_\sim \vdash E = \underline{{\scriptstyle Y:=}\, E'} = E'\{E'/Y\} = E'_1\{E'_1/Y\} = E_1.$$

The second property requires some standard properties of nested substitution, as follows, completing the proof.

$$\begin{array}{lll}
\mathcal{A}_\sim \vdash E_i & = & (\text{definition of } E_i) \\
\quad E'_i\{E'_1/Y\} & = & (\text{by assumption } \mathcal{A}_\sim \vdash E'_i = F'_i\{\boldsymbol{E}'/\boldsymbol{X}\}) \\
\quad F'_i\{\boldsymbol{E}'/\boldsymbol{X}\}\{E'_1/Y\} & = & (\mathcal{A}_\sim \vdash F'_i = F'_i\{F'_1/Y\}) \\
\quad F'_i\{F'_1/Y\}\{\boldsymbol{E}'/\boldsymbol{X}\}\{E'_1/Y\} & = & (Y \text{ not contained in } \boldsymbol{X}) \\
\quad F'_i\{F'_1/Y\}\{\boldsymbol{E}'\{E'_1/Y\}/\boldsymbol{X}\} & = & (\text{definition of } F_i) \\
\quad F_i\{(\boldsymbol{E}'\{E'_1/Y\}/\boldsymbol{X}\} & = & (\text{definition of } \boldsymbol{E}) \\
\quad F_i\{(\boldsymbol{E}/\boldsymbol{X})\}. & &
\end{array}$$

B.3 Theorem 5.3.3

Let E and F be two strongly bisimilar expressions, i.e., $E \sim F$. Furthermore let E $\mathcal{A}_\sim$-provably satisfy the SES S_1 and F $\mathcal{A}_\sim$-provably satisfy the SES S_2, where both S_1 and S_2 are weakly guarded. Then there is some weakly guarded SES S, that both E and F $\mathcal{A}_\sim$-provably satisfy.

Proof. Assume $E \sim F$ and let $S_1 = (\boldsymbol{X} := \boldsymbol{G})$ and $S_2 = (\boldsymbol{Y} := \boldsymbol{H})$ be weakly guarded SES such that there are expressions $\boldsymbol{E} = (E_1, \ldots, E_n)$ and $\boldsymbol{F} = (F_1, \ldots, F_n)$ satisfying

$$\begin{array}{l}
\mathcal{A}_\sim \vdash E = E_1, \\
\mathcal{A}_\sim \vdash F = F_1, \\
\mathcal{A}_\sim \vdash E_i = G_i\{\boldsymbol{E}/\boldsymbol{X}\}, \text{ and} \\
\mathcal{A}_\sim \vdash F_j = H_j\{\boldsymbol{F}/\boldsymbol{Y}\}.
\end{array}$$

Each G_i and H_j is of the following form:

$$G_i \equiv \sum_{k_1=1}^{r_1(i)} (\lambda_{i,k_1}).X_{f_1(i,k_1)} + \sum_{l_1=1}^{s_1(i)} \mathsf{a}_{i,l_1}.X_{g_1(i,l_1)} + \sum_{m_1=1}^{t_1(i)} W_{h_1(i,m_1)} + \sum_{n_1=1}^{u_1(i)} \bot\,,$$

$$H_j \equiv \sum_{k_2=1}^{r_2(j)} (\mu_{j,k_2}).Y_{f_2(j,k_2)} + \sum_{l_2=1}^{s_2(j)} \mathsf{b}_{j,l_2}.Y_{g_2(j,l_2)} + \sum_{m_2=1}^{t_2(j)} W_{h_2(j,m_2)} + \sum_{n_2=1}^{u_2(j)} \bot\,.$$

From law (N) and law $(I2)$ we derive that $\bot + \bot = \bot$, whence we can assume that $u_1(i),\ u_2(j) \in \{0, 1\}$. Now, define the following sets:

- $I = \{(i,j) \mid E_i \sim F_j\}$,
- $J_{i,j} = \{(k_1,k_2) \in \{1,\dots,r_1(i)\} \times \{1,\dots,r_2(j)\} \mid (f_1(i,k_1), f_2(j,k_2)) \in I\}$,
- $K_{i,j} = \{(l_1,l_2) \in \{1,\dots,s_1(i)\} \times \{1,\dots,s_2(j)\} \mid \mathsf{a}_{i,l_1} = \mathsf{b}_{j,l_2} \wedge (g_1(i,l_1), g_2(j,l_2)) \in I\}$.

Since $E \sim F$ (and soundness of $\mathcal{A}_{\sim}$) we have $(1,1) \in I$. In addition, $(i,j) \in I$ implies, by assumption (and soundness), $G_i\{\boldsymbol{E}/\boldsymbol{X}\} \sim H_j\{\boldsymbol{E}/\boldsymbol{Y}\}$.

In the sequel we use $X_i\surd$, $X_i\not\surd$, $X_i\downarrow$, $X_i\uparrow$ to denote $G_i\surd$, $G_i\not\surd$, $G_i\downarrow$, $G_i\uparrow$, respectively, and similar with Y_j and H_j. With this notation we are able to deduce that the following important properties:

If $(i,j) \in I$, then for all $\mathsf{a}, \mathsf{b} \in Act$ and $C \in \mathsf{IML}/\sim$,

(i) $X_i \xrightarrow{\mathsf{a}}_{S_1} X_k$ implies that there is some l such that $Y_j \xrightarrow{\mathsf{a}}_{S_2} Y_l$ and $(k,l) \in I$,
(ii) $Y_j \xrightarrow{\mathsf{b}}_{S_2} Y_l$ implies that there is some k such that $X_i \xrightarrow{\mathsf{b}}_{S_1} X_k$ and $(k,l) \in I$,
(iii) $X_i\surd\downarrow$ implies $Y_j\surd\downarrow$ and $\gamma_{\mathrm{M}}(G_i\{\boldsymbol{E}/\boldsymbol{X}\}, C) = \gamma_{\mathrm{M}}(H_j\{\boldsymbol{F}/\boldsymbol{Y}\}, C))$,
(iv) $Y_j\surd\downarrow$ implies $X_i\surd\downarrow$ and $\gamma_{\mathrm{M}}(G_i\{\boldsymbol{E}/\boldsymbol{X}\}, C) = \gamma_{\mathrm{M}}(H_j\{\boldsymbol{F}/\boldsymbol{Y}\}, C))$,
(v) $\{W_{h_1(i,1)}, \dots, W_{h_1(i,t_1(i))}\} = \{W_{h_2(j,1)}, \dots, W_{h_2(j,t_2(j))}\}$. In addition, if $X_i\surd\downarrow$ (and $Y_j\surd\downarrow$), then
$\{\!|W_{h_1(i,1)}, \dots, W_{h_1(i,t_1(i))}|\!\} = \{\!|W_{h_2(j,1)}, \dots, W_{h_2(j,t_2(j))}|\!\}$.

Assuming $(i,j) \in I$, property (i) implies that for each $l_1 \in \{1,\dots,s_1(i)\}$ there is some $l_2 \in \{1,\dots,s_2(j)\}$ such that $\mathsf{a}_{i,l_1} = \mathsf{b}_{j,l_2}$ and $(g_1(i,l_1), g_2(j,l_2)) \in I)$, and symmetrically by property (ii) with l_2 and l_1 reversed. In other words, $K_{i,j}$ is a total and surjective relation.

In addition to $(i,j) \in I$, let us now assume $X_i\surd\downarrow$ and $Y_j\surd\downarrow$. Property (iii) (or (iv)) implies

$$(\forall C \in \mathsf{IML}/\sim)\ \gamma_{\mathrm{M}}(G_i\{\boldsymbol{E}/\boldsymbol{X}\}, C) = \gamma_{\mathrm{M}}(H_j\{\boldsymbol{F}/\boldsymbol{Y}\}, C).$$

Let $(k_1,k_2) \in J_{i,j}$. i.e., $E_{f_1(i,k_1)} \sim F_{f_2(j,k_2)}$. Then we know, by the definition of $J_{i,j}$, that

$$E_{f_1(i,k_1)} \sim E_{f_1(i,u)} \quad \text{iff} \quad (u,k_2) \in J_{i,j},$$

and symmetrically

$$E_{f_2(j,k_2)} \sim E_{f_2(j,v)} \quad \text{iff} \quad (k_1,v) \in J_{i,j}.$$

Hence, we are able to deduce:

$$\sum_{\substack{u=1\\(u,k_2)\in J_{i,j}}}^{r_1(i)} \lambda_{i,u} = \gamma_{\mathrm{M}}(G_i\{\boldsymbol{E}/\boldsymbol{X}\},[E_{f_1(i,k_1)}]_\sim) = \gamma_{\mathrm{M}}(H_j\{\boldsymbol{F}/\boldsymbol{Y}\},[F_{f_2(j,k_2)}]_\sim) = \sum_{\substack{v=1\\(k_1,v)\in J_{i,j}}}^{r_2(j)} \mu_{j,u}.$$

We abbreviate this sum with $\gamma^{i,j}_{k_1,k_2}$ (though this suggests a dependence from both k_1 and k_2 that obviously is not existing). We are now in the position to create a standard equation set $S = \{Z_{i,j} = L_{i,j} \mid (i,j) \in I\}$ with new formal variables $Z_{i,j}$ (for each pair $(i,j) \in I$) that serves our purposes. Let

$$L_{i,j} \equiv \sum_{(l_1,l_2)\in K_{i,j}} \mathsf{b}_{j,l_2}.Z_{g_1(i,l_1),g_2(j,l_2)} + \sum_{m_2=1}^{t_2(j)} W_{h_2(j,m_2)} + \begin{cases} \displaystyle\sum_{(k_1,k_2)\in J_{i,j}} \left(\frac{\lambda_{i,k_1}\cdot\mu_{j,k_2}}{\gamma^{i,j}_{k_1,k_2}}\right).Z_{f_1(i,k_1),f_2(j,k_2)} & \text{if } X_i\surd\!\downarrow \text{ and } Y_j\surd\!\downarrow \\ \bot & \text{if } X_i\uparrow \text{ or } Y_j\uparrow \\ 0 & \text{else.} \end{cases}$$

Obviously, each $Z_{i,j}$ is weakly guarded in each $L_{i',j'}$ and hence S is weakly guarded. Our claim is that E $\mathcal{A}_\sim$-provably satisfies the SES S (where E will be equated with $Z_{1,1}$). The proof that F $\mathcal{A}_\sim$-provably satisfies S is completely symmetric. In order to show the former, we will prove that $\mathcal{A}_\sim \vdash C_{i,j} = L_{i,j}\{\boldsymbol{C}/\boldsymbol{Z}\}$, where $C_{i,j} \equiv E_i$. Note that $\mathcal{A}_\sim \vdash E = C_{1,1}$, since $\mathcal{A}_\sim \vdash E = E_1$.

We will transform $L_{i,j}\{\boldsymbol{C}/\boldsymbol{Z}\}$ stepwise into $C_{i,j}$, i.e., E_1, starting from

$$L_{i,j}\{\boldsymbol{C}/\boldsymbol{Z}\} \equiv \sum_{(l_1,l_2)\in K_{i,j}} \mathsf{a}_{i,l_1}.E_{g_1(i,l_1)} + \sum_{m_2=1}^{t_2(j)} W_{h_2(j,m_2)} + \begin{cases} \displaystyle\sum_{(k_1,k_2)\in J_{i,j}} \left(\frac{\lambda_{i,k_1}\cdot\mu_{j,k_2}}{\gamma^{i,j}_{k_1,k_2}}\right).E_{f_1(i,k_1)} & \text{if } X_i\surd\!\downarrow \text{ and } Y_j\surd\!\downarrow \\ \bot & \text{if } X_i\uparrow \text{ or } Y_j\uparrow \\ 0 & \text{else.} \end{cases}$$

First, the sum of free variables

$$\sum_{m_2=1}^{t_2(j)} W_{h_2(j,m_2)}$$

has to be transformed into

$$\sum_{m_1=1}^{t_1(i)} W_{h_1(i,m_1)}.$$

If $X_i\surd\!\downarrow$ holds, no transformation is required at all, as an immediate consequence of property (v). If, on the other hand, $X_i\!\uparrow$ or $X_i \,\not\!\!\!\downarrow$ holds, then property (v) ensures $\{W_{h_1(i,1)},\ldots,W_{h_1(i,t_1(i))}\} = \{W_{h_2(j,1)},\ldots,W_{h_2(j,t_2(j))}\}$. In the case of $X_i\!\uparrow$ we can use $\mathcal{A}_\sim \vdash E + E{+}\bot{=}\, E{+}\bot$ to add or remove as many variables from the former sum in order to match the latter. The above transformation is derivable from law $(I3)$ and law (N):

$$\mathcal{A}_\sim \vdash E + E{+}\bot{=}\, E + E + 0{+}\bot{=}\, E + E + 0 + 0{+}\bot{+}\bot{=}\, E{+}\bot$$

In the case of $X_i \,\not\!\!\!\downarrow$, we use $\mathcal{A}_\sim \vdash E + E + \tau.F = E + \tau.F$ instead to match both sums. This transformation can be derived from the former by two applications of law $(\bot 2)$.

Having equated the sums of free variables, we now turn our attention to the sum of delay prefixes. Assuming $X_i\surd\!\downarrow$ and $Y_j\surd\!\downarrow$ we can transform this sum, using law $(I1)$:

$$\begin{aligned}\mathcal{A}_\sim \vdash \sum_{(k_1,k_2)\in J_{i,j}} \left(\frac{\lambda_{i,k_1}\cdot\mu_{j,k_2}}{\gamma^{i,j}_{k_1,k_2}}\right).E_{f_1(i,k_1)} &= \sum_{k_1=1}^{r_1(i)} \sum_{\substack{k_2=1\\(k_1,k_2)\in J_{i,j}}}^{r_2(j)} \left(\frac{\lambda_{i,k_1}\cdot\mu_{j,k_2}}{\gamma^{i,j}_{k_1,k_2}}\right).E_{f_1(i,k_1)}\\ &= \sum_{k_1=1}^{r_1(i)} \left(\sum_{\substack{k_2=1\\(k_1,k_2)\in J_{i,j}}}^{r_2(j)} \frac{\lambda_{i,k_1}\cdot\mu_{j,k_2}}{\gamma^{i,j}_{k_1,k_2}}\right).E_{f_1(i,k_1)}\end{aligned}$$

Note that in the last line, the right symbol '$\sum$' is a sum of real values (inside a delay prefix) rather than a choice operator. This sum can be drastically simplified:

$$\sum_{\substack{k_2=1\\(k_1,k_2)\in J_{i,j}}}^{r_2(j)} \frac{\lambda_{i,k_1}\cdot\mu_{j,k_2}}{\gamma^{i,j}_{k_1,k_2}} = \lambda_{i,k_1}\cdot \sum_{\substack{k_2=1\\(k_1,k_2)\in J_{i,j}}}^{r_2(j)} \frac{\mu_{j,k_2}}{\sum\limits_{\substack{u=1\\(k_1,u)\in J_{i,j}}}^{r_2(j)} \mu_{j,u}} = \lambda_{i,k_1}\cdot \frac{\sum\limits_{\substack{k_2=1\\(k_1,k_2)\in J_{i,j}}}^{r_2(j)} \mu_{j,k_2}}{\sum\limits_{\substack{u=1\\(k_1,u)\in J_{i,j}}}^{r_2(j)} \mu_{j,u}} = \lambda_{i,k_1}$$

Putting both transformations together we obtain

$$\mathcal{A}_\sim \vdash \sum_{(k_1,k_2)\in J_{i,j}} \left(\frac{\lambda_{i,k_1}\cdot\mu_{j,k_2}}{\gamma^{i,j}_{k_1,k_2}}\right).E_{f_1(i,k_1)} = \sum_{k_1=1}^{r_1(i)} (\lambda_{i,k_1}).E_{f_1(i,k_1)}.$$

In the remainder of this proof we use the following abbreviations.

$$D_1 \equiv \sum_{l_1=1}^{s_1(i)} \mathsf{a}_{i,l_1}.E_{g_1(i,l_1)} + \sum_{m_1=1}^{t_1(i)} W_{h_1(i,m_1)} \quad \text{and} \quad D_2 \equiv \sum_{k_1=1}^{r_1(i)} (\lambda_{i,k_1}).E_{f_1(i,k_1)}.$$

With this notation we have, in summary, achieved so far:

$$\mathcal{A}_\sim \vdash L_{i,j}\{\boldsymbol{C}/\boldsymbol{Z}\} = D_1 + \begin{cases} D_2 & \text{if } X_i\surd\downarrow \text{ and } Y_j\surd\downarrow \\ \bot & \text{if } X_i\uparrow \text{ or } Y_j\uparrow \\ 0 & \text{else.} \end{cases} \tag{B.6}$$

On the other hand, we derive from our assumption $\mathcal{A}_\sim \vdash E_i = G_i\{\boldsymbol{E}/\boldsymbol{X}\}$ that

- $X_i\downarrow$ implies $\mathcal{A} \vdash E_i = D_1 + D_2$, and
- $X_i\uparrow$ implies $\mathcal{A} \vdash E_i = D_1 + D_2 {+} \bot$.

Still, $\mathcal{A}_\sim \vdash L_{i,j}\{\boldsymbol{C}/\boldsymbol{Z}\} = C_{i,j}$ has to be checked in 16 different cases, resulting as a combination of $X_i\surd$, $X_i\uparrow$, $Y_j\surd$, $Y_j\uparrow$, and their respective negations. Fortunately six cases can be ruled out immediately and other cases can be grouped together.

- Case $X_i\surd \wedge X_i\downarrow \wedge Y_j\surd \wedge Y_j\uparrow$:
  This combination is impossible because of property (iii).
- Case $X_i\surd \wedge X_i\uparrow \wedge Y_j\surd \wedge Y_j\downarrow$:
  This is ruled out by property (iv).
- Case $X_i\surd \wedge Y_j \not\surd$ (covering four cases):
  These cases are also impossible, because $K_{i,j}$ is total and surjective, and therefore for each l_2 with $\mathsf{b}_{j,l_2} = \tau$ the existence of some l_1 such that $\mathsf{a}_{i,l_1} = \tau$ is guaranteed.
- Case $X_i\surd \wedge X_i\uparrow \wedge Y_j\surd \wedge Y_j\uparrow$:
  Then we have
  $$\mathcal{A}_\sim \vdash L_{i,j}\{\boldsymbol{C}/\boldsymbol{Z}\} = D_1{+}\bot$$
  which can be transformed to the desired result $C_{i,j}$ with the help of $r_1(i)$ applications of law $(\bot 1)$:
  $$\mathcal{A}_\sim \vdash D_1{+}\bot = D_1 + D_2{+}\bot = E_i \equiv C_{i,j}.$$
- Case $X_i\surd \wedge X_i\downarrow \wedge Y_j\surd \wedge Y_j\downarrow$:
  In this case, $\mathcal{A}_\sim \vdash L_{i,j}\{\boldsymbol{C}/\boldsymbol{Z}\} = D_1 + D_2 = E_i \equiv C_{i,j}$ holds immediately.
- Case $X_i \not\surd \wedge X_i\downarrow \wedge Y_j\downarrow$ (covering two cases):
  Then
  $$\mathcal{A}_\sim \vdash L_{i,j}\{\boldsymbol{C}/\boldsymbol{Z}\} = D_1.$$
  Since $X_i \not\surd$ holds, E_1 has to contain a summand of the form $\tau.E'$. We can therefore apply Lemma (5.3.1 $r_1(i)$ times) and obtain
  $$\mathcal{A}_\sim \vdash D_1 = D_1 + D_2 = E_i \equiv C_{i,j}.$$
- Case $X_i \not\surd \wedge X_i\downarrow \wedge Y_j\uparrow$ (covering two cases):
  We have
  $$\mathcal{A}_\sim \vdash L_{i,j}\{\boldsymbol{C}/\boldsymbol{Z}\} = D_1{+}\bot.$$

Again, E_1 has to contain a summand of the form $\tau.E'$. We can therefore use Lemma 5.3.1 as in the preceding case and eventually apply law ($\perp 2$).

$$\mathcal{A}_{\sim} \vdash D_1 + \perp = D_1 + D_2 + \perp = D_1 + D_2 = E_i \equiv C_{i,j}.$$

- Case $X_i \not\Downarrow \wedge X_i \Uparrow$ (covering four cases):
  Then $\mathcal{A}_{\sim} \vdash L_{i,j}\{\boldsymbol{C}/\boldsymbol{Z}\} = D_1 + \perp$ and we can obtain the required result by applying law ($\perp 1$) $r_1(i)$ times.

$$\mathcal{A}_{\sim} \vdash D_1 + \perp = D_1 + D_2 + \perp = E_i \equiv C_{i,j}.$$

We have thus shown that E $\mathcal{A}_{\sim}$-provably satisfies the strongly guarded SES S. As mentioned above, the proof that F $\mathcal{A}_{\sim}$-provably satisfies S is completely symmetric and therefore omitted. This completes the proof of Theorem 5.3.3.

B.4 Theorem 5.3.6

For arbitrary expressions E and F it holds that

$$\mathcal{A}_{\simeq} \vdash E = F \qquad \textit{implies} \qquad E \simeq F.$$

For most of the laws, Definition 5.2.4 is easily verified. We will only discuss the laws for recursion. Law ($:=1$) states that bound variables can be renamed if no additional bindings are introduced, and $\underline{X{:=}\,E} \simeq \underline{Y{:=}\,E\{Y/X\}}$ obviously holds. Law ($:=2$) is immediate from the structure of the operational rules for recursion (Table 5.1) and the definition of the predicate $\downarrow$, Definition 5.1.4. A central law, on the other hand, is ($:=4'$), as discussed in Section 5.3. We develop a detailed soundness proof for ($:=4'$) closely following the proof of congruence with respect to recursion (Appendix B.1). Corollary B.4.1 will be the final point to show soundness of this law. Afterwards we focus on the remaining laws for recursion (($:=5$), ($:=6$), ($:=8$)) that are specific for the treatment of divergence in the presence of maximal progress. Compared to Appendix B.1we make use of a similar 'up to' technique but we use a slightly different definition of 'weak congruence up to $\approx$' than the one given in Definition B.1.1. Instead of introducing a different name we abuse notation for the remainder of this appendix and redefine 'weak congruence up to $\approx$' as follows.

Definition B.4.1. *Let $\mathcal{S}$ be a symmetric relation on IMC and $\mathcal{T} = (\mathcal{S} \cup \approx)^*$* [1]. *$\mathcal{S}$ is a weak congruence up to $\approx$ iff P $\mathcal{S}$ Q implies for all $\mathsf{a} \in Act$, $C \in$ IMC$/\mathcal{T}$ and $P' \in$ IMC that*

[1] Thus, $\mathcal{T}$ is an equivalence relation.

1. $P \overset{a}{\Longrightarrow} P'$ *implies* $Q \overset{a}{\Longrightarrow} Q'$ *and* $P' \approx R_1\ \mathcal{S}\ R_2 \approx Q'$ *for some* $R_1, R_2, Q' \in$ IMC,
2. $P\surd\downarrow$ *(or* $Q\surd\downarrow$*) implies* $\gamma_M(P,C) = \gamma_M(Q,C)$,
3. *If* $\mathcal{P}$ *is a* $(\mathcal{S} \cup \approx)$*-path and* $\mathcal{P}_1\surd\downarrow$*, then there is some* $(\mathcal{S} \cup \approx)$*-path* $\mathcal{P}'$ *such that* $\mathcal{P}'_1 \equiv \mathcal{P}_1$, $\mathcal{P}'_i\surd\downarrow$ *for all* $i \in \{1, \ldots, |\mathcal{P}'|\}$*, and* $\mathcal{P}_{|\mathcal{P}|} \overset{\tau}{\Longrightarrow} \mathcal{P}'_{|\mathcal{P}'|}$,
4. $P\surd\downarrow$ *iff* $Q\surd\downarrow$.

Lemma B.4.1. *If* $\mathcal{S}$ *is a weak congruence up to* $\approx$*, then* $\mathcal{S} \subseteq \simeq$.

Proof. The strategy is analogous to that in Appendix B.1. Apart from proving that $\mathcal{T} = (\mathcal{S} \cup \approx)^*$ is a weak bisimulation the proof follows the lines of that for Lemma B.1.7.

Let $P\ \mathcal{T}\ Q$. Condition (1) of Lemma B.1.1 directly follows from the fact that if $P \overset{a}{\Longrightarrow} P'$ and $\mathcal{P}$ is an $(\mathcal{S} \cup \approx)$–path with $\mathcal{P}_1 \equiv P$ and $\mathcal{P}_{|\mathcal{P}|} \equiv Q$, then there is some $(\mathcal{S} \cup \approx)$–path $\mathcal{Q}$ such that $\mathcal{Q}_1 \equiv P'$ and $Q \overset{a}{\Longrightarrow} \mathcal{Q}_{|\mathcal{Q}|}$. To prove this observation requires a simple induction on $|\mathcal{P}|$ using condition (1) of Definition B.4.1.

In order to show condition (2) of Lemma B.1.1 we use properties (2) and (3) of Definition B.4.1 as follows. Let $P\,\mathcal{T}\,Q$ and $P\surd\downarrow$ and let $\mathcal{P}$ be an $(\mathcal{S} \cup \approx)$–path satisfying $\mathcal{P}_1 \equiv P$ and $\mathcal{P}_{|\mathcal{P}|} \equiv Q$. (3) implies the existence of some $(\mathcal{S} \cup \approx)$–path $\mathcal{P}'$ such that $\mathcal{P}'_1 \equiv P$, $\mathcal{P}'_i\surd\downarrow$ for all $i \in \{1, \ldots, |\mathcal{P}'|\}$, and $Q \overset{\tau}{\Longrightarrow} \mathcal{P}'_{|\mathcal{P}'|}$. Using (2) of Definition B.4.1 we can show $\gamma_M(P,C) = \gamma_M(\mathcal{P}'_i, C)$ for every $i \in \{1, \ldots, |\mathcal{P}'|\}$ and $C \in \mathrm{IMC}/\mathcal{T}$.

Lemma B.4.2. *Let* $E \in$ IML *and* $P, Q \in$ IMC*, with* X *strongly guarded in* E *and* $\mathcal{V}(E) \subseteq \{X\}$*. Furthermore, suppose* $P \simeq E\{P/X\}$ *and* $Q \simeq E\{Q/X\}$ *and define the relation* $\mathcal{R} = \{(G\{P/X\}, G\{Q/X\}) \mid \mathcal{V}(G) \subseteq \{X\}\}$*. Then the relation* $\mathcal{S} = (\mathcal{R} \cup \mathcal{R}^{-1})$ *is a weak congruence up to* $\approx$.

Proof. We have to check conditions (1) - (4) of Definition B.4.1. For symmetry reasons it is sufficient to restrict ourselves to pairs contained in $\mathcal{R}$. We let $\mathcal{T}$ denote $(\mathcal{S} \cup \approx)^*$.

(1): The proof is identical to the proof of Proposition 13 in [145] (only for (1), the fact that X is strongly guarded in E is needed).

(2): Let $\mathcal{V}(G) \subseteq \{X\}$, $G\{P/X\}\surd\downarrow$, and $G\{Q/X\}\surd\downarrow$. We have to show $\gamma_M(G\{P/X\}, C) = \gamma_M(G\{Q/X\}, C)$ for all $C \in \mathrm{IMC}/\mathcal{T}$. We fix some $C \in \mathrm{IMC}/\mathcal{T}$. We first prove the following weaker property (2').

If X is weakly guarded in $G\downarrow$ and $\mathcal{V}(G) \subseteq \{X\}$,
then $\gamma_M(G\{P/X\}, C) = \gamma_M(G\{Q/X\}, C)$.

Assuming weak guardedness of X in G, (i.e., absence of fully unguarded occurrences), Lemma B.1.12 implies that for each transition $G\{P/X\} \overset{\lambda}{\dashrightarrow} G'\{P/X\}$ there is a unique corresponding transition $G\{Q/X\} \overset{\lambda}{\dashrightarrow} G'\{Q/X\}$ and vice versa. Since $G'\{P/X\}\ \mathcal{S}\ G'\{Q/X\}$, we obtain $\gamma_M(G\{P/X\}, D) = \gamma_M(G\{Q/X\}, D)$ for every $D \in \mathrm{IMC}/\mathcal{S}^*$. This in turn implies (2'), because of $\mathcal{S}^* \subseteq \mathcal{T}$ and Lemma B.1.2.

Concerning (2) we can derive from $P \simeq E\{P/X\}$, $Q \simeq E\{Q/X\}$ that $G\{P/X\} \simeq G\{E/X\}\{P/X\}$ as well as $G\{Q/X\} \simeq G\{E/X\}\{Q/X\}$. Because of $\approx \subseteq \mathcal{T}$ and Lemma B.1.2 we have $\gamma_{\mathrm{M}}(G\{P/X\},C) = \gamma_{\mathrm{M}}(G\{E/X\}\{P/X\},C)$ and $P\gamma_{\mathrm{M}}(G\{Q/X\},C) = \gamma_{\mathrm{M}}(G\{E/X\}\{Q/X\},C)$. Since X is strongly guarded in $G\{E/X\}$ (in fact, we need only need weak guardedness) we can apply (2') to obtain $\gamma_{\mathrm{M}}(G\{E/X\}\{P/X\},C) = \gamma_{\mathrm{M}}(G\{E/X\}\{Q/X\},C)$ implying (2).

(3): We show the following generalisation (3').

Let $\mathcal{P}$ be a $(\mathcal{S} \cup \approx)$–path such that $\mathcal{P}_1 \stackrel{\tau}{\Longrightarrow} P'\surd\!\!\downarrow$. Then there is some $(\mathcal{S} \cup \approx)$–path $\mathcal{P}'$ such that $\mathcal{P}'_1 \equiv P'$, $\mathcal{P}_{|\mathcal{P}|} \stackrel{\tau}{\Longrightarrow} \mathcal{P}'_{|\mathcal{P}'|}$ and $\mathcal{P}'_i\surd\!\!\downarrow$ for every $i \in \{1,\ldots,|\mathcal{P}'|\}$.

W.l.o.g., assume that for every $i \in \{1,\ldots,|\mathcal{P}|\}$ it holds that if $\mathcal{P}_i \not\approx \mathcal{P}_{i+1}$ there is some H satisfying

$$\mathcal{P}_i \equiv H\{P/X\} \wedge \mathcal{P}_{i+1} \equiv H\{Q/X\} \wedge X \text{ strongly guarded in } H. \quad \text{(B.7)}$$

(This can be achieved by replacing all subpaths in $\mathcal{P}$ of the form $(G\{P/X\}, G\{Q/X\})$ satisfying $G\{P/X\} \;/\!\!\approx\; G\{Q/X\}$ but where X appears not strongly guarded in G by subpaths of the form $(G\{P/X\}, G\{E/X\}\{P/X\}, G\{E/X\}\{Q/X\}, G\{Q/X\})$.) We prove (3') by an induction on $|\mathcal{P}|$. The base case $|\mathcal{P}| = 0$ is trivial. We distinguish two cases.

- If $\mathcal{P}_1 \approx \mathcal{P}_2$, then $\mathcal{P}_1 \stackrel{\tau}{\Longrightarrow} P'$ implies that there is some R'' such that $\mathcal{P}_2 \stackrel{\tau}{\Longrightarrow} R''$ and $P' \approx R''$. From $P'\surd\!\!\downarrow$ and $P' \approx R''$ we obtain that there is some R' such that $R'' \stackrel{\tau}{\Longrightarrow} R'\surd\!\!\downarrow$. Now, $P' \approx R''$ and $P'\surd$ imply $P' \approx R'$. We have thus obtained $\mathcal{P}_2 \stackrel{\tau}{\Longrightarrow} R'\surd\!\!\downarrow$. Applying the induction hypothesis to the path $\mathcal{P}^{(2)}$ and the chain of transitions $\mathcal{P}_2 \stackrel{\tau}{\Longrightarrow} R'$ returns a path $\mathcal{P}''$ satisfying the conditions of (3'). Thus $\mathcal{P}' = (P', R')\,\mathcal{P}''$ is a path as required in (3').
- If, otherwise, $\mathcal{P}_1 \not\approx \mathcal{P}_2$ holds, (B.7) says that there is some $H \in \mathrm{IML}$ such that $\mathcal{P}_1 \equiv H\{P/X\}$, $\mathcal{P}_2 \equiv H\{Q/X\}$ and X strongly guarded in H. Iterative applications of Lemma B.1.9 and B.1.11 to the chain of transitions $\mathcal{P}_1 \stackrel{\tau}{\Longrightarrow} P'$ leads to some R' such that $\mathcal{P}_2 \stackrel{\tau}{\Longrightarrow} R'$ and $P' \;\mathcal{S}\; R'$. Anticipating that we will show below that (4) of Definition B.4.1 holds, $P'\surd\!\!\downarrow$ and $P' \;\mathcal{S}\; R'$ implies $R'\surd\!\!\downarrow$. We can therefore apply the induction hypothesis to the path $\mathcal{P}^{(2)}$ and the chain of transitions $\mathcal{P}_2 \stackrel{\tau}{\Longrightarrow} R'\surd\!\!\downarrow$. We obtain a path $\mathcal{P}''$ such that $\mathcal{P}' = (P', R')\,\mathcal{P}''$ satisfies the requirements of (3').

(4): It holds

$$\begin{aligned} G\{P/X\}\surd\!\!\downarrow \Leftrightarrow\; & (G\{P/X\} \simeq G\{E/X\}\{P/X\}) \\ G\{E/X\}\{P/X\}\surd\!\!\downarrow \Leftrightarrow\; & (X \text{ strongly guarded in } G\{E/X\}, \\ & \quad \text{using (B.5) on page 181}) \end{aligned}$$

$$\begin{array}{ll} G\{E/X\}\{Q/X\}\surd\!\!\downarrow \Leftrightarrow & (G\{E/X\}\{Q/X\} \simeq G\{Q/X\}) \\ \quad G\{Q/X\}\surd\!\!\downarrow & \end{array}$$

This completes the proof of Lemma B.4.2.

Lemma B.4.3. *Let $E \in \mathsf{IML}$ and $P, Q \in \mathsf{IMC}$, X strongly guarded in E and $\mathcal{V}(E) \subseteq \{X\}$. If $P \simeq E\{P/X\}$ and $Q \simeq E\{Q/X\}$, then $P \simeq Q$.*

Proof. We obtain $P \simeq Q$ by applying Lemma B.4.1 to the relation $\mathcal{R}$ appearing in Lemma B.4.2 where we choose G to be the variable X.

Lemma B.4.4. *Let $E \in \mathsf{IML}$ and $P \in \mathsf{IMC}$, X strongly guarded in E and $\mathcal{V}(E) \subseteq \{X\}$. If $P \simeq E\{P/X\}$, then $P \simeq \underline{{}_{X:=}E}$.*

Proof. $\mathcal{V}(E) \subseteq \{X\}$ implies $\underline{{}_{X:=}E} \in \mathsf{IMC}$ and soundness of $(_{:=}2)$ assures $\underline{{}_{X:=}E} \simeq E\{\underline{{}_{X:=}E}/X\}$. We can therefore use Lemma B.4.3 (with $Q \equiv \underline{{}_{X:=}E}$) to deduce $P \simeq \underline{{}_{X:=}E}$.

Definition 5.2.3 implies that this lemma holds also for arbitrary expressions in IML:

Corollary B.4.1. *$F \simeq E\{F/X\}$ implies $F \simeq \underline{{}_{X:=}E}$ provided X is* strongly *guarded in E.*

After having shown soundness of law $(_{:=}4')$ we now turn our attention towards the remaining laws of recursion that are specific for the treatment of divergence. We will discuss law $(_{:=}8)$ in some detail in order to give some insight into the proof. The proofs for law $(_{:=}5)$ and law $(_{:=}6)$ are instances of the same proof technique and will therefore be only sketched.

In the sequel let $E \in \mathsf{IML}$. To prove correctness of law $(_{:=}8)$ we have to show that $P \simeq Q$ where $P \equiv \underline{{}_{X:=}\tau.X + E}$ and $Q \equiv \underline{{}_{X:=}\tau.(\bot + E)}$. Definition 5.2.3 enables us to restrict ourselves to the case where $\mathcal{V}(E) \subseteq \{X\}$. We now define the following relations $\mathcal{B}_0$, $\mathcal{B}_1$, and $\mathcal{B}$

- $\mathcal{B}_0 = \{(H\{P/X\}, H\{Q/X\}) \mid \mathcal{V}(H) \subseteq \{X\}\}$, $\quad \mathcal{B}_1 = \{(P, \bot + E\{Q/X\})\}$
- $\mathcal{B} = \mathcal{B}_0 \cup \mathcal{B}_0^{-1} \cup \mathcal{B}_1 \cup \mathcal{B}_1^{-1}$

In order to show that $\mathcal{B}^*$ is a weak bisimulation we use the following lemma.

Lemma B.4.5. *Assume $R_1\ \mathcal{B}\ R_2$. Then*

1. *$R_1\ (\mathcal{B}_0 \cup \mathcal{B}_0^{-1})\ R_2$ and $R_1 \xrightarrow{\mathsf{a}} R_1'$ implies $R_2 \overset{\mathsf{a}}{\Longrightarrow} R_2'$ and $R_1'\ \mathcal{B}\ R_2'$ for some $R_2' \in \mathsf{IMC}$,*
2. *$R_1\ (\mathcal{B}_1 \cup \mathcal{B}_1^{-1})\ R_2$ and $R_1 \xrightarrow{\mathsf{a}} R_1'$ implies $R_2 \overset{\mathsf{a}}{\Longrightarrow} R_2'$ and $R_1'\ \mathcal{B}\ R_2'$ for some $R_2' \in \mathsf{IMC}$,*
3. *$R_1\surd\!\!\downarrow$ and $R_2\surd\!\!\downarrow$ imply $(\forall C \in \mathsf{IMC}/\mathcal{B}^*)\ \gamma_{\mathrm{M}}(R_1, C) = \gamma_{\mathrm{M}}(R_2, C)$,*
4. *$F\surd\!\!\downarrow$ iff $G\surd\!\!\downarrow$.*

Sketch of Proof. (1) can be proven by an induction on the height of the proof tree for the transition $F \xrightarrow{a} F'$. The induction base is the case $H \equiv X$. (2) follows easily from (1) by taking $H \equiv E$ in $\mathcal{B}_0$.

(3) can be proven by an induction on the sum of the heights of all proof trees for $\dashrightarrow$ transitions emanating F. Note that the case $R_1\ (\mathcal{B}_1 \cup \mathcal{B}_1^{-1})\ R_2$ is trivial since the premise $R_1 \surd\downarrow$ and $R_2 \surd\downarrow$ is not satisfied in this case.

Finally (4) can be easily proven by using fact (B.5) on page 181. Note that (4) holds for if $R_1\ (\mathcal{B}_1 \cup \mathcal{B}_1^{-1})\ R_2$ since this implies $R_1 \not\surd$, $R_2 \not\surd$.

Now we are able to show soundness of law (:=8).

Lemma B.4.6. $P \equiv \underline{X := \tau.X + E} \simeq \underline{X := \tau.(\perp + E)} \equiv Q$.

Proof. First note that Lemma B.1.3 and Lemma B.4.5 imply that $\mathcal{B}^*$ is a weak bisimulation, i.e., $\mathcal{B} \subseteq \approx$. We have to show that condition (1) to (4) of Definition 5.2.4 are satisfied.

To show condition (1), assume $P \xrightarrow{a} P'$. Since $P\,\mathcal{B}_0\,Q$ (take $H \equiv X$) it follows from Lemma B.4.5 that $Q \stackrel{a}{\Longrightarrow} Q'$ and $P'\,\mathcal{B}\,Q'$ for some $Q' \in$ IMC, thus $P' {\approx} Q'$. (2) of Definition 5.2.4 follows analogously. This completes the proof, because $P \not\surd$ and $Q \not\surd$ directly imply (3) and (4) of Definition 5.2.4.

The proof of soundness for law (:=6) follows the same lines but uses the relation $\mathcal{B}_1 = \{(P + E\{P/X\},\ Q)\}$, where $P \equiv \underline{X := \tau.(X + E) + F}$ and $Q \equiv \underline{X := \tau.X + E + F}$. In addition, the proof of law (:=8) can also be directly adopted to prove soundness of law (:=5). In this case $\mathcal{R}_1 = \{(P + E\{P/X\}, \perp + E\{Q/X\})\}$ where $P \equiv \underline{X := X + E}$ and $Q \equiv \underline{X := \perp + E}$.

B.5 Lemma 5.3.3

For each expression E there exists a strongly guarded expression F such that $\mathcal{A}_{\simeq} \vdash E = F$.

Proof. By induction on the structure of E. The only interesting case is recursion. We show the following stronger property.

If $E \in$ IML, then there is some strongly guarded $F \in$ IML such that
- X is strongly guarded in F,
- Each not strongly guarded occurrence of any variable $Y \in \mathcal{V}(F)$ does not lie within the scope of a subexpression $\underline{Z := G}$ of F,
- $\mathcal{A}_{\simeq} \vdash \underline{X := E} = \underline{X := F}$.

We only consider the case that E already satisfies the second condition. The general case can be reduced to this special case by an induction of the number of nested recursions in E in complete analogy to the proof of Theorem 5.2. in [144].

Thus, it remains to remove not strongly guarded occurrences of X where none of them appears inside the scope of a recursion. First, consider a not strongly guarded occurrence of X that appears in E on the topmost summation–level, i.e., $E = \tau.\tau.\ldots\tau.X + \ldots$. Because of law ($\tau 1$) we can assume that X is guarded by at most one τ–prefix. Such occurrences can be eliminated by means of law ($_{:=}5$) or law ($_{:=}8$) (by transformation into $\bot$). All other not strongly guarded occurrences of X can be lifted to the topmost summation–level by the following iterative procedure.

Case 1: $E \equiv \tau.(X + E') + F'$

Law ($_{:=}6$) implies

$$\mathcal{A}_{\simeq} \vdash \underline{{}_{X:=}E} \equiv \underline{{}_{X:=}\tau.(X + E') + F'} = \underline{{}_{X:=}\tau.X + E' + F'}\,.$$

We continue with expression $\tau.X + E' + F'$.

Case 2: $E \equiv \tau.E' + F'$, where X occurs partially guarded in E'.

By assumption, this occurrence does not lie within the scope of a recursion. E' must hence be of the form $G + \tau.E''$. Law ($\bot 2$) then gives

$$\mathcal{A}_{\simeq} \vdash E' = G + \tau.E'' {+}\bot{=}\ E' {+}\bot\,. \tag{B.8}$$

Taking into account that X appears not strongly guarded in E' we are aiming to establish that

$$\mathcal{A}_{\simeq} \vdash E' = X + E'$$

for some E'' where X occurs strongly guarded. This allows us to proceed with $\tau.(X + E') + F'$ by resorting to Case 1. We know already that

$$\mathcal{A}_{\simeq} \vdash E' = E' {+}\bot\,,$$

and, also from (B.8), that

$$\mathcal{A}_{\simeq} \vdash X + E' {+}\bot{=}\ X + E'.$$

In order to complete this case we have to show that $\mathcal{A}_{\simeq} \vdash E' {+}\ \bot{=}\ X {+} E' {+}\bot$. The verification requires structural induction on E' taking into account that X appears not strongly guarded in E'. The only interesting case is $E' \equiv X$, where $X + X {+}\bot{=}\ X {+}\bot$ is used (which can be derived from ($I3$) and (N)).

Iterating the above two steps will eventually lead to an expression F where all not strongly guarded occurrences of X appear on topmost summation–level, completing the proof of Lemma 5.3.3.

B.6 Theorem 5.3.8

Let E and F be two weakly congruent expressions, i.e., $E \simeq F$. Furthermore let E $\mathcal{A}_{\simeq}$-provably satisfy the SES S_1 and F $\mathcal{A}_{\simeq}$-provably satisfy the SES S_2, where both S_1 and S_2 are strongly guarded and saturated. Then there is some guarded SES S, that both E and F $\mathcal{A}_{\simeq}$-provably satisfy.

Proof. The proof proceeds along the lines of the strong bisimilarity case (Appendix B.3), but is more involved. The problems faced during the proof are essentially the same as in [142, Theorem 3.2], but their solution differs significantly.

Assume $E \simeq F$ and let $S_1 = (\boldsymbol{X} := \boldsymbol{G})$ and $S_2 = (\boldsymbol{Y} := \boldsymbol{H})$ be weakly guarded SES such that there are expressions $\boldsymbol{E} = (E_1, \dots, E_n)$ and $\boldsymbol{F} = (F_1, \dots, F_n)$ satisfying

$$\mathcal{A}_{\simeq} \vdash E = E_1,$$
$$\mathcal{A}_{\simeq} \vdash F = F_1,$$
$$\mathcal{A}_{\simeq} \vdash E_i = G_i\{\boldsymbol{E}/\boldsymbol{X}\}, \text{ and}$$
$$\mathcal{A}_{\simeq} \vdash F_j = H_j\{\boldsymbol{F}/\boldsymbol{Y}\}.$$

We use the sets I, $J_{i,j}$, $K_{i,j}$, and the same notation as in Appendix B.3. Since $E \simeq F$ we have $(1,1) \in I$. In addition, $(i,j) \in I$ implies that, by assumption, $G_i\{\boldsymbol{E}/\boldsymbol{X}\} \approx H_j\{\boldsymbol{E}/\boldsymbol{Y}\}$. We obtain the following properties:

If $(i,j) \in I$, then for all $\mathsf{a}, \mathsf{b} \in \mathit{Act}$ and $C \in \mathsf{IML}/\simeq$,

(i) $X_i \xrightarrow{\mathsf{a}}\!\!\!\!\!+_{S_1} X_{l_1}$ implies that there is some l_2 such that
- $Y_j \xrightarrow{\mathsf{a}}\!\!\!\!\!+_{S_2} Y_{l_2}$ and $(l_1, l_2) \in I$, or
- $\mathsf{a} = \tau$ and $(l_1, j) \in I$.

(ii) $Y_j \xrightarrow{\mathsf{b}}\!\!\!\!\!+_{S_2} Y_{l_2}$ implies that there is some l_1 such that
- $X_i \xrightarrow{\mathsf{b}}\!\!\!\!\!+_{S_1} X_{l_2}$ and $(l_1, l_2) \in I$, or
- $\mathsf{b} = \tau$ and $(i, l_2) \in I$.

(iii) $X_i \surd\!\downarrow$ implies
- $Y_j \not\surd$, or
- $Y_j \surd\!\downarrow$ and $\gamma_{\mathrm{M}}(G_i\{\boldsymbol{E}/\boldsymbol{X}\}, C) = \gamma_{\mathrm{M}}(H_j\{\boldsymbol{F}/\boldsymbol{Y}\}, C))$,

(iv) $Y_j \surd\!\downarrow$ implies
- $X_i \not\surd$, or
- $X_i \surd\!\downarrow$ and $\gamma_{\mathrm{M}}(G_i\{\boldsymbol{E}/\boldsymbol{X}\}, C) = \gamma_{\mathrm{M}}(H_j\{\boldsymbol{F}/\boldsymbol{Y}\}, C))$,

(v) $\{W_{h_1(i,1)}, \dots, W_{h_1(i,t_1(i))}\} = \{W_{h_2(j,1)}, \dots, W_{h_2(j,t_2(j))}\}$.
In addition, if $X_i \surd\!\downarrow$ (and $Y_j \surd\!\downarrow$), then $\{\!|W_{h_1(i,1)}, \dots, W_{h_1(i,t_1(i))}|\!\} = \{\!|W_{h_2(j,1)}, \dots, W_{h_2(j,t_2(j))}|\!\}$.

The proof of these properties requires that both SES are saturated, together with a characterisation of $\approx$ on open expressions, see [104] for details. Notice that the last property does not differ from the strong case.

Indeed, assuming $(i,j) \in I$, properties (i) and (ii) do not directly imply that $K_{i,j}$ is a total and surjective relation, as it was the case in Appendix B.3, except if $(i,j) = (1,1)$. In this case we can use $E_1 = E \simeq F = F_1$ (together with a characterisation of $\simeq$ on open expressions) to ensure (via $G_1\{\boldsymbol{E}/\boldsymbol{X}\} \simeq F_1\{\boldsymbol{H}/\boldsymbol{Y}\}$) that the first alternative of property (i) and (ii) has to be fulfilled. In other words $K_{1,1}$ is total and surjective.

To clarify the situation when (i,j) is different from $(1,1)$, we introduce for arbitrary sets M, N of indices and $O \subseteq M \times N$ the projections on M, respectively N:

$$\begin{aligned}\pi^{(1)}(O) &= \{m \in M \mid \exists n \in N : (m,n) \in O\}, \text{ and}\\ \pi^{(2)}(O) &= \{n \in N \mid \exists m \in M : (m,n) \in O\}.\end{aligned}$$

Properties (i) and (ii) imply that for each $l_1 \in \{1,\ldots,s_1(i)\}$ there is some $l_2 \in \{1,\ldots,s_2(j)\}$ such that

- $(l_1,l_2) \in K_{i,j}$, or
- $\mathsf{a}_{i,l_1} = \tau$ and $(g_1(i,l_1),j) \in I$,

as well as that for each $l_2 \in \{1,\ldots,s_2(j)\}$ there is some $l_1 \in \{1,\ldots,s_1(j)\}$ such that

- $(l_1,l_2) \in K_{i,j}$, or
- $\mathsf{b}_{j,l_2} = \tau$ and $(i,g_2(j,l_2)) \in I$.

Stated differently, it holds for each $l_1 \in \{1,\ldots,s_1(i)\}$ that

$$l_1 \notin \pi^{(1)}(K_{i,j}) \quad \text{implies} \quad \mathsf{a}_{i,l_1} = \tau \text{ and } (g_1(i,l_1),j) \in I \tag{B.9}$$

as well as that for each $l_2 \in \{1,\ldots,s_2(j)\}$

$$l_2 \notin \pi^{(2)}(K_{i,j}) \quad \text{implies} \quad \mathsf{b}_{j,l_2} = \tau \text{ and } (i,g_2(j,l_2)) \in I. \tag{B.10}$$

Assuming $X_i\surd\!\downarrow$ and $Y_j\surd\!\downarrow$, in addition to $(i,j) \in I$, we are able to deduce from property (iii) or (iv), in complete analogy to the strong bisimilarity case:

$$\begin{aligned}\sum_{\substack{u=1\\(u,k_2)\in J_{i,j}}}^{r_1(i)} \lambda_{i,u} = {} & \gamma_{\mathrm{M}}(G_i\{\boldsymbol{E}/\boldsymbol{X}\},[E_{f_1(i,k_1)}]_{\approx}) &=\\ & \gamma_{\mathrm{M}}(H_j\{\boldsymbol{F}/\boldsymbol{Y}\},[F_{f_2(j,k_2)}]_{\approx}) &= \sum_{\substack{v=1\\(k_1,v)\in J_{i,j}}}^{r_2(j)} \mu_{j,u}.\end{aligned}$$

In the remainder of this proof we use $\gamma^{i,j}_{k_1,k_2}$ to denote this sum.

We will now create a standard equation set $S = \{Z_{i,j} := L_{i,j} \mid (i,j) \in I\}$ with new formal variables $Z_{i,j}$ (for each pair $(i,j) \in I$). It differs from the SES used in Appendix B.3 only in its second line. Let

$$L_{i,j} \equiv \sum_{(l_1,l_2)\in K_{i,j}} \mathsf{b}_{j,l_2}.Z_{g_1(i,l_1),g_2(j,l_2)} + \sum_{m_2=1}^{t_2(j)} W_{h_2(j,m_2)}$$

$$+ \sum_{l_1 \notin \pi^{(1)}(K_{i,j})} \tau.Z_{g_1(i,l_1),j} + \sum_{l_2 \notin \pi^{(2)}(K_{i,j})} \tau.Z_{i,g_2(j,l_2)}$$

$$+ \begin{cases} \sum_{(k_1,k_2)\in J_{i,j}} \left(\frac{\lambda_{i,k_1} \cdot \mu_{j,k_2}}{\gamma^{i,j}_{k_1,k_2}} \right) .Z_{f_1(i,k_1),f_2(j,k_2)} & \text{if } X_i\surd\downarrow \text{ and } Y_j\surd\downarrow \\ \bot & \text{if } X_i\uparrow \text{ or } Y_j\uparrow \\ 0 & \text{else.} \end{cases}$$

Our claim is that E $\mathcal{A}_{\simeq}$-provably satisfies the SES S (where E will be equated with $Z_{1,1}$). The proof that F $\mathcal{A}_{\simeq}$-provably satisfies S is completely symmetric. In order to show the former we define terms $C_{i,j}$ (for $(i,j) \in I$) as follows:

$$C_{i,j} \equiv \begin{cases} \tau.E_i & \text{if } \pi^{(2)}(K_{i,j}) \neq \{1,\ldots,s_2(j)\} \\ E_i & \text{else.} \end{cases}$$

We will prove that $\mathcal{A}_{\simeq} \vdash C_{i,j} = L_{i,j}\{C/Z\}$. As a consequence of the fact that $K_{1,1}$ is total and surjective we have that $C_{1,1} \equiv E_1 = E$.

We proceed and construct the term $L_{i,j}\{C/Z\}$. Besides of variables $W_{h_2(i,m_1)}$ and $\bot$, summands of the form $\mathsf{a}.E_i$, $\mathsf{a}.\tau.E_i$, $(\lambda).E_i$ and $(\lambda).\tau.E_i$ may occur in this term. Because of law $(\tau 1)$ and $(\tau 4)$ we can transform $L_{i,j}\{C/Z\}$ to

$$\mathcal{A}_{\simeq} \vdash L_{i,j} = \sum_{(l_1,l_2)\in K_{i,j}} \mathsf{a}_{i,l_1}.E_{g_1(i,l_1)} + \sum_{m_2=1}^{t_2(j)} W_{h_2(j,m_2)}$$

$$+ \sum_{l_1 \notin \pi^{(1)}(K_{i,j})} \tau.E_{g_1(i,l_1)} + \sum_{l_2 \notin \pi^{(2)}(K_{i,j})} \tau.E_i$$

$$+ \begin{cases} \sum_{(k_1,k_2)\in J_{i,j}} \left(\frac{\lambda_{i,k_1} \cdot \mu_{j,k_2}}{\gamma^{i,j}_{k_1,k_2}} \right) .E_{f_1(i,k_1)} & \text{if } X_i\surd\downarrow \text{ and } Y_j\surd\downarrow \\ \bot & \text{if } X_i\uparrow \text{ or } Y_j\uparrow \\ 0 & \text{else.} \end{cases}$$

To prove that $\mathcal{A}_{\simeq} \vdash C_{i,j} = L_{i,j}\{C/Z\}$ we proceed exactly as in the strong bisimilarity case, where we transformed free variables as well as delay prefixes using the laws (N), $(I1)$, $(I3)$, and $(\bot 2)$. As a result, we have that

$$\mathcal{A}_{\simeq} \vdash \sum_{m_2=1}^{t_2(j)} W_{h_2(j,m_2)} = \sum_{m_1=1}^{t_1(i)} W_{h_1(i,m_1)},$$

and

$$\mathcal{A}_{\simeq} \vdash \sum_{(k_1,k_2)\in J_{i,j}} \left(\frac{\lambda_{i,k_1} \cdot \mu_{j,k_2}}{\gamma^{i,j}_{k_1,k_2}} \right) .E_{f_1(i,k_1)} = \sum_{k_1=1}^{r_1(i)} (\lambda_{i,k_1}).E_{f_1(i,k_1)}.$$

In the remainder of this proof we use the following abbreviations (two of them appeared also in Appendix B.3).

$$\begin{aligned}
D_1 &\equiv \sum_{l_1=1}^{s_1(i)} \mathsf{a}_{i,l_1}.E_{g_1(i,l_1)} + \sum_{m_1=1}^{t_1(i)} W_{h_1(i,m_1)},\\
D_2 &\equiv \sum_{k_1=1}^{r_1(i)} (\lambda_{i,k_1}).E_{f_1(i,k_1)},\\
D_3 &\equiv \sum_{l_1 \notin \pi^{(1)}(K_{i,j})} \tau.E_{g_1(i,l_1)}, \text{ and}\\
D_4 &\equiv \sum_{l_2 \notin \pi^{(2)}(K_{i,j})} \tau.E_i
\end{aligned}$$

With this notation we have achieved so far that

$$\mathcal{A}_{\simeq} \vdash L_{i,j}\{\boldsymbol{C}/\boldsymbol{Z}\} = D_1 + D_3 + D_4 + \begin{cases} D_2 & \text{if } X_i \surd\!\downarrow \text{ and } Y_j \surd\!\downarrow \\ \bot & \text{if } X_i\uparrow \text{ or } Y_j\uparrow \\ 0 & \text{else.} \end{cases}$$

A close look into D_3 reveals that we can apply (B.10): Whenever $l_1 \notin \pi^{(1)}(K_{i,j})$, then $\mathsf{a}_{i,l_1} = \tau$. Hence D_3 is nothing else than

$$\sum_{l_1 \notin \pi^{(1)}(K_{i,j})} \mathsf{a}_{i,l_1}.E_{g_1(i,l_1)}.$$

As a consequence, observe that all the summands of D_3 also appear somewhere in the first sum of D_1. We can therefore apply law $(I2)$ summand by summand in order to erase D_3:

$$\mathcal{A}_{\simeq} \vdash D_1 + D_3 = D_1.$$

So, we have

$$\mathcal{A}_{\simeq} \vdash L_{i,j}\{\boldsymbol{C}/\boldsymbol{Z}\} = D_1 + D_4 + \begin{cases} D_2 & \text{if } X_i \surd\!\downarrow \text{ and } Y_j \surd\!\downarrow \\ \bot & \text{if } X_i\uparrow \text{ or } Y_j\uparrow \\ 0 & \text{else.} \end{cases}$$

On the other hand, we have, by means of our assumption $\mathcal{A}_{\sim} \vdash E_i = G_i\{\boldsymbol{E}/\boldsymbol{X}\}$, that

- $X_i\downarrow$ implies $\mathcal{A} \vdash E_i = D_1 + D_2$, and
- $X_i\uparrow$ implies $\mathcal{A} \vdash E_i = D_1 + D_2 + \bot$.

Now, different from the strong bisimilarity case, we have to distinguish two cases, dependent on the fact whether $\pi^{(2)}(K_{i,j})$ is equal or different from $\{1, \ldots, s_2(j)\}$.

If $\pi^{(2)}(K_{i,j}) = \{1, \ldots, s_2(j)\}$, then $C_{i,j} \equiv E_i$, and D_4 denotes an empty sum. As a consequence, the situation is exactly as in Appendix B.3 where we analysed 16 different cases in order to complete the proof. Indeed, this case analysis can be directly adopted to complete the proof of the particular case $\pi^{(2)}(K_{i,j}) = \{1, \ldots, s_2(j)\}$.

If, on the other hand $\pi^{(2)}(K_{i,j}) \neq \{1, \ldots, s_2(j)\}$, then $C_{i,j} \equiv \tau.E_i$, and $L_{i,j}\{\boldsymbol{C}/\boldsymbol{Z}\}$ contains a summand $\tau.E_i$ at least once, resulting from D_4. After multiple applications of law ($I2$) we have

$$\mathcal{A}_{\simeq} \vdash D_1 + \tau.E_i + \begin{cases} D_2 & \text{if } X_i\surd\downarrow \text{ and } Y_j\surd\downarrow \\ \bot & \text{if } X_i\uparrow \text{ or } Y_j\uparrow \\ 0 & \text{else.} \end{cases}$$

If $X_i\surd\downarrow$ does not hold, or $Y_j\surd\downarrow$ does not hold we require an intermediate step, in which we add D_2 by means of $r_1(i)$ applications of Lemma 5.3.1. Hence

$$\mathcal{A}_{\simeq} \vdash D_1 + D_2 + \tau.E_i + \begin{cases} \bot & \text{if } X_i\uparrow \text{ or } Y_j\uparrow \\ 0 & \text{else.} \end{cases}$$

Since either $E_i = D_1 + D_2$ or $E_i = D_1 + D_2 + \bot$ we possibly have to apply law ($\bot 2$) to obtain

$$\mathcal{A}_{\simeq} \vdash E_i + \tau.E_i.$$

Law ($\tau 2$) eventually produces the desired result

$$\mathcal{A}_{\simeq} \vdash L_{i,j}\{\boldsymbol{C}/\boldsymbol{Z}\} = E_i + \tau.E_i = \tau.E_i \equiv C_{i,j}.$$

We have thus shown that E $\mathcal{A}_{\simeq}$-provably satisfies the weakly guarded SES S. As mentioned above, the proof that F $\mathcal{A}_{\simeq}$-provably satisfies S is completely symmetric and therefore omitted. Obviously, each $Z_{i,j}$ is weakly guarded in each $L_{i',j'}$ and hence S is weakly guarded. However, in order to serve as an SES that satisfies Theorem 5.3.8, S has to be ***strongly*** guarded.

Assume that S is not strongly guarded. Hence, there is a cycle of τ-prefixes in $\boldsymbol{L}$. But, according to the construction of $\boldsymbol{L}$ this implies a cycle of τ-prefixes in either $\boldsymbol{G}$ or $\boldsymbol{H}$. This contradicts our assumption that both S_1 and S_2 are strongly guarded, and the proof is complete.

B.7 Theorem 5.5.2

If P is minimal with respect to strong bisimilarity, then $\{n\,!\,P\}A$ is minimal with respect to strong bisimilarity provided that P is linear or that A is empty.

Sketch of Proof. Let us first consider the case that P is linear, but A is not necessarily empty. Since synchronising activities have to be performed in the same order by all replicas, $\{n!P\}A$ terminates (with $\{n\,!\,0\}A$) if and only if P terminates (with 0). Let $P_1, \ldots .P_m$ enumerate the finite number of different processes reachable from P, (totally) ordered by their distance from P and let P_0 denote P. Note that by assumption none of these P_i are pairwise bisimilar. It requires a detailed case analysis to check that for all $0 < i \leq m$, $\{n\,!\,P_i\}A$ is reachable from $\{n\,!\,P\}A$ and that the behaviour of $\{n\,!\,P_i\}A$ is unique (with respect to strong bisimilarity) among the behaviours reachable from $\{n\,!\,P\}A$.

Let us now assume that there are two distinct expressions, say $\mathcal{M}_1 A$ and $\mathcal{M}_2 A$, reachable from $\{n\,!\,P\}A$ that are bisimilar. We know that $\sum_i \mathcal{M}_1(P_i) = \sum_i \mathcal{M}_2(P_i) = n$, and, since $\mathcal{M}_1$ and $\mathcal{M}_2$ are distinct, there are (more than one) $P_{k_1}, \ldots, P_{k_r}$, such that $\mathcal{M}_1(P_{k_i}) \neq \mathcal{M}_2(P_{k_i})$, again ordered by distance from P. Let l be the maximum number such that $\mathcal{M}_1(P_l) > 0$ or $\mathcal{M}_2(P_l) > 0$, and let j be the respective minimum. $l - j$ will be denoted g. The expression $\{n!P_l\}A$ is reachable from both, $\mathcal{M}_1 A$ and $\mathcal{M}_2 A$ without further synchronisation. We now have that $\mathcal{M}_1 A$ as well as $\mathcal{M}_2 A$ possess traces to $\{n\,!\,P_l\}A$ which is the unique representative of this class of behaviours. Strong bisimilarity of $\mathcal{M}_1 A$ and $\mathcal{M}_2 A$ implies that their set of traces leading to $\{n\,!\,P_l\}A$ have to coincide. An induction on g, the difference between j and l, shows that this is not the case.

This completes the proof of the case that P is linear, but A is not empty. If A is empty, the result follows from similar (but simpler) arguments.

B.8 Theorem 5.5.3

Strong bisimilarity and weak congruence are congruences with respect to all operators of $\mathsf{IMC}_{\mathsf{XXL}}$.

Sketch of Proof. Substitutivity of the operators of $\mathsf{IMC}_{\mathsf{L}}$ is covered by Theorem 5.2.1 and Theorem 5.2.2. Proofs for parallel composition and abstraction have appeared in Appendix A.2, respectively Appendix A.5. Note that it is negligible that the definitions of strong bisimilarity and weak congruence in Chapter 5 account for ill-definedness while those of Chapter 4 do not. This is because all expressions of $\mathsf{IMC}_{\mathsf{XXL}}$ are well-defined by definition, since those of $\mathsf{IMC}_{\mathsf{L}}$ are. So, we only have to discuss substitutivity of the elapse operator and of symmetric composition.

We first turn our attention towards symmetric composition. Let $A \subseteq Act$, $P, P' \in \mathsf{IMC}_{\mathsf{XXL}}$, and let $\mathcal{M}$ be an arbitrary multiset of elements from $\mathsf{IMC}_{\mathsf{XXL}}$. In order to show substitutivity of strong bisimilarity, we have to show that $(\mathcal{M} \oplus P)A \sim (\mathcal{M} \oplus Q)A$ holds whenever $P \sim Q$.

We construct an equivalence relation $\mathcal{E} = \mathcal{R} \cup \mathcal{R}^{-1} \cup \mathcal{Id}_{\mathsf{XXL}}$ (the identity relation on $\mathsf{IMC}_{\mathsf{XXL}}$) where

$$\mathcal{R} = \left\{ ((\mathcal{M}' \oplus P')A, (\mathcal{M}' \oplus Q')A) \mid P' \sim Q' \ \wedge\ \mathcal{M}'A \in \mathsf{IMC}_{\mathsf{XXL}} \right\}.$$

In order to verify substitutivity, it is sufficient to show that $\mathcal{E}$ is a strong bisimulation according to Definition 5.2.1. For symmetry reasons, we can restrict to elements of $\mathcal{R}$. Assume $(\mathcal{M} \oplus P)A \ \mathcal{R}\ (\mathcal{M} \oplus Q)A$. In order to verify the first clause of Definition 5.2.1 we distinguish two cases.

Case 1: $(\mathcal{M} \oplus P)A \xrightarrow{\mathsf{a}} (\mathcal{M}' \oplus P')A$ and $\mathsf{a} \in A$.
Then, $\mathcal{M}A \xrightarrow{\mathsf{a}} \mathcal{M}'A$ and $P \xrightarrow{\mathsf{a}} P'$, according to the second rule in Table 5.10. Since $P \sim Q$, there is some Q' such that $Q \xrightarrow{\mathsf{a}} Q'$ and $P' \sim Q'$ (Definition 5.2.1). As a consequence, $(\mathcal{M} \oplus Q)A \xrightarrow{\mathsf{a}} (\mathcal{M}' \oplus Q')A$ and $(\mathcal{M}' \oplus P')A \ \mathcal{E}\ (\mathcal{M}' \oplus Q')A$, as required.

Case 2: $(\mathcal{M} \oplus P)A \xrightarrow{\mathsf{a}} (\mathcal{M}' \oplus P')$ and $\mathsf{a} \notin A$.
By the structure of the third rule in Table 5.10 we either have $\mathcal{M} \equiv \mathcal{M}'$ and $P \xrightarrow{\mathsf{a}} P'$, or we have that $\mathcal{M}A \xrightarrow{\mathsf{a}} \mathcal{M}'A$ while P has not moved, i.e., $P \equiv P'$. In the latter case, we directly obtain $(\mathcal{M} \oplus Q)A \xrightarrow{\mathsf{a}} (\mathcal{M}' \oplus Q)A$, and $(\mathcal{M}' \oplus P)A \ \mathcal{E}\ (\mathcal{M}' \oplus Q)A$. In the former case, use $P \sim Q$ and Definition 5.2.1 to obtain the existence of some Q' such that $Q \xrightarrow{\mathsf{a}} Q'$ and $P' \sim Q'$. Thus, $(\mathcal{M} \oplus Q)A \xrightarrow{\mathsf{a}} (\mathcal{M}' \oplus Q')A$ and $(\mathcal{M}' \oplus P')A \ \mathcal{E}\ (\mathcal{M}' \oplus Q')A$, as required.

For verification of the second clause of Definition 5.2.1, assume $(\mathcal{M} \oplus P)A \not\xrightarrow{\tau}$ and $\gamma_M((\mathcal{M} \oplus P)A, C) = \mu$ for some equivalence class C of $\mathcal{E}$. The elements of C are either all of the form $(\mathcal{M}'' \oplus R)A$, or C is a singleton set (induced by $\mathcal{Id}_{\mathsf{XXL}}$). The latter case is trivial, since $\mu = 0$. In the former case, the equivalence class is basically a crossproduct of some singleton set $\{\mathcal{M}''\}$ and an equivalence class of $\sim$. Let $(\mathcal{M}'' \oplus R)A$ be a representative of this class and let C_R denote the equivalence class of R, with respect to $\sim$. We have from the structure of the third rule in Table 5.10 that $P \not\xrightarrow{\tau}$. Hence, $\gamma_M(P, C_R) = \gamma_M(Q, C_R)$, by assumption. Due to the structure of C, this implies $\gamma_M(P, C) = \gamma_M(Q, C)$.

We have to show $\gamma_M((\mathcal{M} \oplus Q)A, C) = \mu$. Now assume $\mathcal{M}(P) = l \geq 0$ and $\mathcal{M}(Q) = k \geq 0$. We split the value μ in three parts, i.e., $\gamma_M((\mathcal{M} \oplus P)A, C) = \mu = \mu_P + \mu_Q + \mu'$. μ_P is the rate contributed by Markovian transitions of P and μ_Q is the same for those transitions caused by Q. μ' subsumes the remaining rates. By the last rule in Table 5.10 we derive $\mu_P = (l+1)\gamma_M(P, C)$ and $\mu_Q = k\gamma_M(Q, C)$. Using $\gamma_M(P, C) = \gamma_M(Q, C)$ we obtain after some straightforward transformations that $\mu = l\,\gamma_M(P, C) + (k+1)\,\gamma_M(Q, C) + \mu'$. By the last rule in Table 5.10, we finally derive that $\gamma_M((\mathcal{M} \oplus Q)A, C) = \mu$, as required.

Hence, $\mathcal{E}$ is a strong bisimulation. So, strong bisimilarity is substitutive with respect to symmetric composition. In order to show substitutivity for weak congruence, the proof proceeds essentially in the same way. Assuming $P \simeq Q$, it is shown that the pair $(\mathcal{M} \oplus P)A, (\mathcal{M} \oplus Q))A$ satisfies the requirements of Lemma B.1.4. We omit the details.

We now sketch substitutivity of the elapse operator. Let us consider the strong bisimilarity case. Choose $S, I, D \subseteq Act$, $P, P', Q, Q', R, R' \in \mathsf{IMC_{XXL}}$, and assume $P \sim P'$, $Q \sim Q'$ and $R \sim R'$. We have to show that

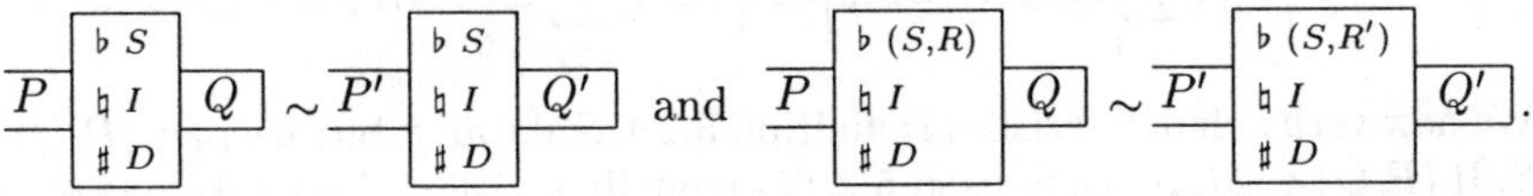

$$
P \left[\begin{array}{c} \flat\, S \\ \natural\, I \\ \sharp\, D \end{array}\right] Q \;\sim\; P' \left[\begin{array}{c} \flat\, S \\ \natural\, I \\ \sharp\, D \end{array}\right] Q' \quad \text{and} \quad P \left[\begin{array}{c} \flat\, (S,R) \\ \natural\, I \\ \sharp\, D \end{array}\right] Q \;\sim\; P' \left[\begin{array}{c} \flat\, (S,R') \\ \natural\, I \\ \sharp\, D \end{array}\right] Q' \, .
$$

$$(\text{B.11})$$

For this purpose, define an equivalence relation $\mathcal{E}$ as

$$
\mathit{Id}_{\mathsf{XXL}} \;\cup\; \left\{ \left(P \left[\begin{array}{c} \flat\, S \\ \natural\, I \\ \sharp\, D \end{array}\right] Q \,,\; P' \left[\begin{array}{c} \flat\, S \\ \natural\, I \\ \sharp\, D \end{array}\right] Q' \right) \;\Big|\; P \sim P', Q \sim Q', R \sim R' \right\} \;\cup
$$

$$
\left\{ \left(P \left[\begin{array}{c} \flat\, (S,R) \\ \natural\, I \\ \sharp\, D \end{array}\right] Q \,,\; P' \left[\begin{array}{c} \flat\, (S,R') \\ \natural\, I \\ \sharp\, D \end{array}\right] Q' \right) \;\Big|\; P \sim P', Q \sim Q', R \sim R' \right\}.
$$

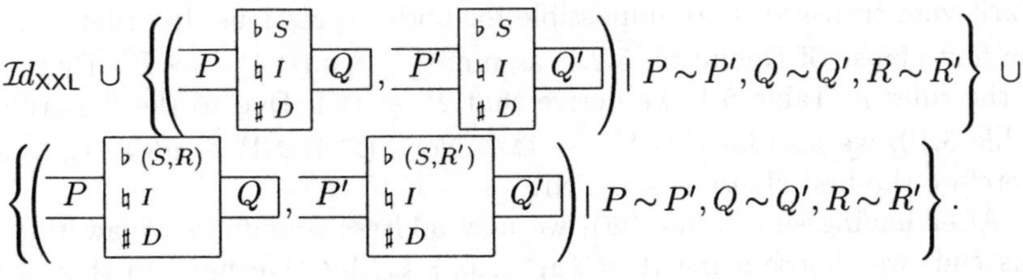

It requires a tedious case analysis of Table 5.7 and 5.8 to verify that equivalence $\mathcal{E}$ satisfies Definition 5.2.1, similar to the symmetric composition case. We do not work out the details. $\mathcal{E}$ is hence a strong bisimulation, and it contains the above two pairs by definition. Thus, strong bisimilarity is indeed a congruence for the elapse operator. Congruence with respect to $\simeq$ follows similar lines, verifying Lemma B.1.4 for any pair of the above form (B.11), but replacing $\sim$ by $\simeq$ in the hypotheses concerning P, P', Q, Q', R, R'.

B.9 Theorem 5.5.4

For arbitrary expressions $P \in \mathsf{IMC_{XXL}}$ and $Q \in \mathsf{IMC_{XXL}}$ it holds that

1. $P \sim Q$ *if and only if* $(\mathcal{A}_\sim \cup \mathcal{A}_X \cup \mathcal{A}_\sharp \cup \mathcal{A}_!) \vdash P = Q$, *and*
2. $P \simeq Q$ *if and only if* $(\mathcal{A}_\simeq \cup \mathcal{A}_X \cup \mathcal{A}_\sharp \cup \mathcal{A}_!) \vdash P = Q$.

We will first show soundness of the laws with respect to strong bisimilarity (Lemma B.9.1). Soundness with respect to weak congruence follows from an application of Lemma 5.2.2. Then we show completeness of the laws. For this purpose, we only have to consider the elapse operator and symmetric composition. The other operators are covered by Theorem 5.4.1. Corollary B.9.1 will summarise soundness and completeness, proving Theorem 5.5.4.

Lemma B.9.1. *For* $P, Q \in \mathsf{IMC_{XXL}}$ *it holds that* $P \sim Q$ *if* $(\mathcal{A}_\sharp \cup \mathcal{A}_!) \vdash P = Q$.

Sketch of Proof. We only tackle Table 5.11, i.e., the two laws for symmetric composition. Verification of the laws of $\mathcal{A}_\sharp$ (Table 5.9) proceeds in a similar way, but requires a tedious distinction of many case, dependent on the rules of Table 5.7 and 5.8 used to derive a transition.

In order to show that law ($\varnothing$) is sound we fix a set $A \subseteq Act$, and construct an equivalence relation $\mathcal{E}$ as follows:

$$\mathcal{E} = \Big\{\big(\sum_{\mathsf{a}\in A} \mathsf{a}.(\varnothing A), \varnothing A\big)\Big\} \cup \Big\{\big(\varnothing A, \sum_{\mathsf{a}\in A} \mathsf{a}.(\varnothing A)\big)\Big\} \cup \mathcal{Id}_{\mathsf{XXL}}$$

We now verify that $\mathcal{E}$ satisfies Definition 5.2.1. Take an arbitrary pair $(P, Q) \in \mathcal{E}$. If $(P, Q) \in \mathcal{Id}_{\mathsf{XXL}}$, Definition 5.2.1 is trivially satisfied. Now take one of the remaining pairs, say the first. For the other pair, the situation is symmetric. To verify Definition 5.2.1, note that the second clause is irrelevant, because Markovian transitions are impossible for both expressions. In order to show the first clause of Definition 5.2.1, assume $\sum_{\mathsf{a}\in A} \mathsf{a}.(\varnothing A) \xrightarrow{\mathsf{a}} P'$. By means of the rules in Table 5.1, we derive that $P' \equiv \varnothing A$. Due to the first rule of Table 5.10, we also have $\varnothing A \xrightarrow{\mathsf{a}} \varnothing A$. Now, $(\varnothing A, \varnothing A) \in \mathcal{Id}_{\mathsf{XXL}} \subseteq \mathcal{E}$, and therefore the first clause is satisfied.

After having shown law ($\varnothing$), we now address soundness of law ($\oplus$). To this end, we choose some $A = \{\mathsf{a}_1 \ldots \mathsf{a}_m\} \subseteq Act$. We have to show that, for arbitrary P and $\mathcal{M}$ it holds that $(\mathcal{M} \oplus P)A \sim \mathcal{M}A \overset{\mathsf{a}_1\ldots\mathsf{a}_m}{=\!=\!=} P$. For this purpose, define an equivalence relation $\mathcal{E} = (\mathcal{R} \cup \mathcal{R}^{-1})^*$ with

$$\mathcal{R} = \Big\{\big((\widetilde{\mathcal{M}} \oplus \widetilde{P})A, \widetilde{\mathcal{M}}A \overset{\mathsf{a}_1\ldots\mathsf{a}_m}{=\!=\!=} \widetilde{P} \mid \widetilde{P} \in \mathsf{IMC}_{\mathsf{XXL}} \wedge \widetilde{\mathcal{M}}A \in \mathsf{IMC}_{\mathsf{XXL}}\big)\Big\}$$

Obviously, $\mathcal{E}$ contains the above pair. In order to show that $\mathcal{E}$ is a strong bisimulation we consider an arbitrary pair from (w.l.o.g.) $\mathcal{R}$, say $((\mathcal{M} \oplus P)A, \mathcal{M}A \overset{\mathsf{a}_1\ldots\mathsf{a}_m}{=\!=\!=} P) \in \mathcal{R}$. In order to show the first clause of Definition 5.2.1, we distinguish two cases.

Case 1: $(\mathcal{M} \oplus P)A \xrightarrow{\mathsf{a}} (\mathcal{M}' \oplus P')A$ and $\mathsf{a} \in A$.
Then, $\mathcal{M}A \xrightarrow{\mathsf{a}} \mathcal{M}'A$ and $P \xrightarrow{\mathsf{a}} P'$, according to the second rule in Table 5.10. This implies (third rule of Table 2.1), that $\mathcal{M}A \overset{\mathsf{a}_1\ldots\mathsf{a}_m}{=\!=\!=} P \xrightarrow{\mathsf{a}} \mathcal{M}'A \overset{\mathsf{a}_1\ldots\mathsf{a}_m}{=\!=\!=} P'$, as well. Furthermore, $(\mathcal{M}' \oplus P')A \;\mathcal{E}\; \mathcal{M}'A \overset{\mathsf{a}_1\ldots\mathsf{a}_m}{=\!=\!=} P'$, as required.

Case 2: $(\mathcal{M} \oplus P)A \xrightarrow{\mathsf{a}} (\mathcal{M}' \oplus P')$ and $\mathsf{a} \notin A$.
By the structure of the third rule in Table 5.10 we either have $\mathcal{M} \equiv \mathcal{M}'$ and $P \xrightarrow{\mathsf{a}} P'$, or $\mathcal{M}A \xrightarrow{\mathsf{a}} \mathcal{M}'A$ and $P \equiv P'$. In the latter case, we obtain $\mathcal{M}A \overset{\mathsf{a}_1\ldots\mathsf{a}_m}{=\!=\!=} P \xrightarrow{\mathsf{a}} \mathcal{M}'A \overset{\mathsf{a}_1\ldots\mathsf{a}_m}{=\!=\!=} P$. In the former case, the second rule of Table 2.1 gives $\mathcal{M}A \overset{\mathsf{a}_1\ldots\mathsf{a}_m}{=\!=\!=} P \xrightarrow{\mathsf{a}} \mathcal{M}'A \overset{\mathsf{a}_1\ldots\mathsf{a}_m}{=\!=\!=} P'$, as well. In either of the above two cases, $(\mathcal{M}' \oplus P')A \;\mathcal{E}\; \mathcal{M}'A \overset{\mathsf{a}_1\ldots\mathsf{a}_m}{=\!=\!=} P'$, as required.

In order to show the second clause of Definition 5.2.1, let us assume that for an arbitrary equivalence class C of $\mathcal{E}$, $\gamma_M((\mathcal{M}A \oplus P), C) = \mu$. We have to show that also $\gamma_M(\mathcal{M}A \overset{\mathsf{a}_1\ldots\mathsf{a}_m}{=\!=\!=} P, C) = \mu$. Recall that γ_M builds the sum of rates leading into class C. By the structure of the rules, we can assume that this sum is identical for both expressions except for those rates caused by a Markovian transition of P. So, suppose that $P \xrightarrow{\lambda} P'$. If $\mathcal{M}(P) = 0$, we have

$$(\mathcal{M} \oplus P)A \xrightarrow{\lambda} (\mathcal{M} \oplus P')A,$$

as well as

$$\mathcal{M}A \,\underline{\underline{\overline{\overline{\;a_1 \ldots a_m\;}}}}\, P \xrightarrow{\lambda} \mathcal{M}A \,\underline{\underline{\overline{\overline{\;a_1 \ldots a_m\;}}}}\, P',$$

and $(\mathcal{M} \oplus P') \;\mathcal{E}\; \mathcal{M}A \,\underline{\underline{\overline{\overline{\;a_1 \ldots a_m\;}}}}\, P'$. If, otherwise, $\mathcal{M}(P) = n > 0$, we obtain

$$(\mathcal{M} \oplus P)A \xrightarrow{(n+1)\lambda} (\mathcal{M} \oplus P')A.$$

On the other hand side,[2]

$$\mathcal{M}A \,\underline{\underline{\overline{\overline{\;a_1 \ldots a_m\;}}}}\, P \xrightarrow{\lambda} \mathcal{M}A \,\underline{\underline{\overline{\overline{\;a_1 \ldots a_m\;}}}}\, P', \text{ and}$$

$$\mathcal{M}A \,\underline{\underline{\overline{\overline{\;a_1 \ldots a_m\;}}}}\, P \xrightarrow{n\lambda} ((\mathcal{M}A \ominus P) \oplus P') \,\underline{\underline{\overline{\overline{\;a_1 \ldots a_m\;}}}}\, P.$$

Since the chains $(\mathcal{M} \oplus P')A$, $\mathcal{M}A \,\underline{\underline{\overline{\overline{\;a_1 \ldots a_m\;}}}}\, P'$ and $((\mathcal{M}A \ominus P) \oplus P') \,\underline{\underline{\overline{\overline{\;a_1 \ldots a_m\;}}}}\, P$ are pairwise contained in $\mathcal{E}$ (the latter pair due to transitive closure), they belong to the same equivalence class C of $\mathcal{E}$. Therefore, the transition $P \xrightarrow{\lambda} P$ contributes the value $(n+1)\lambda$ to both $\gamma_M((\mathcal{M}A \oplus P), C)$ and $\gamma_M(\mathcal{M}A \,\underline{\underline{\overline{\overline{\;a_1 \ldots a_m\;}}}}\, P, C)$, completing the proof.

Lemma B.9.2. *For each expression $P \in \mathsf{IMC_{XXL}}$ there is some expression $P' \in \mathsf{IMC_{XXL}}$ not containing symmetric composition, such that $\mathcal{A}_! \vdash P = P'$.*

Proof. The proof is by structural induction on P. The only interesting case is symmetric composition. Assume that P is of the form $\mathcal{M}A$, and that the multiset $\mathcal{M}$ contains a total number of n elements (each of them not containing symmetric composition, by our induction hypothesis). By means of n applications of law ($\oplus$) and a final application of law ($\varnothing$), we transform P into P' not containing symmetric composition.

Lemma B.9.3. *For each expression $P \in \mathsf{IMC_{XXL}}$ there is some expression $P' \in \mathsf{IMC_{L}}$ such that $(\mathcal{A}_\sim \cup \mathcal{A}_X \cup \mathcal{A}_\sharp \cup \mathcal{A}_!) \vdash P = P'$.*

Sketch of Proof. The proof is by structural induction on P. Due to Theorem 5.4.1 and Lemma B.9.2 we can assume that apart from the operators of IML at most elapse operators appear in P. This is thus the only nontrivial induction step. So, let us assume that P is of the following form (covering the case that the time constraint is inactive; the converse case that P has the form of an active time constraint proceeds analogously)

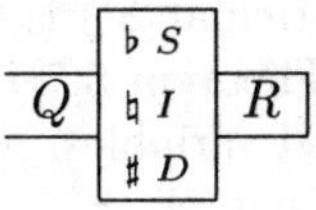

,

[2] Define $\mathcal{M} \ominus P := \mathcal{M}'$ with $\mathcal{M}'(P') :=$ **if** $(P' \equiv P)$ **then** $\max(0, \mathcal{M}(P') - 1)$ **else** $\mathcal{M}(P')$.

where $Q \in \mathsf{IMC_L}$ and $R \in \mathsf{IMC_L}$ holds by our induction hypothesis. Due to Theorem 5.3.2 there are two weakly guarded SES $S_1 = (\boldsymbol{X} := \boldsymbol{G})$ and $S_2 = (\boldsymbol{Y} := \boldsymbol{H})$ such that there are expressions $\boldsymbol{E} = (E_1, \ldots, E_n)$ and $\boldsymbol{F} = (F_1, \ldots, F_m)$ satisfying

$$\mathcal{A}_\sim \vdash Q = E_1,\ \mathcal{A}_\sim \vdash R = F_1,\ \mathcal{A}_\sim \vdash E_i = G_i\{\boldsymbol{E}/\boldsymbol{X}\},\ \mathcal{A}_\sim \vdash F_j = H_j\{\boldsymbol{F}/\boldsymbol{Y}\}. \tag{B.12}$$

Since $Q, R \in \mathsf{IMC_L}$ we can assume that each G_i and H_j is of the following form:

$$G_i \equiv \sum_{k_1=1}^{r_1(i)} (\lambda_{i,k_1}).X_{f_1(i,k_1)} + \sum_{l_1=1}^{s_1(i)} \mathsf{a}_{i,l_1}.X_{g_1(i,l_1)}, \text{ and}$$

$$H_j \equiv \sum_{k_2=1}^{r_2(j)} (\mu_{j,k_2}).Y_{f_2(j,k_2)} + \sum_{l_2=1}^{s_2(j)} \mathsf{b}_{j,l_2}.Y_{g_2(j,l_2)}.$$

Now define a new (weakly guarded) SES $S = (\boldsymbol{Z} := \boldsymbol{L})$ by setting

$$L_{i,1} \equiv G_i \boxed{\begin{array}{l}\flat\, S\\ \natural\, I\\ \sharp\, D\end{array}} H_1, \text{ and } L_{i,j+1} \equiv G_i \boxed{\begin{array}{l}\flat\, (S,H_1)\\ \natural\, I\\ \sharp\, D\end{array}} H_j.$$

Obviously, we can derive from the above properties (B.12) that $\mathcal{A}_\sim \vdash P = L_{1,1}\{\boldsymbol{E}/\boldsymbol{X}\}\{\boldsymbol{F}/\boldsymbol{Y}\}$. In addition, we get by application of law ($\natural$),

$$\begin{aligned}
L_{i,1} &\equiv G_i \boxed{\begin{array}{l}\flat\, S\\ \natural\, I\\ \sharp\, D\end{array}} H_1\\
&= \sum(\lambda_{k_1}).\ X_{f(i,k_1)} \boxed{\begin{array}{l}\flat\, S\\ \natural\, I\\ \sharp\, D\end{array}} H_1 +\\
&\quad \sum_{\mathsf{a}_{l_1} \notin S} \mathsf{a}_{l_1}.\ X_{g(i,l_1)} \boxed{\begin{array}{l}\flat\, S\\ \natural\, I\\ \sharp\, D\end{array}} H_1 + \sum_{\mathsf{a}_{l_1} \in S} \mathsf{a}_{l_1}.\ X_{g(i,l_1)} \boxed{\begin{array}{l}\flat\, (S,H_1)\\ \natural\, I\\ \sharp\, D\end{array}} H_1\\
&\equiv \sum(\lambda_{k_1}).Z_{f(i,k_1),1} + \sum_{\mathsf{a}_{l_1} \notin S} \mathsf{a}_{l_1}.Z_{g(i,l_1),1} + \sum_{\mathsf{a}_{l_1} \in S} \mathsf{a}_{l_1}.Z_{g(i,l_1),2}.
\end{aligned}$$

Similar transformations for the remaining $L_{i,j+1}$ show that S can be equated to a (weakly guarded) SES $S' = (\boldsymbol{Z} := \boldsymbol{L}')$ where $\boldsymbol{L}'$ just contains expressions from IML. Now, a slight variation of Theorem 5.3.4 is needed, showing that any weakly guarded SES (in particular S') is $\mathcal{A}_\sim$-provably satisfied by some expression P' in IML (see [142, Theorem 5.7]). Since the above SES S' does neither contain $\perp$, nor unguarded variables, we can assume that $P' \in \mathsf{IMC_L}$. In summary, we have achieved that $\mathcal{A}_\sim \vdash P = L_{1,1}\{\boldsymbol{E}/\boldsymbol{X}\}\{\boldsymbol{F}/\boldsymbol{Y}\}$, and we have that $\mathcal{A}_\sharp \vdash L_{1,1} = L'_{1,1}$. Since P' $\mathcal{A}_\sim$-provably satisfies S', we obtain, as a consequence of Theorem 5.3.4 that $(\mathcal{A}_\sim \cup \mathcal{A}_X \cup \mathcal{A}_\sharp \cup \mathcal{A}_!) \vdash P = P'$, completing the proof.

Corollary B.9.1. *Let $P \in \mathsf{IMC}_{\mathsf{XXL}}$ and $Q \in \mathsf{IMC}_{\mathsf{XXL}}$. Then we have that*

1. $P \sim Q$ *if and only if* $(\mathcal{A}_{\sim} \cup \mathcal{A}_X \cup \mathcal{A}_{\sharp} \cup \mathcal{A}_{!}) \vdash P = Q$, *and*
2. $P \simeq Q$ *if and only if* $(\mathcal{A}_{\simeq} \cup \mathcal{A}_X \cup \mathcal{A}_{\sharp} \cup \mathcal{A}_{!}) \vdash P = Q$.

Proof. The 'if' part of the first clause (soundness with respect to strong bisimilarity) is a consequence of Theorem 5.3.1 and Lemma B.9.1. For weak congruence, the same direction follows from an application of Lemma 5.2.1 to Lemma B.9.1, together with Theorem 5.3.6. The converse directions (completeness) are covered by Lemma B.9.3. For strong bisimilarity the lemma can be applied directly, while weak congruence requires a straightforward variation of the lemma, replacing $\sim$ by $\simeq$ and using Theorem 5.3.7, respectively Theorem 5.3.9 instead of Theorem 5.3.2, respectively Theorem 5.3.4.

Bibliography

1. L. Aceto and A. Jeffrey. A complete axiomatization of timed bisimulation for a class of timed regular behaviours. *Theoretical Computer Science*, 152:251–268, 1995.
2. A.V. Aho, J.E. Hopcroft, and J.D. Ullman. *The design and analysis of computer algorithms.* Addison-Wesley, 1994.
3. M. Ajmone Marsan, G. Balbo, G. Conte, S. Donatelli, and G. Franceschinis. *Modelling with generalized stochastic Petri nets.* John Wiley and Sons, 1995.
4. M. Ajmone Marsan, G. Conte, and G. Balbo. A class of generalised stochastic Petri nets for the performance evaluation of multiprocessor systems. *ACM Transactions on Computer Systems*, 2(2):93–122, May 1984.
5. L. de Alfaro. How to specify and verify the long-run average behaviour of probabilistic systems. In *13th Annual Symposium on Logic in Computer Science (LICS)*, pages 454–465. IEEE, Computer Society Press, July 1998.
6. L. de Alfaro. Stochastic transition systems. In Sangiorgi and de Simone [168], pages 423–438.
7. R. Alur and T.A. Henzinger, editors. *CAV'96: Computer Aided Verification*, volume 1102 of *Lecture Notes in Computer Science.* Springer-Verlag, 1996.
8. ATM Forum. *ATM User-Network Interface Specification V3.1.* ATM Forum, 1994.
9. A. Aziz, K. Sanwal, V. Singhal, and R.K.Brayton. Verifying continuous time Markov chains. In Alur and Henzinger [7], pages 269–276.
10. F. Baccelli, A. Jean-Marie, and Z. Liu. A survey on solution methods for task graph models. In Götz et al. [80], pages 163–183.
11. F. Baccelli, A.J. Marie, and I. Mitrani, editors. *Quantitative Methods in Parallel Systems*, Esprit Basic Research Series. Springer-Verlag, 1995.
12. J.C.M. Baeten, J.A. Bergstra, and S.A. Smolka. Axiomatizing probabilistic processes: ACP with generative probabilities. *Information and Computation*, 121(2):234–255, 1995.
13. J.C.M. Baeten and J.W. Klop, editors. *Theories of Concurrency: Unification and Extension (CONCUR'90)*, volume 458 of *Lecture Notes in Computer Science.* Springer-Verlag, 1990.
14. J.C.M. Baeten and P. Weijland. *Process Algebra*, volume 18 of *Cambridge Tracts in Computer Science.* Cambridge University Press, 1990.
15. C. Baier, B. Haverkort, H. Hermanns, and J.-P. Katoen. Model checking continuous time Markov chains by transient analysis. In E.A. Emerson and A.P. Sistla, editors, *Computer Aided Verification*, volume 1855 of *Lecture Notes in Computer Science*, pages 358–372. Springer-Verlag, 2000.
16. C. Baier and H. Hermanns. Weak bisimulation for fully probabilistic processes. In O. Grumberg, editor, *CAV'97: Computer Aided Verification*, volume 1254 of *Lecture Notes in Computer Science*, pages 119–130. Springer-Verlag, 1997.

17. C. Baier, J.-P. Katoen, and H. Hermanns. Approximate symbolic model checking of continuous-time Markov chains. In J.C.M. Baeten and S. Mauw, editors, *Concurrency Theory*, volume 1664 of *Lecture Notes in Computer Science*, pages 146–162. Springer-Verlag, 1999.
18. J.W. de Bakker and E.P. de Vink. *Control Flow Semantics.* MIT Press, 1996.
19. C. Baier. *On the Algorithmic Verification of Probabilistic Systems.* Universität Mannheim, 1996. Habilitiation thesis.
20. G. Balbo. On the success of stochastic Petri nets. In *Proc. of the 6th Int. Workshop on Petri Nets and Performance Models*, pages 2–9, Durham, 1995. IEEE Computer Society Press.
21. E. Bandini and R. Segala. Axiomatizations for probabilistic bisimulation. In F. Orejas, P.G. Spirakis, and J. van Leeuwen, editors, *Proceedings of the 28th Colloquium on Automata, Languages and Programming (ICALP)*, volume 2076 of *Lecture Notes in Computer Science*, pages 370–381. Springer-Verlag, 2001.
22. T. Basten. Branching bisimilarity is an equivalence indeed! *Information Processing Letters*, 58(3):141–147, 1996.
23. J.A. Bergstra, J.W. Klop, and E.-R. Olderog. Failures without chaos: A new process semantics for fair abstraction. In *Formal Description of Programming Concepts - III.* Elsevier Science Publishers, 1987.
24. M. Bernardo. An algebra-based method to associate rewards with EMPA terms. In P. Degano, R. Gorrieri, and A. Marchetti, editors, *Proceedings of the 24th Colloquium on Automata, Languages and Programming (ICALP)*, volume 1256 of *Lecture Notes in Computer Science*, pages 358–368. Springer-Verlag, 1997.
25. M. Bernardo and R. Gorrieri. A tutorial on EMPA: A theory of processes with nondeterminism, priorities, probabilities and time. *Theoretical Computer Science*, 202(1-2):1–54, 1998.
26. M. Bernardo and R. Gorrieri. Corrigendum to "A tutorial on EMPA: A theory of processes with nondeterminism, priorities, probabilities, and time". *Theoretical Computer Science*, 254(1-2):691–694, 2001.
27. T. Bolognesi and E. Brinksma. Introduction to the ISO specification language LOTOS. *Computer Networks and ISDN Systems*, 14:25–59, 1987.
28. H. Bohnenkamp. *Compositional Solution of Stochastic Process Algebra Models.* PhD thesis, Department of Computer Science, RWTH Aachen, 2002.
29. A. Bouajjani, J.C. Fernandez, N. Halbwachs, P. Raymond, and C. Ratel. Minimal state graph generation. *Science of Computer Programming*, 18(3):247–271, 1992.
30. A. Bouali. Weak and branching bisimulation in FCTOOL. Technical Report 1575, INRIA Sophia Antipolis, Valbonne Cedex, France, 1992.
31. M. Bravetti. Revisiting Interactive Markov Chains. Technical Report 2002-07, University of Bologna, 2002.
32. M. Bravetti. *Specification and Analysis of Stochastic Real-Time Systems.* PhD thesis, University of Bologna, 2002.
33. M. Bravetti and M. Bernardo. Compositional asymmetric cooperations for process algebras with probabilities, priorities, and time. In *Proc. of the 1st Int. Workshop on Models for Time-Critical Systems*, volume 39(3) of *Electronic Notes in Theoretical Computer Science.* Elsevier Science Publishers, 2000.
34. M. Bravetti, M. Bernardo, and R. Gorrieri. Towards performance evaluation with general distributions in process algebras. In Sangiorgi and de Simone [168], pages 405–422.
35. M. Bravetti and R. Gorrieri. A complete axiomatization for observational congruence of prioritized finite-state behaviors. In U. Montanari, J.D.P. Rolim,

and E. Welzl, editors, *Proceedings of the 27th Colloquium on Automata, Languages and Programming (ICALP)*, volume 1853 of *Lecture Notes in Computer Science*, pages 744–755. Springer-Verlag, 2000.
36. M. Bravetti and R. Gorrieri. The theory of interactive generalized semi-markov processes. *Theoretical Computer Science*, 282(1):5–32, 2002.
37. G. Brebner. A CCS-based Investigation of deadlock in a multi-process electronic mail system. *Formal Aspects of Computing*, 5(5):467–479, 1993.
38. E. Brinksma, J.-P. Katoen, R. Langerak, and D. Latella. A stochastic causality-based process algebra. In Gilmore and Hillston [70].
39. E. Brinksma, A. Rensink, and W. Vogler. Fair testing. In Insup Lee and Scott Smolka, editors, *Proceedings of 6th International Conference on Concurrency Theory (CONCUR'95)*, volume 962 of *Lecture Notes in Computer Science*. Springer-Verlag, 1995.
40. E. Brinksma, G. Scollo, and C. Steenbergen. Lotos specifications, their implementations and their tests. In *International Workshop on Protocol Specification, Testing and Verification VI*, pages 349–360. North-Holland, 1986. IFIP WG 6.1 International Workshop.
41. S.D. Brookes, C.A.R. Hoare, and A.W. Roscoe. A theory of communicating sequential processes. *Journal of the ACM*, 31(3):560–599, 1984.
42. P. Buchholz. Markovian process algebra: Composition and equivalence. In Herzog and Rettelbach [112], pages 11–30.
43. P. Buchholz. *A framework for the hierarchical analysis of discrete event dynamic systems.* Universität Dortmund, 1996. Habilitiation thesis.
44. G. Chehaivbar, H. Garavel, L. Mounier, N. Tawbi, and F. Zulian. Specification and verification of the Powerscale bus arbitration protocol: an industrial experiment with LOTOS. In Gotzhein and Bredereke [81], pages 435–450.
45. G. Chiola. Petri Nets versus queueing networks: Similarities and differences. In M. Silva and G. Balbo, editors, *Performance Models for Discrete Event Systems with Synchronisations: Formalisms and Analysis Techniques*, pages 121–134. Prensas Universitarias de Zaragoza, 1998.
46. G. Chiola, C. Dutheillet, G. Fraceschinis, and S. Haddad. Stochastic well-formed coloured nets for symmetric modelling applications. *IEEE Transactions on Computers*, 42(11):1343–1360, 1993.
47. I. Christoff. Testing equivalences and fully abstract models for probabilistic processes. In Baeten and Klop [13], pages 126–140.
48. E.M. Clarke, O. Grumberg, and D. Peled. *Model Checking.* MIT Press, 2000.
49. R. Cleaveland, G. Lüttgen, and M. Mendler. An algebraic theory of multiple clocks. In A. Mazurkievicz and J. Winkowski, editors, *CONCUR'97: Concurrency Theory*, volume 1243 of *Lecture Notes in Computer Science*, pages 166–180. Springer-Verlag, 1996.
50. R. Cleaveland, S. Smolka, and A. Zwarico. Testing preorders for probabilistic processes. In W. Kuich, editor, *19th Colloquium on Automata, Languages and Programming (ICALP)*, volume 623 of *Lecture Notes in Computer Science*, pages 708–719. Springer-Verlag, 1992.
51. J. Conlisk. Interactive Markov Chains. *Journal of Mathematical Sociology*, 4:157–185, 1976.
52. D. Coppersmith and S. Winograd. Matrix multiplication via arithmetic progressions. In *Proc. 19th ACM Symposium on Theory of Computing*, pages 1–6, 1987.
53. D.R. Cox. A use of complex probabilities in the theory of stochastic processes. *Cambridge Philosophical Society*, 51:313–319, 1955.
54. P.R. D'Argenio. *Algebras and Automata for Timed and Stochastic Systems.* PhD thesis, University of Twente, 1999.

55. P.R. D'Argenio, H. Hermanns, and J.-P. Katoen. On asynchronous generative parallel composition. In *Proc. of the 1st Int. Workshop on Probabilistic Methods in Verification*, volume 22 of *Electronic Notes in Theoretical Computer Science.* Elsevier Science Publishers, 1999.
56. P.R. D'Argenio, J.-P. Katoen, and E. Brinksma. An algebraic approach to the specification of stochastic systems (extended abstract). In D. Gries and W.-P. de Roever, editors, *Programming Concepts and Methods*, New York, USA, 1998. Chapman and Hall.
57. P. Darondeau. An enlarged definition and complete axiomatisation of observational congruence of finite processes. In M. Dezani-Ciancaglini and U. Montanari, editors, *International Symposium on Programming*, volume 137 of *Lecture Notes in Computer Science*, pages 47–61. Springer-Verlag, 1982.
58. R. De Nicola and M. Hennessy. Testing equivalences for processes. *Theoretical Computer Science*, 34:83–133, 1984.
59. S. Derisavi, H. Hermanns, and W.H. Sanders. Optimal state space lumping in Markov chains. submitted for publication, 2002.
60. A.K. Erlang. Lösung einger Probleme der Wahrscheinlichkeitsrechnung von Bedeutung für die selbsttätigen Fernsprechämter. *Elektrotechnische Zeitschrift*, 51:504–508, 1918.
61. J.-C. Fernandez, H. Garavel, A. Kerbrat amd R. Mateescu, and L. Mounier. CADP (Caesar/Aldebaran Development Package): A protocol validation and verification toolbox. In Alur and Henzinger [7], pages 437–440.
62. J.-C. Fernandez and L. Mounier. A tool set for deciding behavioral equivalences. In J.C.M. Baeten and J.F. Groote, editors, *2nd International Conference on Concurrency Theory (CONCUR'91)*, volume 527 of *Lecture Notes in Computer Science*, pages 23–42. Springer-Verlag, 1991.
63. J.C. Fernandez. An implementation of an efficient algorithm for bisimulation equivalence. *Science of Computer Programming*, 13:219–236, 1989.
64. H. Garavel. An overview of the Eucalyptus toolbox. In Z. Brezočnik and T. Kapus, editors, *Proceedings of the COST 247 International Workshop on Applied Formal Methods in System Design*, pages 76–88. University of Maribor, Slovenia, June 1996.
65. H. Garavel and H. Hermanns. On combining functional verification and performance evaluation using CADP. In L.-H. Eriksson and P.A. Lindsay, editor, *Proceedings of FME 2002: Formal Methods Europe*, volume 2391 of *Lecture Notes in Computer Science.* Springer-Verlag, June 2002.
66. H. Garavel, F. Lang, and R. Mateescu. An overview of CADP 2001. Technical Report RT-254, INRIA Rhone-Alpes, 2001.
67. H. Garavel and J. Sifakis. Compilation and verification of LOTOS specifications. In L. Logrippo, R. Probert, and H. Ural, editors, *International Workshop on Protocol Specification, Testing and Verification X*, pages 379–394. North-Holland, 1990. IFIP TC-6 International Workshop.
68. H. Garavel and M. Sighireanu. On the introduction of exceptions in E-LOTOS. In Gotzhein and Bredereke [81], pages 469–484.
69. A. Giacalone, C.-C. Jou, and S.A. Smolka. Algebraic reasoning for probabilistic concurrent systems. In M. Broy and C.B. Jones, editors, *Proc. PROCOMET Working Conference*, pages 443–458. IFIP TC2, North-Holland, 1990.
70. S. Gilmore and J. Hillston, editors. *Proc. of the 3rd Int. Workshop on Process Algebras and Performance Modelling*, volume 38(7) of *The Computer Journal.* Oxford University Press, 1995.

71. S. Gilmore, J. Hillston, R. Holton, and M. Rettelbach. Specifications in Stochastic Process Algebra for a Robot Control Problem. *International Journal of Production Research*, 34(4):1065–1080, 1996.
72. S. Gilmore, J. Hillston, and M. Ribaudo. An efficient algorithm for aggregating PEPA models. *IEEE Transactions on Software Engineering*, 27(5):449–464, 2001.
73. R.J. van Glabbeek and W. Weijland. Branching time and abstraction in bisimulation semantics. *Journal of the ACM*, 43(3):555–600, 1996.
74. R.J. van Glabbeek. The linear time – Branching time spectrum II: The semantics of sequential systems with silent moves (extended abstract). In E. Best, editor, *4th International Conference on Concurrency Theory, CONCUR'93*, volume 715 of *Lecture Notes in Computer Science*, pages 66–81. Springer-Verlag, 1993.
75. R.J. van Glabbeek. *The linear time – Branching time spectrum I*, pages 3–99. Elsevier Science Publishers, 2001.
76. R.J. van Glabbeek, S.A. Smolka, and B. Steffen. Reactive, generative, and stratified models of probabilistic processes. *Information and Computation*, 121:59–80, 1995.
77. N. Götz. *Stochastische Prozeßalgebren – Integration von funktionalem Entwurf und Leistungsbewertung Verteilter Systeme.* PhD thesis, Universität Erlangen-Nürnberg, 1994.
78. N. Götz, H. Hermanns, U. Herzog, V. Mertsiotakis, and M. Rettelbach. Stochastic process algebras – Constructive specification techniques integrating functional, performance and dependability aspects. In Baccelli et al. [11], pages 3–17.
79. N. Götz, U. Herzog, and M. Rettelbach. Multiprocessor and distributed system design: The integration of functional specification and performance analysis using stochastic process algebras. In *Tutorial Proc. of the 16th Int. Symposium on Computer Performance Modelling, Measurement and Evaluation, PERFORMANCE'93*, volume 729 of *Lecture Notes in Computer Science*, pages 121–146. Springer-Verlag, 1993.
80. N. Götz, U. Herzog, and M. Rettelbach, editors. *Proc. of the QMIPS-Workshop on Formalism, Principles and State-of-the-art*, volume 26(14) of *Arbeitsberichte des IMMD*. Universität Erlangen-Nürnberg, September 1993.
81. R. Gotzhein and J. Bredereke, editors. *Formal Description Techniques IX*. Chapman and Hall, 1996.
82. S. Graf and P. Loiseaux. Programm verification using compositional abstraction. In *TAPSOFT'93: Theory and Practice of Software Development*, 1993.
83. S. Graf, B. Steffen, and G. Luettgen. Compositional minimization of finite state systems. *Formal Aspects of Computing*, 8(5):607–616, 1996.
84. C. Gregorio-Rodriguez, L. Llana-Díaz, Manuel Núñez, and P. Palao-Gostanza. Testing semantics for a probabilistic-timed process algebra. In M. Bertran and T. Rus, editors, *Proceedings of the 4th AMAST Workshop on Real-Time System Development*, volume 1231 of *Lecture Notes in Computer Science*, pages 353–367. Springer-Verlag, 1997.
85. J.F. Groote. Transition system specifications with negative premises. *Theoretical Computer Science*, 118:263–299, 1993.
86. J.F. Groote and F.W. Vaandrager. An efficient algorithm for branching bisimulation and stuttering equivalence. In M.S. Paterson, editor, *17th Colloquium on Automata, Languages and Programming (ICALP)*, volume 443 of *Lecture Notes in Computer Science*, pages 626–638. Springer-Verlag, 1990.
87. H.A. Hansson. *Time and Probability in Formal Design of Distributed Systems*. Elsevier Science Publishers, 1994.

88. H.A. Hansson and B. Jonsson. A logic for reasoning about time and probability. *Formal Aspects of Computing*, 6:512–535, 1994.
89. P. Harrison and B. Strulo. Process algebra for discrete event simulation. In Baccelli et al. [11], pages 18–37.
90. F. Hartleb. Stochastic graph models for performance evaluation of parallel programs and the evaluation tool PEPP. In Götz et al. [80], pages 207–224.
91. B.R. Haverkort. *Performance of Computer Communication Systems: A Model-Based Approach*. John Wiley and Sons, 1998.
92. B.R. Haverkort and K.S. Trivedi. Specification techniques for Markov reward models. *Discrete Event Systems: Theory and Applications*, 3:219–247, 1993.
93. M. Hennessy and T. Regan. A process algebra for timed systems. *Information and Computation*, 117:221–239, 1995.
94. H. Hermanns. Semantik für Prozeßsprachen zur Leistungsbewertung. Master's thesis, Universität Erlangen-Nürnberg, IMMD 7, November 1993.
95. H. Hermanns. An operator for symmetry representation and exploitation in stochastic process algebras. In E. Brinksma and A. Nymeyer, editors, *Proc. of 5th Workshop on Process Algebras and Performance Modelling*, number 97–14 in CTIT Technical Report series. University of Twente, June 1997.
96. H. Hermanns. *Interactive Markov Chains*. PhD thesis, Universität Erlangen-Nürnberg, 1998.
97. H. Hermanns, U. Herzog, and J.-P. Katoen. Process algebra for performance evaluation. *Theoretical Computer Science*, 274(1-2):43–87, 2002.
98. H. Hermanns, U. Herzog, U. Klehmet, M.Siegle, and V. Mertsiotakis. Compositional performance modelling with the TIPPtool. *Performance Evaluation*, 39(1-4):5–35, 2000.
99. H. Hermanns, U. Herzog, and V. Mertsiotakis. Stochastic process algebras as a tool for performance and dependability modelling. In *Proc. of IEEE Int. Computer Performance and Dependability Symposium*, pages 102–113, Erlangen, April 1995. IEEE Computer Society Press.
100. H. Hermanns, U. Herzog, and V. Mertsiotakis. Stochastic process algebras - Between LOTOS and Markov chains. *Computer Networks and ISDN Systems*, 30(9-10):901–924, 1998.
101. H. Hermanns and J.-P. Katoen. Automated compositional Markov chain generation for a plain-old telephony system. *Science of Computer Programming*, 36(1):97–127, 2000.
102. H. Hermanns, J.-P. Katoen, J. Meyer-Kayser, and M. Siegle. A Markov chain model checker. In S. Graf and M. Schwartzbach, editors, *Tools and Algorithms for the Construction and Analysis of Systems (TACAS 2000)*, volume 1785 of *Lecture Notes in Computer Science*, pages 347–362, Berlin, 2000. Springer-Verlag.
103. H. Hermanns, J.-P. Katoen, J. Meyer-Kayser, and M. Siegle. Towards model checking stochastic process algebra. In W. Grieskamp, T. Santen, and B. Stoddart, editors, *2nd Int. Conference on Integrated Formal Methods*, volume 1945 of *Lecture Notes in Computer Science*, pages 420–439, Dagstuhl, November 2000. Springer-Verlag.
104. H. Hermanns and M. Lohrey. Observational congruence in a stochastic timed calculus with maximal progress. Technical Report IMMD-VII-7/97, Universität Erlangen-Nürnberg, 1997.
105. H. Hermanns and M. Lohrey. Priority and maximal progress are completely axiomatisable. In Sangiorgi and de Simone [168], pages 237–252.
106. H. Hermanns, V. Mertsiotakis, and M. Rettelbach. Performance Analysis of Distributed Systems Using TIPP – A Case Study. In J. Hillston and R. Pooley, editors, *Proc. of the 10th U.K. Performance Engineering Workshop for*

Computer and Telecommunication Systems. Department of Computer Science, University of Edinburgh, 1994.
107. H. Hermanns and M. Rettelbach. Syntax, semantics, equivalences, and axioms for MTIPP. In Herzog and Rettelbach [112], pages 71–87.
108. H. Hermanns, M. Rettelbach, and T. Weiß. Formal characterisation of immediate actions in SPA with nondeterministic branching. In Gilmore and Hillston [70].
109. H. Hermanns and M. Ribaudo. Exploiting symmetries in stochastic process algebras. In *Proc. of the 12th European Simulation Multiconference (ESM)*, pages 763–770. SCS Europe, 1998.
110. U. Herzog. Formal description, time, and performance analysis: A framework. In T. Härder, H. Wedekind, and G. Zimmermann, editors, *Entwurf und Betrieb Verteilter Systeme*, volume 264 of *Informatik-Fachberichte*, pages 172–190. Springer-Verlag, 1990.
111. U. Herzog. A concept for graph-based stochastic process algebras, generally distributed activity times and hierarchical modelling. In Ribaudo [164].
112. U. Herzog and M. Rettelbach, editors. *Proc. of the 2nd Int. Workshop on Process Algebras and Performance Modelling*, volume 27(4) of *Arbeitsberichte des IMMD.* Universität Erlangen-Nürnberg, 1994.
113. J. Hillston. The nature of synchronisation. In Herzog and Rettelbach [112], pages 51–70.
114. J. Hillston. *A Compositional Approach to Performance Modelling.* Cambridge University Press, 1996.
115. J. Hillston and V. Mertsiotakis. A simple time scale decomposition technique for stochastic process algebras. In Gilmore and Hillston [70], pages 566–577.
116. C.A.R. Hoare. *Communicating Sequential Processes.* Prentice-Hall, 1985.
117. K. Honda and M. Tokoro. On asynchronous communication semantics. In M. Tokoro, O. Nierstrasz, and P. Wegner, editors, *Object-Based Concurrent Computing*, volume 612 of *Lecture Notes in Computer Science*, pages 21–51. Springer-Verlag, 1992.
118. R. Howard. *Dynamic Probabilistic Systems*, volume 2: Semi Markov and Decision Processes. John Wiley and Sons, 1971.
119. R. Howard. *Dynamic Probabilistic Systems*, volume 1: Markov Chains. John Wiley and Sons, 1971.
120. T. Huynh and L. Tian. On some equivalence relations for probabilistic processes. *Fundamenta Informaticae*, 17:211–234, 1992.
121. P. Inverardi and M. Nesi. Deciding observational congruence of finite state CCS expressions by rewriting. *Theoretical Computer Science*, 139:315–354, 1995.
122. ISO. *LOTOS: A Formal Description Technique Based on the Temporal Ordering of Observational Behaviour*, 1989. International standard IS 8807.
123. B. Jacobs and J. Rutten. A tutorial on (co)algebras and (co)induction. *Bulletin of the EATCS*, 62:222–259, 1997.
124. B. Jonsson and W. Yi. Compositional testing preorders for probabilistic processes. In *10th Annual Symposium on Logic in Computer Science (LICS)*, pages 431–443. IEEE, Computer Society Press, June 1995.
125. P. Kanellakis and S. Smolka. CCS expressions, finite state processes, and three problems of equivalence. *Information and Computation*, 86:43–68, 1990.
126. J.-P. Katoen. *Quantitative and Qualitative Extensions of Event Structures.* PhD thesis, University of Twente, 1996.
127. J.-P. Katoen, R. Langerak, and D. Latella. Modeling systems by probabilistic process algebra: An event structures approach. In *6th International*

Conference on Formal Description Techniques FORTE'93, pages 253–268. North-Holland, 1993.

128. J.G. Kemeny and J.L. Snell. *Finite Markov Chains.* Springer, 1976.
129. L. Kleinrock. *Queueing Systems*, volume 1: Theory. John Wiley and Sons, 1975.
130. L. Kloul. *Méthodes d' Évaluation des Performances pour les Résaux ATM.* PhD thesis, Université de Versailles, 1996.
131. C.J. Koomen. *The Design of Communicating Systems: A System Engineering Approach.* Kluwer, 1991.
132. J.-P. Krimm and L. Mounier. Compositional state space generation from Lotos programs. In *Tools and Algorithms for the Construction and Analysis of Systems (TACAS'97)*, volume 1217 of *Lecture Notes in Computer Science*, Twente, April 1997. Springer-Verlag.
133. K. Larsen and A. Skou. Bisimulation through probabilistic testing. *Information and Computation*, 94(1):1–28, September 1991.
134. M. Lohrey. Vollständige Axiomatisierung für PIM-TIPP. Master's thesis, Universität Erlangen-Nürnberg, 1997.
135. M. Lohrey, P.R. D'Argenio, and H. Hermanns. Axiomatising divergence. In *Proceedings of the 29th Colloquium on Automata, Languages, and Programming (ICALP)*, volume 2380 of *Lecture Notes in Computer Science*, pages 585–596. Springer-Verlag, 2002.
136. N.A. Lynch and M.R. Tuttle. Hierarchical correctness proofs for distributed algorithms. In *6th Annual Symposium on Principles of Distributed Computing (PODC)*, pages 137–151. ACM, ACM Press, August 1987.
137. A.A. Markov. Extension of the limit theorems of probability theory to a sum of variables connected in a chain. reprinted in Appendix B of [119], 1907.
138. D.E. McDysan and D.L. Spohn. *ATM Theory and Application.* McGraw-Hill, 1994.
139. V. Mertsiotakis. *Approximate Analysis Methods for Stochastic Process Algebras.* PhD thesis, Universität Erlangen-Nürnberg, 1998.
140. V. Mertsiotakis and M. Silva. Throughput approximation of decision free processes using decomposition. In *Proc. of the 7th Int. Workshop on Petri Nets and Performance Models*, St. Malo, June 1997. IEEE Computer Society Press.
141. R. Milner. Calculi for Synchrony and Asynchrony. *Theoretical Computer Science*, 25:269–310, 1983.
142. R. Milner. A complete inference system for a class of regular behaviours. *Journal of Computer and System Science*, 28:439–466, 1984.
143. R. Milner. Process constructors and interpretations. In *Proc. IFIP-WG Information Processing.* North-Holland, 1986.
144. R. Milner. A complete axiomatization for observational congruence of finite-state behaviours. *Information and Computation*, 91:227–247, 1989.
145. R. Milner. *Communication and Concurrency.* Prentice-Hall, London, 1989.
146. F. Moller and C. Tofts. A temporal calculus for communicating systems. In Baeten and Klop [13], pages 401–415.
147. U. Montanari and V. Sassone. Dynamic congruence vs. progressing bisimulation for CCS. *Fundamenta Informaticae*, XVI(2):171–199, 1992.
148. V. Natarajan, I. Christoff, L. Christoff, and R. Cleaveland. Priorities and abstraction in process algebra. In P. S. Thiagarajan, editor, *Foundations of Software Technology and Theoretical Computer Science*, volume 880 of *Lecture Notes in Computer Science*, pages 217–230. Springer-Verlag, 1994.

149. V. Natarjan and R. Cleaveland. Divergence and fair testing. In Zoltán Fülöp and Ferenc Gécseg, editors, *22nd Colloquium on Automata, Languages and Programming (ICALP)*, volume 944 of *Lecture Notes in Computer Science*, pages 648–659. Springer-Verlag, 1995.
150. M.F. Neuts. *Matrix-geometric solutions in stochastic models - An algorithmic approach.* The Johns Hopkins University Press, 1981.
151. X. Nicollin and J. Sifakis. An overview and synthesis on timed process algebras. In J.W. de Bakker, K. Huizing, and W.-P. de Roever, editors, *Real-Time: Theory in Practice (REX Workshop)*, volume 600 of *Lecture Notes in Computer Science*, pages 526–548. Springer-Verlag, 1990.
152. M. Núñez and D. de Frutos. Testing semantics for probabilistic LOTOS. In *Proceedings of the 8th International Conference on Formal Description Techniques, FORTE'95*, pages 365–380. North-Holland, 1995.
153. E.-R. Olderog and C.A.R. Hoare. Specification oriented semantics for communicating processes. *Acta Informatica*, 23:9–66, 1986.
154. R. Paige and R. Tarjan. Three partition refinement algorithms. *SIAM Journal of Computing*, 16(6):973–989, 1987.
155. D. Park. Concurrency and automata on infinite sequences. In P. Deussen, editor, *5th GI Conference on Theoretical Computer Science*, volume 104 of *Lecture Notes in Computer Science.* Springer-Verlag, 1981.
156. Joachim Parrow and Peter Sjödin. The complete axiomatization of cs-congruence. In P. Enjalbert, E.W. Mayr, and K.W. Wagner, editors, *STACS'94*, volume 775 of *Lecture Notes in Computer Science*, pages 557–568. Springer-Verlag, 1994.
157. G.D. Plotkin. A structured approach to operational semantics. Technical Report DAIMI FM-19, Computer Science Department, Aarhus University, 1981.
158. A. Pnueli and L. Zuck. Verification of multiprocess probabilistic protocols. *Distributed Computing*, 1(1):53–72, 1986.
159. C. Priami. Stochastic π-calculus. In Gilmore and Hillston [70], pages 578–689.
160. C. Priami. Stochastic π-calculus with general distributions. In Ribaudo [164], pages 41–57.
161. M.L. Puterman. *Markov Decision Processes: Discrete Stochastic Dynamic Programming.* John Wiley and Sons, 1994.
162. W. Reisig. *Petrinetze — Eine Einführung.* Springer-Verlag, Berlin, Heidelberg, New York, 1982.
163. M. Rettelbach. *Stochastische Prozeßalgebren mit zeitlosen Aktivitäten und probabilistischen Verzweigungen.* PhD thesis, Universität Erlangen-Nürnberg, 1996.
164. M. Ribaudo, editor. *Proc. of the 4th Workshop on Process Algebras and Performance Modelling.* Universita di Torino, CLUT, 1996.
165. T.C. Ruys, R. Langerak, J.-P. Katoen, D. Latella, and M. Massink. First passage time analysis of stochastic process algebra using partial orders. In *Tools and Algorithms for the Construction and Analysis of Systems (TACAS 2001)*, volume 2031 of *Lecture Notes in Computer Science*, pages 220 - 235. Springer-Verlag, 2001.
166. A. Sahner and K.S. Trivedi. Performance and reliability analysis using directed acyclic raphs. *IEEE Transactions on Software Engineering*, 10:1105–1114, 1987.
167. W.H. Sanders and J.F. Meyer. Reduced base model construction methods for stochastic activity networks. *IEEE Journal on Selected Areas in Communications*, 9(1):25–36, 1991.

168. D. Sangiorgi and R. de Simone, editors. *CONCUR'98: Concurrency Theory*, volume 1466 of *Lecture Notes in Computer Science*. Springer-Verlag, September 1998.
169. D. Sangiorgi and R. Milner. The problem of 'weak bisimulation up to'. In Rance Cleaveland, editor, *3rd International Conference on Concurrency Theory (CONCUR'92)*, volume 630 of *Lecture Notes in Computer Science*, pages 32–46. Springer-Verlag, 1992.
170. D.A. Schmidt. *Denotational Semantics: A Methodology for Language Development*. Allyn and Bacon, 1986.
171. R. Segala. *Modeling and Verification of Randomized Distributed Real-Time Systems*. PhD thesis, MIT, Dept. of Electrical Engineering and Computer Science, 1995.
172. R. Segala. Verification of randomized distributed algorithms. In E. Brinksma, H. Hermanns, and J.-P. Katoen, editors, *Lectures on Formal Methods and Performance Analysis*, volume 2090 of *Lecture Notes in Computer Science*, pages 232–260. Springer-Verlag, 2001.
173. R. Segala and N. Lynch. Probabilistic simulations for probabilistic processes. *Nordic Journal of Computing*, 2(2):250–273, 1995.
174. G.S. Shedler. *Regenerative Stochastic Simulation*. Academic Press, 1993.
175. M. Siegle. Reduced Markov models of parallel programs with replicated processes. In *Proc. of 2nd. EUROMICRO Workshop on Parallel and Distributed Processing*, pages 126–133, Malaga, Spain, January 1994.
176. M. Siegle. *Beschreibung und Analyse von Markovmodellen mit großem Zustandsraum*. PhD thesis, Universität Erlangen-Nürnberg, 1995.
177. M. Siegle, B. Wentz, A. Klingler, and M. Simon. Neue Ansätze zur Planung von Klinikkommunikationssystemen mittels stochastischer Leistungsmodellierung. In R. Muche, G. Büchele, D. Harder, and W. Gaus, editors, *42. Jahrestagung der Deutschen Gesellschaft für Medizinische Informatik, Biometrie und Epidemiologie (GMDS)*, pages 188 – 192, Ulm, September 1997. MMV Medizin Verlag.
178. R. Sisto, L. Ciminiera, and A. Valenzano. Probabilistic characterization of algebraic protocol specifications. In *12th International Conference on Distributed Computing Systems*, pages 260–268. IEEE Computer Society Press, 1992.
179. D.D. Sleator and R. Tarjan. Self-adjusting binary search trees. *Journal of the ACM*, 32(3):652–686, 1985.
180. T. Speed. Probabilistic risk assessment in the nuclear industry: WASH 1400 and beyond. In L. LeCam and R.A. Olshen, editors, *Proceedings of the Berkeley Conference in Honor of Jerzy Neyman and Jack Kiefer*, pages 173–200. Wadsworth Inc, 1985.
181. W.J. Stewart. *Introduction to the numerical solution of Markov chains*. Princeton University Press, 1994.
182. A. Valmari and M. Tienari. Compositional failure-based semantic models for basic LOTOS. *Formal Aspects of Computing*, 7:440–468, 1995.
183. M. Vardi. Automatic verification of probabilistic concurrent finite-state programs. In *Proc. 26th Annual Symposium on Foundations of Computer Science*, pages 327–338, 1985.
184. C.A. Vissers, G. Scollo, M. van Sinderen, and E. Brinksma. Specification styles in distributed systems design and verification. *Theoretical Computer Science*, 89(1):179–206, 1991.
185. D.J. Walker. Bisimulation and divergence. *Information and Computation*, 85:202–241, 1990.

186. S. Wu, S.A. Smolka, and E.W. Stark. Composition and behaviors of probabilistic I/O automata. *Theoretical Computer Science*, 176(1-2):1–38, 1997.
187. W. Yi. CCS + Time = An Interleaving Model for Real Time Systems. In J. Leach Albert, B. Monien, and M. Rodríguez, editors, *18th Colloquium on Automata, Languages and Programming (ICALP)*, volume 510 of *Lecture Notes in Computer Science*, pages 217–228. Springer-Verlag, 1991.
188. S. Yuen, R. Cleaveland, Z. Dayar, and S. Smolka. Fully abstract characterizations of testing preorders for probabilistic processes. In B. Jonsson and J. Parrow, editors, *CONCUR'94: Concurrency Theory*, volume 836 of *Lecture Notes in Computer Science*, pages 497–512. Springer-Verlag, August 1994.

Lecture Notes in Computer Science

For information about Vols. 1–2404
please contact your bookseller or Springer-Verlag

Vol. 2343: I. Chong (Ed.), Information Networking. Proceedings, Part I, 2002. XIX, 729 pages. 2002.

Vol. 2344: I. Chong (Ed.), Information Networking. Proceedings, Part II, 2002. XX, 825 pages. 2002.

Vol. 2376: E. Gregori, L. Cherkasova, G. Cugola, F. Panzieri, G.P. Picco (Eds.), Web Engineering and Peer-to-Peer Computing. Proceedings, 2002. XV, 352 pages. 2002.

Vol. 2405: B. Eaglestone, S. North, A. Poulovassilis (Eds.), Advances in Databases. Proceedings, 2002. XII, 199 pages. 2002.

Vol. 2406: C. Peters, M. Braschler, J. Gonzalo, M. Kluck (Eds.), Evaluation of Cross-Language Information Retrieval Systems. Proceedings, 2001. X, 601 pages. 2002.

Vol. 2407: A.C. Kakas, F. Sadri (Eds.), Computational Logic: Logic Programming and Beyond. Part I. XII, 678 pages. 2002. (Subseries LNAI).

Vol. 2408: A.C. Kakas, F. Sadri (Eds.), Computational Logic: Logic Programming and Beyond. Part II. XII, 628 pages. 2002. (Subseries LNAI).

Vol. 2409: D.M. Mount, C. Stein (Eds.), Algorithm Engineering and Experiments. Proceedings, 2002. VIII, 207 pages. 2002.

Vol. 2410: V.A. Carreño, C.A. Muñoz, S. Tahar (Eds.), Theorem Proving in Higher Order Logics. Proceedings, 2002. X, 349 pages. 2002.

Vol. 2411: F. Paternò (Ed.), Mobile Human-Computer Interaction. Proceeedings, 2002. XIII, 432 pages. 2002.

Vol. 2412: H. Yin, N. Allinson, R. Freeman, J. Keane, S. Hubbard (Eds.), Intelligent Data Engineering and Automated Learning – IDEAL 2002. Proceedings, 2002. XV, 597 pages. 2002.

Vol. 2413: K. Kuwabara, J. Lee (Eds.), Intelligent Agents and Multi-Agent Systems. Proceedings, 2002. X, 221 pages. 2002. (Subseries LNAI).

Vol. 2414: F. Mattern, M. Naghshineh (Eds.), Pervasive Computing. Proceedings, 2002. XI, 298 pages. 2002.

Vol. 2415: J.R. Dorronsoro (Ed.), Artificial Neural Networks – ICANN 2002. Proceedings, 2002. XXVIII, 1382 pages. 2002.

Vol. 2416: S. Craw, A. Preece (Eds.), Advances in Case-Based Reasoning. Proceedings, 2002. XII, 656 pages. 2002. (Subseries LNAI).

Vol. 2417: M. Ishizuka, A. Sattar (Eds.), PRICAI 2002: Trends in Artificial Intelligence. Proceedings, 2002. XX, 623 pages. 2002. (Subseries LNAI).

Vol. 2418: D. Wells, L. Williams (Eds.), Extreme Programming and Agile Methods – XP/Agile Universe 2002. Proceedings, 2002. XII, 292 pages. 2002.

Vol. 2419: X. Meng, J. Su, Y. Wang (Eds.), Advances in Web-Age Information Management. Proceedings, 2002. XV, 446 pages. 2002.

Vol. 2420: K. Diks, W. Rytter (Eds.), Mathematical Foundations of Computer Science 2002. Proceedings, 2002. XII, 652 pages. 2002.

Vol. 2421: L. Brim, P. Jančar, M. Křetínský, A. Kučera (Eds.), CONCUR 2002 – Concurrency Theory. Proceedings, 2002. XII, 611 pages. 2002.

Vol. 2422: H. Kirchner, Ch. Ringeissen (Eds.), Algebraic Methodology and Software Technology. Proceedings, 2002. XI, 503 pages. 2002.

Vol. 2423: D. Lopresti, J. Hu, R. Kashi (Eds.), Document Analysis Systems V. Proceedings, 2002. XIII, 570 pages. 2002.

Vol. 2424: S. Flesca, G. Ianni (Eds.), Logics in Artificial Intelligence. Proceedings, 2002. XIII, 572 pages. 2002. (Subseries LNAI).

Vol. 2425: Z. Bellahsène, D. Patel, C. Rolland (Eds.), Object-Oriented Information Systems. Proceedings, 2002. XIII, 550 pages. 2002.

Vol. 2426: J.-M. Bruel, Z. Bellahsène (Eds.), Advances in Object-Oriented Information Systems.Procedings, 2002. IX, 314 pages. 2002.

Vol. 2428: H. Hermanns, Interactive Markov Chains. XII, 217 pages. 2002.

Vol. 2430: T. Elomaa, H. Mannila, H. Toivonen (Eds.), Machine Learning: ECML 2002. Proceedings, 2002. XIII, 532 pages. 2002. (Subseries LNAI).

Vol. 2431: T. Elomaa, H. Mannila, H. Toivonen (Eds.), Principles of Data Mining and Knowledge Discovery. Proceedings, 2002. XIV, 514 pages. 2002. (Subseries LNAI).

Vol. 2432: R. Bergmann, Experience Management. XXI, 393 pages. 2002. (Subseries LNAI).

Vol. 2433: A.H. Chan, V. Gligor (Eds.), Information Security. Proceedings, 2002. XII, 502 pages. 2002.

Vol. 2434: S. Anderson, S. Bologna, M. Felici (Eds.), Computer Safety, Reliability and Security. Proceedings, 2002. XX, 347 pages. 2002.

Vol. 2435: Y. Manolopoulos, P. Návrat (Eds.), Advances in Databases and Information Systems. Proceedings, 2002. XIII, 415 pages. 2002.

Vol. 2436: J. Fong, C.T. Cheung, H.V. Leong, Q. Li (Eds.), Advances in Web-Based Learning. Proceedings, 2002. XIII, 434 pages. 2002.

Vol. 2438: M. Glesner, P. Zipf, M. Renovell (Eds.), Field-Programmable Logic and Applications. Proceedings, 2002. XXII, 1187 pages. 2002.

Vol. 2439: J.J. Merelo Guervós, P. Adamidis, H.-G. Beyer, J.-L. Fernández-Villacañas, H.-P. Schwefel (Eds.), Parallel Problem Solving from Nature – PPSN VII. Proceedings, 2002. XXII, 947 pages. 2002.

Vol. 2440: J.M. Haake, J.A. Pino (Eds.), Groupware: Design, Implementation and Use. Proceedings, 2002. XII, 285 pages. 2002.

Vol. 2441: Z. Hu, M. Rodríguez-Artalejo (Eds.), Functional and Logic Programming. Proceedings, 2002. X, 305 pages. 2002.

Vol. 2442: M. Yung (Ed.), Advances in Cryptology – CRYPTO 2002. Proceedings, 2002. XIV, 627 pages. 2002.

Vol. 2443: D. Scott (Ed.), Artificial Intelligence: Methodology, Systems, and Applications. Proceedings, 2002. X, 279 pages. 2002. (Subseries LNAI).

Vol. 2444: A. Buchmann, F. Casati, L. Fiege, M.-C. Hsu, M.-C. Shan (Eds.), Technologies for E-Services. Proceedings, 2002. X, 171 pages. 2002.

Vol. 2445: C. Anagnostopoulou, M. Ferrand, A. Smaill (Eds.), Music and Artificial Intelligence. Proceedings, 2002. VIII, 207 pages. 2002. (Subseries LNAI).

Vol. 2446: M. Klusch, S. Ossowski, O. Shehory (Eds.), Cooperative Information Agents VI. Proceedings, 2002. XI, 321 pages. 2002. (Subseries LNAI).

Vol. 2447: D.J. Hand, N.M. Adams, R.J. Bolton (Eds.), Pattern Detection and Discovery. Proceedings, 2002. XII, 227 pages. 2002. (Subseries LNAI).

Vol. 2448: P. Sojka, I. Kopeček, K. Pala (Eds.), Text, Speech and Dialogue. Proceedings, 2002. XII, 481 pages. 2002. (Subseries LNAI).

Vol. 2449: L. Van Gool (Ed.), Pattern Recognotion. Proceedings, 2002. XVI, 628 pages. 2002.

Vol. 2451: B. Hochet, A.J. Acosta, M.J. Bellido (Eds.), Integrated Circuit Design. Proceedings, 2002. XVI, 496 pages. 2002.

Vol. 2452: R. Guigó, D. Gusfield (Eds.), Algorithms in Bioinformatics. Proceedings, 2002. X, 554 pages. 2002.

Vol. 2453: A. Hameurlain, R. Cicchetti, R. Traunmüller (Eds.), Database and Expert Systems Applications. Proceedings, 2002. XVIII, 951 pages. 2002.

Vol. 2454: Y. Kambayashi, W. Winiwarter, M. Arikawa (Eds.), Data Warehousing and Knowledge Discovery. Proceedings, 2002. XIII, 339 pages. 2002.

Vol. 2455: K. Bauknecht, A M. Tjoa, G. Quirchmayr (Eds.), E-Commerce and Web Technologies. Proceedings, 2002. XIV, 414 pages. 2002.

Vol. 2456: R. Traunmüller, K. Lenk (Eds.), Electronic Government. Proceedings, 2002. XIII, 486 pages. 2002.

Vol. 2458: M. Agosti, C. Thanos (Eds.), Research and Advanced Technology for Digital Libraries. Proceedings, 2002. XVI, 664 pages. 2002.

Vol. 2459: M.C. Calzarossa, S. Tucci (Eds.), Performance Evaluation of Complex Systems: Techniques and Tools. Proceedings, 2002. VIII, 501 pages. 2002.

Vol. 2460: J.-M. Jézéquel, H. Hussmann, S. Cook (Eds.), «UML» 2002 – The Unified Modeling Language. Proceedings, 2002. XII, 449 pages. 2002.

Vol. 2461: R. Möhring, R. Raman (Eds.), Algorithms – ESA 2002. Proceedings, 2002. XIV, 917 pages. 2002.

Vol. 2462: K. Jansen, S. Leonardi, V. Vazirani (Eds.), Approximation Algorithms for Combinatorial Optimization. Proceedings, 2002. VIII, 271 pages. 2002.

Vol. 2463: M. Dorigo, G. Di Caro, M. Sampels (Eds.), Ant Algorithms. Proceedings, 2002. XIII, 305 pages. 2002.

Vol. 2464: M. O'Neill, R.F.E. Sutcliffe, C. Ryan, M. Eaton, N. Griffith (Eds.), Artificial Intelligence and Cognitive Science. Proceedings, 2002. XI, 247 pages. 2002. (Subseries LNAI).

Vol. 2465: H. Arisawa, Y. Kambayashi (Eds.), Conceptual Modeling for New Information Systems Technologies. Proceedings, 2001. XVII, 500 pages. 2002.

Vol. 2469: W. Damm, E.-R. Olderog (Eds.), Formal Techniques in Real-Time and Fault-Tolerant Systems. Proceedings, 2002. X, 455 pages. 2002.

Vol. 2470: P. Van Hentenryck (Ed.), Principles and Practice of Constraint Programming – CP 2002. Proceedings, 2002. XVI, 794 pages. 2002.

Vol. 2471: J. Bradfield (Ed.), Computer Science Logic. Proceedings, 2002. XII, 613 pages. 2002.

Vol. 2475: J.J. Alpigini, J.F. Peters, A. Skowron, N. Zhong (Eds.), Rough Sets and Current Trends in Computing. Proceedings, 2002. XV, 640 pages. 2002. (Subseries LNAI).

Vol. 2476: A.H.F. Laender, A.L. Oliveira (Eds.), String Processing and Information Retrieval. Proceedings, 2002. XI, 337 pages. 2002.

Vol. 2477: M.V. Hermenegildo, G. Puebla (Eds.), Static Analysis. Proceedings, 2002. XI, 527 pages. 2002.

Vol. 2478: M.J. Egenhofer, D.M. Mark (Eds.), Geographic Information Science. Proceedings, 2002. X, 363 pages. 2002.

Vol. 2479: M. Jarke, J. Koehler, G. Lakemeyer (Eds.), KI 2002: Advances in Artificial Intelligence. Proceedings, 2002. XIII, 327 pages. (Subseries LNAI).

Vol. 2480: Y. Han, S. Tai, D. Wikarski (Eds.), Engineering and Deployment of Cooperative Information Systems. Proceedings, 2002. XIII, 564 pages. 2002.

Vol. 2483: J.D.P. Rolim, S. Vadhan (Eds.), Randomization and Approximation Techniques in Computer Science. Proceedings, 2002. VIII, 275 pages. 2002.

Vol. 2484: P. Adriaans, H. Fernau, M. van Zaanen (Eds.), Grammatical Inference: Algorithms and Applications. Proceedings, 2002. IX, 315 pages. 2002. (Subseries LNAI).

Vol. 2488: T. Dohi, R. Kikinis (Eds), Medical Image Computing and Computer-Assisted Intervention – MICCAI 2002. Proceedings, Part I. XXIX, 807 pages. 2002.

Vol. 2489: T. Dohi, R. Kikinis (Eds), Medical Image Computing and Computer-Assisted Intervention – MICCAI 2002. Proceedings, Part II. XXIX, 693 pages. 2002.

Vol. 2496: K.C. Almeroth, M. Hasan (Eds.), Management of Multimedia in the Internet. Proceedings, 2002. XI, 355 pages. 2002.

Vol. 2498: G. Borriello, L.E. Holmquist (Eds.), UbiComp 2002: Ubiquitous Computing. Proceedings, 2002. XV, 380 pages. 2002.